CW01182870

Braking 2002

Collected papers from the International Conference
Braking 2002
From the Driver to the Road
held at the Royal Armouries Museum and Conference Centre, Leeds, UK, 10–12 July 2002.

The Conference was organized by
the Yorkshire Centre, Automobile Division of
the Institution of Mechanical Engineers (IMechE)

In association with
the University of Leeds
the Society of Automotive Engineers
FISITA

Conference Organizing Committee

Professor David Barton (Joint Chairman)
University of Leeds

Mr Brian Shilton (Joint Chairman)
Consultant

Dr Martin Haigh
WABCO Vehicle Control Systems

Mr John Hahn
TMD Friction Limited

Mr Bert Smales
Consultant

Dr Andrew Blackwood
WABCO Vehicle Control Systems

Mr Jim Douglas
IMechE Yorks Admin Officer

Cover design/illustration by Terry Bambrook, Leeds, UK.
Tel: + 44 (0)113 236 2855

International Conference

Braking 2002

From the Driver to the Road

Edited by

David Barton

and

Brian Shilton

Professional
Engineering
Publishing

Professional Engineering Publishing Limited
Bury St Edmunds and London, UK.

First Published 2002

This publication is copyright under the Berne Convention and the International Copyright Convention. All rights reserved. Apart from any fair dealing for the purpose of private study, research, criticism or review, as permitted under the Copyright, Designs and Patents Act, 1988, no part may be reproduced, stored in a retrieval system, or transmitted in any form or by any means, electronic, electrical, chemical, mechanical, photocopying, recording or otherwise, without the prior permission of the copyright owners. *Unlicensed multiple copying of the contents of this publication is illegal.* Inquiries should be addressed to: The Publishing Editor, Professional Engineering Publishing Limited, Northgate Avenue, Bury St. Edmunds, Suffolk, IP32 6BW, UK. Fax: 01284 705271.

© 2002 With Authors

ISBN 1 86058 371 7

A CIP catalogue record for this book is available from the British Library.

Printed and bound in Great Britain by The Cromwell Press Limited, Wiltshire, UK.

The Publishers are not responsible for any statement made in this publication. Data, discussion, and conclusions developed by authors are for information only and are not intended for use without independent substantiating investigation on the part of potential users. Opinions expressed are those of the Author and are not necessarily those of the Institution of Mechanical Engineers or its Publishers.

Related Titles of Interest

Title	Editor/Author	ISBN
Brakes 2000 – Automotive Braking	D Barton and S Earle	1 86058 261 3
Advances in Vehicle Design	J Fenton	1 86058 181 1
Brakes and Friction Materials	G A Harper	1 86058 127 7
Road Vehicle Suspensions	W Matschinsky	1 86058 202 8
IMechE Engineers' Data Book – Second Edition	C Matthews	1 86058 248 6
Multi-body Dynamics – Vehicles, Machines, and Mechanisms	H Rahnejat	1 86058 122 6
Vehicle Systems Integration – The Way Ahead	A V Smith and C Hickman	1 86058 262 1
How Did That Happen? Engineering Safety and Reliability	W Wong	1 86058 359 8
Multi-body Dynamics – New Techniques and Application	IMechE Conference	1 86058 152 8
Vehicle Noise and Vibration 2000	IMechE Conference	1 86058 270 2
Vehicle Safety 2000	IMechE Conference	1 86058 271 0

For the full range of titles published by Professional Engineering Publishing contact:

Marketing Department
Professional Engineering Publishing Limited
Northgate Avenue
Bury St Edmunds
Suffolk
IP32 6BW
UK

Tel: +44 (0)1284 724384
Fax: +44 (0)1284 718692
Website: www.pepublishing.com

Contents

Foreword *xi*

Brake Refinement

One year's experience utilizing the SAE J2521 brake noise test procedure
J K Thompson and C M Fudge ... 3

Towards more accurate brake testing
E J Slevin and H Smales ... 15

Holographic interferometry used to investigate noise from a drum brake mounted on a half vehicle test rig
J D Fieldhouse, C Talbot, C Beveridge, and W Steel ... 25

Modelling and simulation of the vibration and squeal of a car disc brake
H Ouyang, Q Cao, J E Mottershead, T Treyde, and M P Cartmell ... 43

Brake system noise and vibration – a review
P Ioannidis, P C Brooks, D C Barton, and M Nishiwaki ... 53

Evaluation of disc brake pad pressure distribution by multibody dynamic analysis
W Rumold and R A Swift ... 75

Investigation of the time development of disc brakes generating noise using laser holography
C Talbot and J D Fieldhouse ... 85

Stick–slip vibration in a disc brake system
Q Cao, H Ouyang, J E Mottershead, D J Brookfield, and S James ... 101

Brake-squeal measurement using electronic speckle pattern interferometry
R Krupka, T Walz, A Ettemeyer, and R Evans ... 109

Mechatronic Braking Systems

A regulatory approach to complex electronic control systems when sharing life with the braking system
A J Mendelson ... 123

Development of linear solenoid valve for brake-by-wire system
T Miyazaki, A Sakai, A Otomo, E Nakamura, A Shimura, Y Tanaka, T Matsunaga, and S Niwa ... 129

Design and control of an electric parking brake
R Leiter ... 145

The new Range Rover – maximizing customer benefits from a slip control system
P Thompson 163

Regenerative braking for all-electric vehicles
N Schofield, C M Bingham, and D Howe 175

Development of 'ECB' system for hybrid vehicles
M Soga, M Shimada, and Y Obuchi 185

Enhancing road vehicle efficiency by regenerative braking
A M Walker, M U Lamperty, and S Wilkins 197

Braking Systems and Vehicle Performance

Influence of tyre properties to modern vehicle control systems
K-H Hartmann and H Grünberg 211

Brake performance monitoring for commercial vehicles using estimated tire–road friction information
T Dieckmann and A Stenman 223

From tyre contact patch to satellite
A R Williams 235

Intelligent braking management for commercial vehicles
E-C von Glasner, H Marwitz, R Povel, and C Kohrs 245

The intelligent brake system
E Gerum 261

Current position and development trends in air-brake systems for Mercedes-Benz commercial vehicles
T Markovic 271

Wheel movement during braking
J Klaps and A J Day 285

Materials and Modelling

Simulation of brake sequences and corresponding brake wear
R Harju 299

Analysis of automotive disc brake cooling characteristics
G Voller, M Tirovic, R Morris, and P Gibbens 307

Interface temperatures in friction braking
H S Qi, K Noor, and A J Day 319

New development of a wet brake system
S Nowak ... 329

Friction and wear of polymer matrix composite materials for automotive braking industry
P Filip ... 341

The interactions of phenolic polymers in friction materials
B H McCormick ... 355

Effect of friction lining additives on interfaces of friction components
R Holinski ... 369

Authors' Index ... 379

Foreword

Having established itself as a recognized centre of excellence for vehicle braking, and following on from previous seminars and the highly successful two-day International Conference *Brakes 2000*, the Yorkshire Centre of the Automobile Division of the IMechE has organized a further event of major importance to all automotive engineers concerned with the braking of road vehicles.

The scope of this Conference, *Braking 2002*, has been widened to encompass all aspects of vehicle braking 'from the driver to the road', and in this respect it continues in the tradition of the successful and highly acclaimed IMechE AD Conferences of 1976, 1983, and 1993, initiated and organized by the highly respected, late Professor Peter Newcomb.

The first two sessions of the Conference are devoted to brake refinement, with papers from academia and industry presenting the latest work on both modelling and testing techniques aimed at achieving a better understanding of the mechanisms of brake noise and vibration, and means of combating these undesirable phenomena.

The third session focuses on the electronic control and electrical actuation of brakes, including the legislative issues raised by such systems, followed by descriptions of systems for electric and hybrid vehicles incorporating energy saving regenerative braking.

In the fourth session, the influence of road surface and tyre properties on future system design and performance is examined, followed by descriptions of integrated intelligent systems for heavy vehicles and an assessment of vehicle steering and suspension effects.

The fifth and final session describes modelling and analyses of brake temperature, cooling and wear characteristics, and reports on the effects of materials on the tribology of the friction interface.

The editors of this published volume, which is a collection of the papers presented at *Braking 2002*, would like to thank all authors for their contributions to the Conference. Thanks are also due to the many people who kindly refereed the papers, the members of the Conference Organizing Committee, and Susan Lacey and Sheila Moore of the University of Leeds for all their hard work in organizing the event, and finally to Terry Bambrook for the publicity material. We hope that Conference delegates and readers of this publication will find the papers both informative and interesting, and we look forward to meeting many of you at our next event, planned for 2004, by which time we are sure that further significant advances will have been made in road vehicle braking.

David Barton *and* **Brian Shilton**
Joint Chairmen of the *Braking 2002* Organizing Committee

Brake Refinement

One year's experience utilizing the SAE J2521 brake noise test procedure

J K THOMPSON and **C M FUDGE**
Link Engineering Company, Plymouth, Michigan, USA

SYNOPSIS

The SAE J2521 procedure has been employed in over 50 vehicle platform brake noise evaluations. The experience with this procedure and the results obtained are presented. The major issues that arose in the implementation of SAE J2521 are discussed. The required test time, cooling optimization, and special requirements for warm up of the brake to meet test requirements are described. The discussion of the results compares the drag and deceleration test outcomes and their effectiveness in indicating brake squeal. It is shown that, unlike the initial expectations for the test, the drag sequence is not always the most effective in eliciting squeal noise. There are instances where noise from the deceleration segments exceeds that from the drag segments. The noise results obtained before and after fade are also compared. The variations in the changes in noise after fade recovery segment are presented. A comparison of dynamometer test results to those obtained from vehicle testing is provided. Finally, recommendations are made for improvements and revisions to the procedure.

1. INTRODUCTION

In 2000 the US Working Group on Brake Noise proposed the first SAE standard for brake noise measurement. This proposed procedure has been adopted as a standard by SAE and is beginning to see widespread usage around the world. For those unfamiliar with this procedure see References 1 and 2 and Figure 1 below. Link Testing Laboratories was one of the first independent facilities to perform this test procedure. This paper summarizes the results of a year's experience in performing this test and makes suggestions for potential improvements.

Table 1 – SAE J2521 summary

Test Segment	Section Nos.	Description	No of Applicaitons	Speed, km/h	Pressure, bar	Initial Temp., °C	Notes
Initial Conditioning	1	Break-In	30	80-30	30	100	
	2	Bedding	32	80-30	15-51	100	
	3	Friction Characteristic Value After Break-In	6	80-30	30	100	
Evaluation	4, 9, & 14	Drag Module	266	3 & 10	0-30	50-200	
Repeated 3 times	5, 10, & 15	Intermediate Conditioning and Warm-Up	24	50-0	5-30	100 & 150	
	6, 11, & 16	Backward/Forward	50	-3 & 3	0-20	150-50	
	7, 12, & 17	Deceleration Module	108	50-0	5-30	50-250	
	8, 13, & 18	Friction Characteristic Value After Break-In	6	80-30	30	100	
Fade & Recovery	19	Temperature Fade Module	15	100-0		100-550	0.4 g
	20	Recovery	18	80-30	30	100	
	21	Drag Module	266	3 & 10	0-30	50-200	
	22	Intermediate Conditioning and Warm-Up	24	50-0	5-30	100 & 150	
	23	Backward/Forward	50	-3 & 3	0-20	150-50	
	24	Deceleration Module	108	50-0	5-30	50-250	
	25	Friction Characteristic Value After Break-In	6	80-30	30	100	

The results from 105 tests are reported here. The optional fade and recovery segments of the J2521 procedure were included in 24 of these tests. Since Link Testing is an independent laboratory doing tests for a wide range of customers, the reported results are for a broad spectrum of applications. These results cover passenger vehicle and light truck braking systems for the US, South American, European, and Korean markets. They range from subcompact vehicles to heavy-duty light trucks. In many cases, tests represented initial screening of friction material and other components. Final selections and NVH problem resolution tests were also included in the spectrum of evaluations included. A detailed summary of the tests and the results obtained are presented in Tables 1 and 2 of the Appendix.

2. TEST DYNAMOMETER

Nearly all of the tests reported here were performed on the primary NVH dynamometer at Link Testing. This is a LINK Model 1900 dynamometer specifically configured for NVH testing. A gearbox is used to ensure that sufficient torque is available to run the low speed drag procedures found in the SAE J2521 protocol and other NVH procedures. The test section is enclosed in a combined environmental and acoustic enclosure. High transmission loss walls are provided to ensure that the background is kept to acceptable limits and sound absorption material is placed on the inside of the enclosure to better simulate the free field environment of outdoor operation of the brake. A photograph of this enclosure is shown in Figure 1. The entire enclosure is isolated from the rest of the dynamometer. The measurement microphone is located in accordance with SAE J2521. All instrumentation and analysis equipment complies with the requirements of this procedure. Testing has verified the performance of this dynamometer to exceed all the requirements of SAE J2521.

3. RUNNING THE SAE J2521 TEST PROCEDURE

Overall, the implementation of the SAE J2521 procedure was found to be straightforward on this dynamometer. Tests ranging from small passenger vehicles to whole heavy-duty light truck axles have been accommodated on the dynamometer running the SAE procedure. After running over 100 J2521 tests, a few issues with the test procedure were uncovered. First, the Warm-Up procedure specified in the SAE recommended practice resulted in difficulties due to the upper speed limitations of the dynamometer's gearbox. The warm-up is specified to occur at 50 km/h. When one is running the drag or backward/forward sections at 3 and 10 km/h with the gearbox engaged, the gearbox has to be disengaged to increase the speed to 50 km/h. Although this process is easily programmed into the test procedure and is accomplished by the dynamometer controls without difficulties, it adds time to the procedure each time this cycle is repeated. Each cycle of disengagement and engagement adds approximately 50 seconds to the test time. This does not seem like much time until one considers that there are 1401 brake applications in the test procedure, even without the fade and recovery section. Depending on the ambient temperature, brake size, and other factors; the delays due to the gearbox can add several minutes to hours to the required test time.

Figure 1 – Link NVH dynamometer test section

The time required for the test procedure was found to be highly dependent on the cooling air system used. Initially, testing was done using ambient air to cool the brake. This worked fine with low outside air temperatures. However, in the summer months the ambient air temperatures were higher than the required test temperatures. Now air-conditioned cooling air is used. This has resulted in significant reductions in cycle times. Perhaps consideration should be given to specifying cooling air temperatures in future revisions to the standard.

4. DISCUSSION OF RESULTS

The results obtained are summarized in Table 2 below. The maximum sound pressure level is the maximum level recorded throughout the test using a peak hold averaging. The average maximum is simply the average among the 105 tests studied. There was little difference in the maximum levels seen between the test with and without the fade module included. It is interesting to note that there was a 14,866 Hz span in the range of the maximum sound levels measured. The highest occurrence of 71.9% in noise levels above the threshold is quite high. Clearly, brakes with such a high level of noise occurrence would be commercially unacceptable in most applications for passenger cars and light trucks. Note that there was no significant difference in the maximum and average occurrences where the tests with the optional fade section were included or not.

Table 2 – Summary of measured data

Maximum Sound Pressure Level	123.7 dB(A)
Average Maximum Sound Pressure Level	101.2 dB(A)
Highest Frequency for the Maximum Sound Pressure Level	16,244 Hz.
Lowest Frequency for the Maximum Sound Pressure Level	1378 Hz.
Average Frequency for the Maximum Sound Pressure Level	6993 Hz.
Maximum Overall Percentage of Noise Occurrences	71.9 %
Average Overall Percentage of Noise Occurrences	20.9 %

Since the maximum sound pressure level occurred over such a wide range of frequencies, it is helpful to look at the distribution of frequencies at which the maximum sound level occurred as shown in Figure 2. Clearly the majority of maximum sound pressure levels occur in the

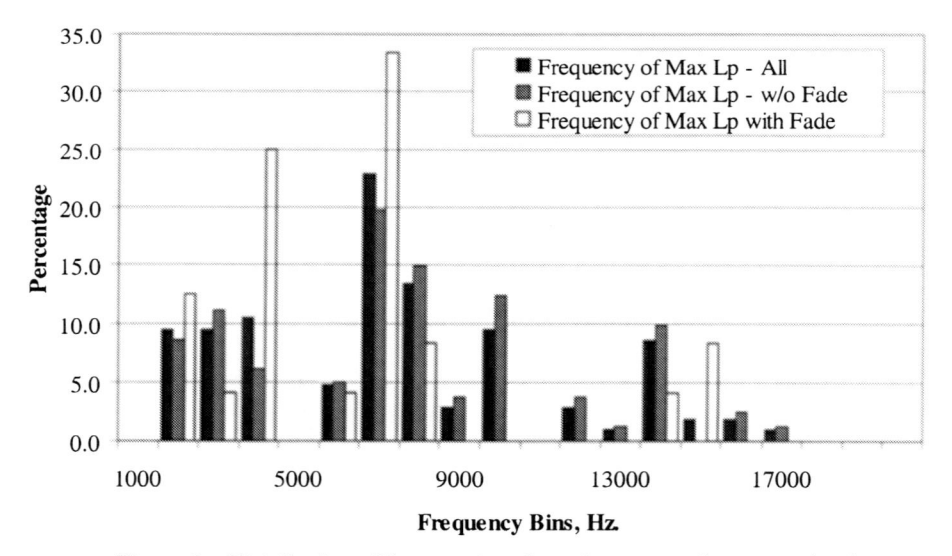

Figure 2 – Distribution of frequencies of maximum sound pressure level

6000-7000 Hz bin. The distributions are modified when the fade and recovery module is included. There is a tendency to the occurrence of higher frequency maximums with the fade section.

The accumulated data also provide a good indication of the time required to run this SAE J2521 procedure. These data are summarized in Table 3.

Table 3 – Summary of test times

	Time, Hours
Maximum Required	159
Minimum Required Without The Fade Section	19
Minimum Required With The Fade Section	26
Average Required Without The Fade Section	39.4
Average Required With The Fade Section	55

The exceptionally high maximum test time is due to a case where the required minimum temperature was nearly the same as the ambient temperature. A great deal of time was required to cool the brake to this temperature each time it was required. From these times it is clear that there is a minimum of 7 additional hours required to run the optional fade and recovery section.

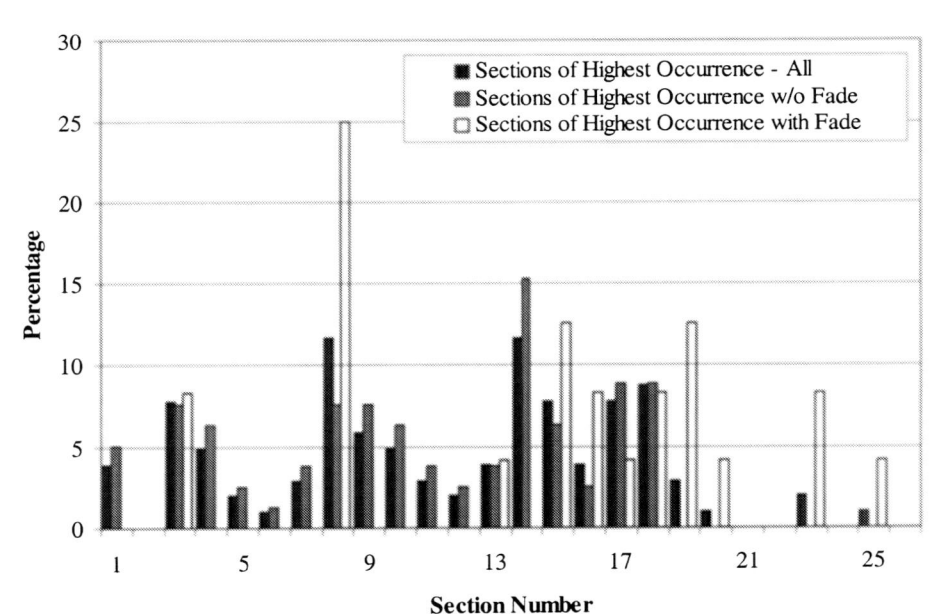

Figure 3 – Distribution of sections of highest relative occurrences

To further understand the implications of these data, the sections where the highest relative occurrences percentage was found were plotted. These data are shown in Figure 3. Clearly there is a tendency for the highest relative occurrence to occur in Section 8. This result was a surprise since Section 8 is the friction characterization segment and was not designed to generate high levels of noise. To better understand the areas of the test where the highest occurrences of noise were being generated, a second analysis was done grouping all the sections that were the same. For example, as shown in Table 1, sections 3, 8, 13, 18, and 25 were grouped together to represent the friction characterization segments of the test. Similarly, sections 4, 9, 14, and 21 were grouped together to represent the drag segments of the test. Using these segments a new distribution of highest relative occurrences was calculated and these results are shown in Figure 4. This chart even more clearly illustrates the tendency for noise to occur during the friction characterization segments of the test. For tests without the fade sections, the drag segments are most likely to produce the highest relative occurrences. The drag, deceleration, and backward/forward sections, which were the sections primary designed to elicit noise, are ranked highly next to the friction characterization section in terms of the percentage of occurrences.

In a few incidences there was considerable "green" brake noise, which can be seen in the break-in section having nearly 5% of the maximum relative occurrences. This was a controversial section of the procedure and these results would seem to validate the decision to measure noise in all segments of the test.

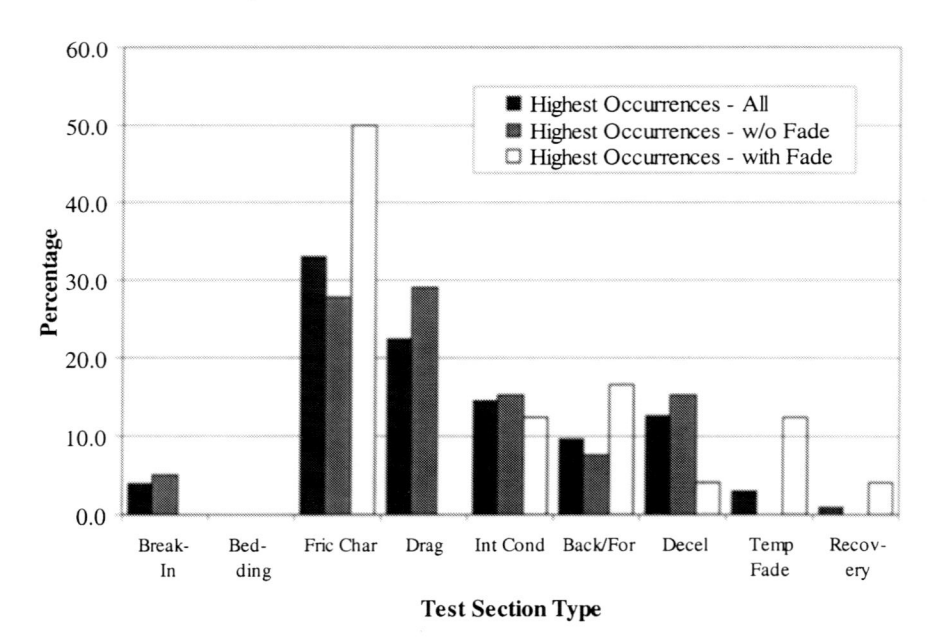

Figure 4 – Distribution of highest relative occurrence by test section type

A key issue not yet addressed in this discussion is the usefulness of the SAE J2521 procedure in predicting vehicle NVH performance. In many cases, there were not comparable vehicle tests conducted to compare to the dynamometer results. In other cases, the data are proprietary and cannot be discussed here. What follows is one case study that illustrates the level of correlation that is possible.

In this particular case, both dynamometer and vehicle tests were performed in a program to resolve a noise issue. SAE J2521 tests were conducted on the dynamometer. Tests are reported for the original brake hardware and the corrected brake hardware. The results of these tests are summarized in Table 4. Clearly, a significant improvement was achieved.

Table 4 – Summary of dynamometer test results for case study

	Original Brake Hardware	**Corrected Brake Hardware**
Peak Sound Level, dB(A)	112 at 5000 Hz.	101 at 5000 Hz.
Noise Occurrences (> 70dB(A))	58%	9%

The same hardware was also tested on the vehicle. Objective measurements and subjective ratings were collected from on-road tests. The results of these tests are summarized below in Table 5. Again, the corrected hardware was shown to be highly effective.

Table 5 – Summary of vehicle test results for case study

	Original Brake Hardware	**Corrected Brake Hardware**
Peak Wheel Well Sound Level, dB(A)	90 at 7800 Hz.	70 at 6800 Hz.
Peak Interior Sound Level, dB(A)	52 at 7800 Hz	45 at 5000 Hz
Peak Caliper Acceleration, g (g = 9.8 m/s^2)	22	1.8
Test Driver Subjective Rating	5	7-8

It is interesting to note that the 11 dB reduction seen on the dynamometer falls between the 20 and 7 dB reductions seen on the vehicle. There is no indication of an absolute correlation between the results from the dynamometer and the vehicle. However, there is an indication of correlation between relative results.

5. SUMMARY

The SAE J2521 procedure has been demonstrated to elicit significant noise from the brake system. The distribution of noise occurrences by section indicates a clear difference in the noise generated by this procedure in comparison to the AK Noise procedure. The high occurrences of noise in the friction characterization and deceleration stop portions of the test demonstrate this difference since these types of sections are minimally represented in the AK Noise procedure.

The propensity for the friction correlation section to generate the highest noise occurrences is a surprising result from this analysis. Since this test is run at the relatively high-speed range of 80 to 50 km/h, it may be that the speed ranges for other portions of SAE J2521 should be reconsidered. It is suggested that this topic be reviewed for future revisions to SAE J2521.

The question remains to be objectively addressed as to how well the results of this procedure correlate with vehicle noise performance. There is not sufficient objective data to evaluate this correlation. The experience from these tests was that the customers felt the results did well in indicating vehicle noise performance. The single case study illustrates there is potential for SAE J2521 tests to accurately represent vehicle trends. However, much more data needs to be accumulated to show a true statistical correlation. There is a need for a standard vehicle test procedure under which objective and subjective data can be accumulated to provide a database for correlation with SAE J2521.

6. REFERENCES

1. "SAE J2521 – Disc Brake Dynamometer Squeal Matrix," *SAE International, Warrendale, PA, 2001*
2. Thompson, James K., "Summary of the Brake Noise Recommended Practice Draft Developed by the US Working Group on Brake Noise," *Proceedings Brakes 2000*, Leeds, England, July 2000.

APPENDIX: SUMMARY OF SAE J2521 TEST RESULTS

Table A1 - Summary of SAE J2521 measurements without the optional fade module

Test No.	Time Required, Hr	Max Lp, dB(A)	Max Lp, Freq, Hz	Max Lp, Section	Overall % Occurrences	Overall Occurences	Highest Rel % Occurrences	Highest Rel % Occurrences, Freq, Hz	Highest Rel % Occurrences, Freq, Hz	Highest Relative Occurrences, Section
1	57	113.1	15625	4	1.4	21	10	2100	15625	11
2		85	5500		2.4	38	20	3000	5500	18
3	19	106.5	7675	13	22.5	220	83.3	7675		8
4	159	121.9	9868	9	36.6	508	100	13715		3
5	34	106.8	9035	9	25.5	356	50	13500		3
6	116.5	112.7	9164	4	16	223	100	9800		1
7	39	114.5	9740	14	22.4	312	66.6	9965		10
8	52.5	116.7	9580	14	51.6	719	100	8395		10
9	57	113.1	6152	9	21.9	305	83.3	8550		18
10	67	115	9868	9	29.6	412	100	9708		18
11	91	117	13649	14	44.5	619	91.9	10000	13700	10
12	32	90.5	1900	11	0.7	10	3.7	1650		7
13	27	102.9	7175	14	4.8	69	9	7150		14
14	29.5	82.4	9475	5	10.7	153	54.1	9100		5
15	29.5	77.9	1450	12	0.6	8	2.7	1450		12
16	73	109.9	6216	14	43.7	626	100	8500		8
17	35	111.1	6216	14	19.6	280	79.1	9000		10
18	32.5	106.2	9708	17	20.9	299	100	9050		3
19	28.5	104.7	9035	9	7.9	113	66.7	9035		3
20	25.5	99.7	1875	6	2.2	31	6	1900		6
21	25	76.3	1650	12	0.1	2	1.8	1700	17800	7
22	31	76	1650	12	0.1	1	1.9	1475		17
23	27.5	87.6	2075	4	1.5	21	7.1	2100		4
24	29.5	113.7	8425	14	54.3	776	100	4400		3
25	27	77	4000	4	0.1	1	0.3	4000		4
26	25.5	93.2	5625	14	2.1	30	6.3	5675		4
27	35	75.2	1650	7	0.1	1	0.3	1950		14
28	25	79.6	13975	7	0.4	5	4.6	14000		7
29	25	91.9	2275	14	1.2	17	6	2300		14
30	37	112.3	5550	4	56.7	811	100	9350		3
31	26	94	11200	14	46.6	666	83.3	11250		13
32	41	100.6	5400	9	8.5	122	17.6	5400		9
33	26.5	92.6	13600	14	0.6	9	1.5	5450		14
34	35.5	109.9	6152	14	42	600	100	8138		13
35	42	109	6100	14	31.5	450	68	7500		14
36	80	123.7	6248	9	51.4	737	100	8500		18
37	36.5	121.9	13361	14	71.9	1028	100	13750		8
38	32	103.8	8266	14	13.8	197	59	8200		5
39	29	121.6	13297	14	68.4	978	100	13500		15
40	39	110	7850	14	38.5	551	77.8	7750		17
41	27	101.9	9932	14	34.5	493	83.3	9800		15
42	67	112	7497	16	25.3	362	68	6500	14500	11
43	29	96.7	13000	4	0.8	12	3.3	13000		1
44	42	87.3	2243	14	5.2	73	14.2	2275		9
45	53	79.3	16244	9	0.1	2	0.3	16400		9
46	29	97.6	15892	16	8	111	40	15900		16
47	29.5	92.8	7337	12	2.7	39	16.6	14600		18

Table A1 - Summary of SAE J2521 measurements without the optional fade module

Test No.	Time Required, Hr	Max Lp, dB(A)	Max Lp, Freq, Hz	Max Lp, Section	Overall % Occurrences	Overall Occurences	Highest Rel % Occurrences	Highest Rel % Occurrences, Freq, Hz	Highest Rel % Occurrences, Freq, Hz	Highest Relative Occurrences, Section
48	32	105.9	7241	11	10.5	150	32	7600		16
49	33	111	7200	9	9.2	131	29.1	13300		15
50	30	111.9	6800	12	51.4	735	70.7	7000		14
51	31	107.4	6550	15	26.6	380	63.9	9000		14
52	30.5	92.3	2050	14	2.3	33	6.7	2075		9
53	23.5	83.1	7305	7	0.9	13	4.6	5575		17
54	34	94.4	6475	9	2.6	37	6.3	6675		9
55	23.5	83.1	7305	7	0.9	13	4.6	5575		17
56	35	72.6	13040	4	0.1	1	0.3	1090		4
57	42	107.4	7369	14	13.2	188	45.8	3500	7433	15
58	25	118.2	6650	4	63.3	905	75.9	6600		9
59	39	111.7	13617	14	63.4	906	91.7	11275		10
60	55	110.9	13265	14	55.5	793	83.3	17878		15
61	29	100	6825	1	11.4	164	36.7	6825		1
62	28.5	115.3	6800	1	25.3	363	100	3800		18
63	41	106.7	2475	9	22.3	319	36.8	2475		11
64	51	106.3	3925	14	21.2	303	40.2	16275		14
65	32	112	2475	11	49.9	713	100	3825		8
66	40.5	101.6	3825	14	42.9	612	100	3825		8
67	19	89.6	8475	9	0.8	11	16.7	1850		18
68	46	104	2025	14	7.1	102	22.1	9700		14
69	20	77.1	3556	16	4.6	66	33.3	3550		8
70	23.5	75.6	2750	16	0.8	12	4.6	1650		17
71	25	74.6	1750	17	0.4	5	2.8	1750		12
72	25	94.8	3621	16	14.3	205	26.9	3714		17
73	25	102	11951	17	25	357	54.1	15900		14
74	25.5	106	11727	17	29.5	422	53			14
75	68	115.1	7890	14	27.5	393	83.3	7722		13
76	33	112.1	2560	15	24.7	353				
77	52	112.4	6150	6	69.1	990				
78	27	102.7	6700	4	5.2	74	11.5	6700	6500	4
79	47	101.8	8000	4	13.5	193	32.4	11575		17
80	38.5	106.4	6150	14	11	158	22.2	5000	6500	14
81	67	104.9	6775	14	23.6	339	88	1500		1

Table A2 - Summary of SAE J2521 Measurements with the Optional Fade

Test No.	Time Required, Hr	Max Lp, dB(A)	Max Lp, Freq, Hz	Max Lp, Section	Overall % Occurrences	Overall Occurences	Highest Rel % Occurrences	Highest Rel % Occurrences, Freq, Hz	Highest Rel % Occurrences, Freq, Hz	Highest Relative Occurrences, Section
1	48	102.9	14975	14	16.9	324	42	10950	14950	16
2	47	111.7	3375	14	41.9	802	100	3350		8
3	45	102.9	14975	14	5.9	111	32	14500		16
4	49.5	112	3375	7	64.7	1242	100	3375		8
5	48.5	107	3375	15	31	595	100	3375		13
6	55.5	102	3350	14	9.7	185	41.6	3325	14900	15
7	68	112	3375	9	60.2	1153	100	14600		18
8	39.5	105.1	13925	25	6.7	125	20	3000	4500	23
9	39	106.1	6650	24	10.8	201	50	14050		25
10	40.5	99.3	5650	23	5.1	95	36	5675		23
11	43	109.8	6250	21	29.6	549	83.3	3425		8
12	45	95.5	3400	25	11.2	208	83.3	3450		8
13	40	108.4	6800	9	22.7	434	100	1350	7100	19
14	33	82	1400	7	2.9	56	100	2900		3
15	56	109.6	6675	14	37.6	720	100	6625	12525	8
16	124	110.2	6675	9	53.6	1027	100	6175		19
17	55.5	97.1	7150	21	2.3	44	93.3	1625		18
18	118	109.7	7275	14	35.8	687	83.3	7125		8
19	48.5	105.7	6925	21	5.3	102	94.4	1375		20
20	131	101	1378	19	0.5	9	26.6	8300		19
21	46	74.8	2115	17	0.3	5	8.3	2200		15
22	26	77	1675	12	0.4	6	1.9	2075		17
23	31	99.9	6300	14	1.4	20	8.3	6350		15
24	43	113	6575	14	58	1112	100	6575		3

Towards more accurate brake testing

E J SLEVIN and **H SMALES**
Microface Limited, Newburgh, UK

SYNOPSIS

Classically, all brake designers have used vehicle and dynamometer testing to evaluate new designs of brake. Often there is no vehicle to fully test the design, and when it does arrive, the brake design is well established and changes are more expensive to implement and re-design quickly.

There are often considerable differences encountered between the initial dynamometer and the actual vehicle performance. Designers are aware of this and invariably learn over years of experience which of the dynamometers they own give the most realistic results and add a margin of safety or they over-engineer the design so that it will enjoy a smoother passage through realistic trials. On the other hand vehicle testing is subject to the variance of the weather, the state of surface and not least the driver.

Neither dynamometer nor vehicle methods therefore are ideal.

Confusingly enough, the poorer the perceived performance of the dynamometer the better the correlation with reality in many cases. For instance, the dynamometer with poor main bearings and gearboxes, giving low run-down times is probably more realistically following the losses in a vehicle than a perfect dynamometer. This was particularly brought to the forefront of consideration when in 1993 the "Friction Free" dynamometer with real time compensations for losses was introduced.

Many users complained that although the stopping distances and times had to be measured in milliseconds and millimetres to detect errors, the brakes required excessive levels of cooling flow to keep the temperature rise during braking similar to the vehicle.

By incorporating precision measuring equipment and software it is possible to overcome the limitations of dynamometer testing by driving/retarding the main motor during testing to add or subtract energy from the braking application making them more accurately repeat the true vehicle test.

In addition the interface between tyre and road can have a significant effect on the work done in decelerating the vehicle. This can also be simulated by the latest equipment.

1 BACKGROUND AND HISTORY

As far as can be established, dynamometer testing of friction materials started in 1897 with a water-powered machine (1)
Over the succeeding years improvements in every aspect of the machines has taken place with important milestones such as

- a) Heenan Eddy Current variable speed drives 1958
- b) Automation to remove human element and drudgery 50's and 60's
- c) Constant Torque Control (2)

None of the above improvements achieved fast enough response rates for the main source of power (usually an electric motor) to be adjusted whilst the brake is applied and thus either subtract or enhance the flywheel energy which is still used as the main source of energy to be dissipated during braking.

All machines at present still use a flywheel as the main source of energy during braking. This is because generally the brakes can stop a vehicle a great deal quicker than the engine can accelerate it, so if you only use a motor to test brakes it has to be inordinately large.

However the motor can make a significant modification to the raw flywheel inertia to improve the vehicle model.

Typically for example the deceleration torque of a passenger vehicle can amount to 5000 Nm and a typical motor used on such a dynamometer would produce at best 1500 Nm Torque. This is a useful amount that can be used to add or subtract from the torque produced by the flywheel. This can be used in various ways as follows:-

- d) To reduce the total number of flywheels required by interpolation between widely spaced steps. It is unusual to require more than three removable flywheels in a new dynamometer design nowadays. This can be +ve or –ve
- e) Compensate for mechanical losses such as belt drives and gearboxes. The use of gearboxes greatly simplifies the design of test rigs by allowing smaller flywheels to be used. Because of the losses in gearboxes their use was frowned on in the 70's and 80's. However now they can be reintroduced because the losses they introduce is compensated for. Always +ve additional energy required
- f) Correct for vehicle rolling resistance including engine drag. Always –ve energy removal

g) Correct for wind resistance on the vehicle. Always –ve, energy removal

h) Correct for work done on a gradient. This is already simulated at constant drag speed and torque but simulation opens the possibility of speed, gradient and deceleration all changing simultaneously for more complete modelling. +ve or –ve.

i) Correct for different states of vehicle loading although if this is large the motor may be incapable of simulating this effect in full. +ve and –ve.

j) Model tyre losses. This is very significant and the new element is considered in detail in this paper. Always –ve.

k) Model other effects in the future.

2 HOW INERTIA SIMULATION WORKS.

The test is undertaken on a conventional dynamometer, this is best achieved with a double-ended machine as this allows the interaction between the front and rear brakes (Load sharing) to be included in the simulation automatically. It is hoped that in the future this effect will also be able to be modelled to allow single ended machines to be used to greater effect. (i.e. in effect k above).

The computer monitors the performance of the brakes in real time and uses the main drive, which originally was used to accelerate the flywheel inertia between tests, to correct the model from a simple dynamometer to a vehicle simulation. This involves adding energy back in to compensate for losses in the dynamometer and any other positive inputs such as simulating braking on a down hill slope and taking out any energy which assists braking such as the wind and the overdriven engine.

2.1 Operation of Inertia Simulation.

A simplistic view of the requirement to simulate inertia on a dynamometer is to measure the total torque being produced at the brake, calculate the speed which would have been achieved with this torque and demand this speed from the D.C. motor. This method produces a very slow acting control loop of circa 1 Hz that is about the same as can be achieved by a manual operator. Anyone who has tried to control a dynamometer to give just constant torque will know how difficult this is with a high deceleration stop. For instance, with a 60 tooth speed measuring wheel it takes a one second gate to measure speed to ±1 rpm. With this kind of long time constant the operation of inertia simulation is impossible with any accuracy.

A radically different approach is required to overcome this problem. Firstly the main torque is measured as above but instead of demanding a speed from the drive, the current required for the motor is calculated and demanded instead. This current loop is much faster acting than the speed loop and the approximate torque appears within 30 milliseconds rather than seconds. Unfortunately the torque produced in the motor is related to current in the armature by complex functions which must take into account:-

(a) Speed to assess iron losses in the motor,

(b) Field current (altered due to field weakening),

(c) Non-linearity in current response

(d) Bearing and other losses in the total machine.

A first approximation is for the software program to include a model of the motor so that it can predict the current required for a given torque. This model is typically circa 3 % accurate at worst but is a good first estimate and allows time for a more accurate calculation to be made.

2.2 Accurate Inertia Simulation.

The speed required of the dynamometer has to be accurately and quickly controlled. The measurement of speed is either inaccurate and fast such as from a tacho generator or accurate and slow from a 60 tooth wheel. Neither of these is suitable for use in a feedback loop to control speed, as the requirement is fast response coupled with high accuracy.

The key to a more accurate calculation is accurate distance measurement with sufficient resolution so that speed can be calculated from rate of change of distance.

This requires at least 1000 pulses per rev from a two phase shaft encoder. These are electrically processed to give 4000 counts (sampled by using several sensing heads) per rev and fed to a dedicated microprocessor which provides the necessary digital signal processing. Once a very accurate speed measurement is made this is compared with the desired speed to achieve an iterative correction to the current demand.

To further improve accuracy, a position control loop is provided in the design. This continually uses the two-phase signal from the pulse generator to provide the absolute position of the flywheels and the number of rotations to date. Comparison is made with the required distance travelled and a servo loop included to control position in real time, providing the final control loop to achieve virtually perfect inertia simulation. Once accomplished other modelling aspects follow naturally.

A simplified diagram of the operation is included as Fig 1.

Figure 1

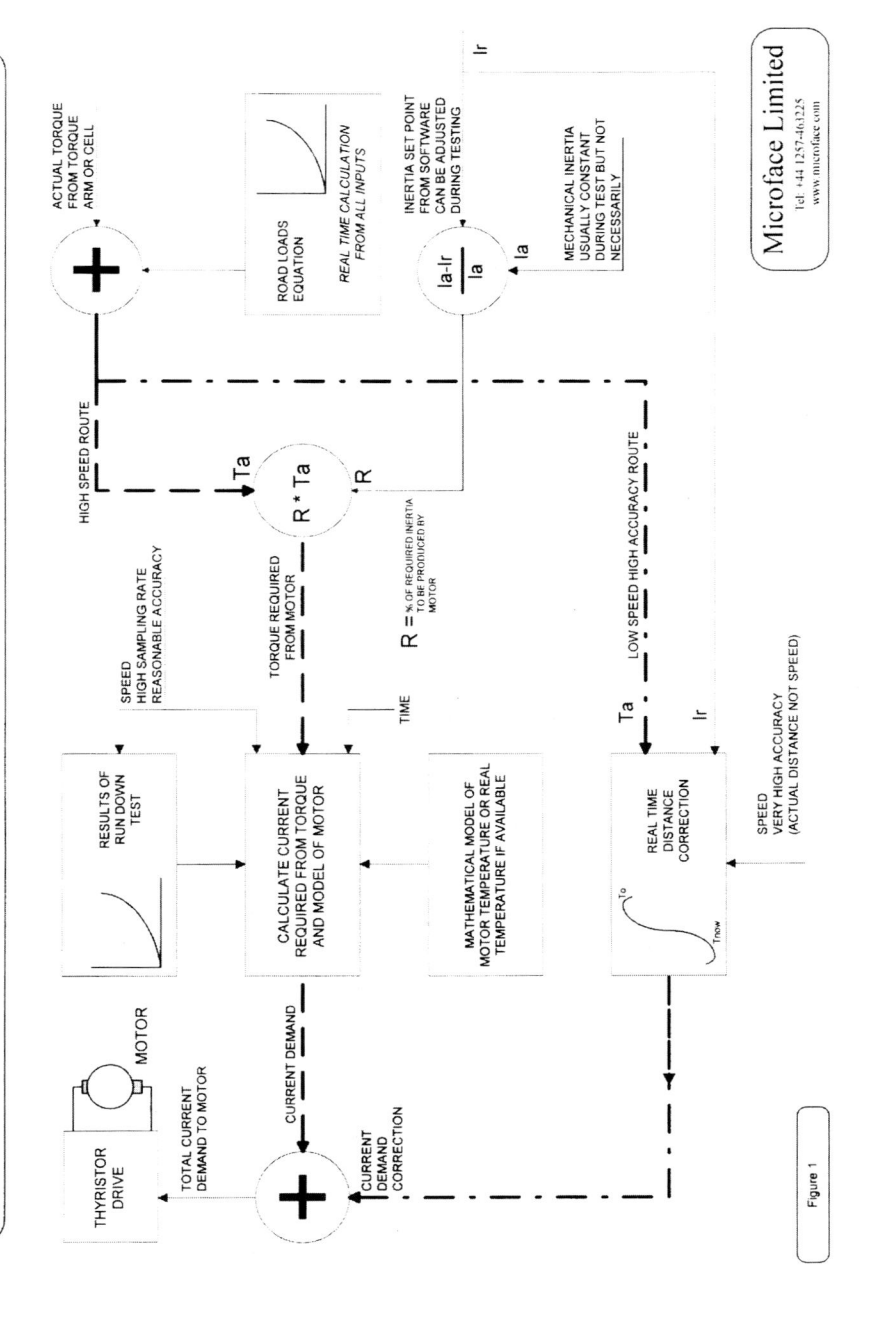

2.3 Operation of other features.

As is shown from Fig 1, once a good inertia simulation has been achieved the simulation of other effects follows on naturally as indicated by the addition of the road load equation to the measured torque. All that is required is to calculate in real time the torque that would be produced by each item you wish to simulate and to add them (Or subtract them) from the measured torque or the torque required from the motor. (not shown on the diagram for simplicity)

For example wind resistance is known and produces a drag related to vehicle speed. When added to the actual measured torque a total deceleration effect is obtained. Similarly road gradient can be taken into account.
The biggest error by far is the work done by the tyres under braking conditions

3 MODELLING PARAMETERS

3.1 Tyre losses when braking

This accounts for by far the largest difference between simple dynamometer and vehicle testing. Even modest braking torque without considering the situation where there is obvious slip between the tyre and the road can cause a noticeable difference.

This extra work done is manifested as reduced torque and speed at the brake. The torque seen by the brake is reduced by rolling resistance which increases with higher braking forces. The speed as seen by the brake is reduced by the slip between the tyre and the road.

As braking occurs, the rubber on the tyre is first pulled towards the rear of the vehicle and then released from tension as the contact rubber comes out of contact with the road. At this point it springs back to its original position. The heat output is clearly understood and datalogged in various rigs and vehicles by the tyre manufacturers. This effect is greater on the front brakes of the vehicle than on the rear as the braking forces are higher. Since most wear occurs on braking components at high braking forces, a difference between the model for front and rear brakes is necessary (3).

Fig.2 is an example. Here the slip between the tyre and the road is plotted at various decelerations. This is somewhat simplistic as the weight transfer to the front tyre changes the load during braking which adds another variable to the model and on high performance cars aerodynamics may add significant down force. However these can all be taken into account in the model of the vehicle in real time.

Braking on poor surfaces is not considered, as the work done is significantly limited and neither causes a problem of heat input nor torque limitation, hence testing for wear and performance of the brakes is only necessary on good surfaces because these are the worst case.

Fig 2. Tyre slip data (4)

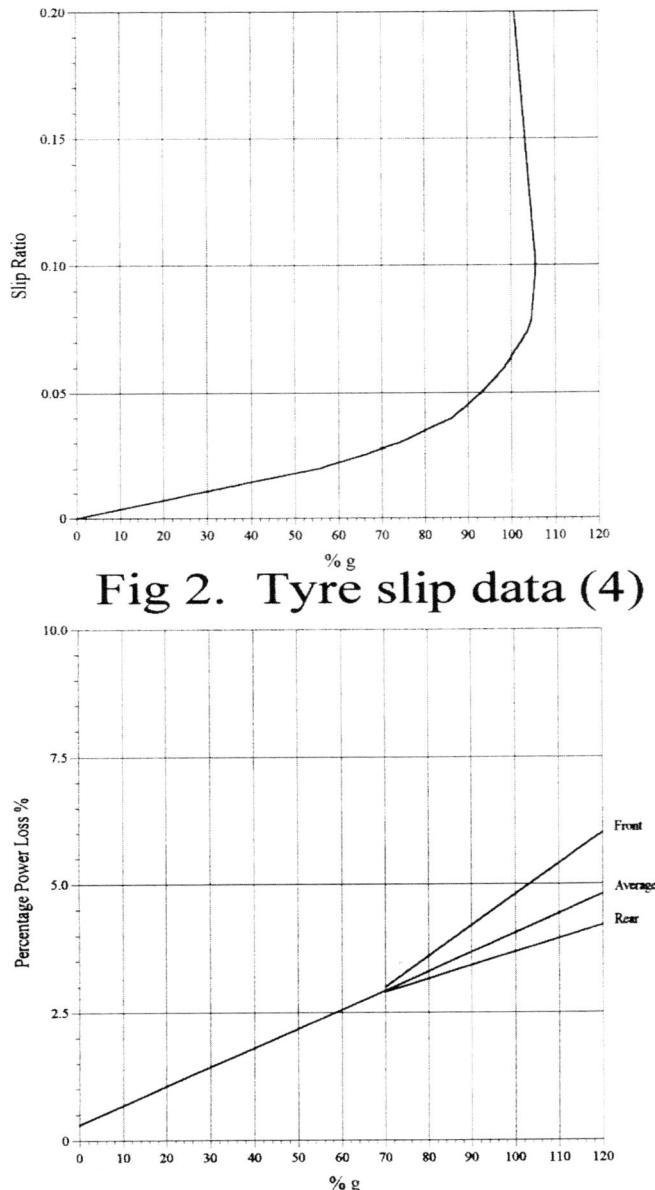

Fig 3. Tyre rolling loss data (5)

Brakes are usually tested at various decelerations from say 10% to 110 % where slip would occur on even the best surface.

The graphs (Fig.2 and 3) show that :-
>At 10% g the work done by the tyre is 0.6% +0.3% of the total work
>At 60% g the work done by the tyre is 2.6 % +2% of the total work
>At 100% g the work done by the tyre is 3.8% +6.4% of total work.

This is the work done in both rolling resistance and slip added together to produce a total work done by the tyre.

At high decelerations the load transfer to the front axle is greater allowing even greater work to be done by the tyre. Ideally in a totally correct model the work done by the tyre should be included as a drop in speed of the test brake. Since the test brake is rigidly fastened to the flywheels that cannot quickly be changed in speed in response to torque change from the brake, we have to do the next best thing. This is to simulate the work done by the tyre by taking extra energy out of the flywheels with the main motor after first reducing the initial speed in line with a prediction as to the likely losses. Unless the speed sensitivity of the brake is particularly acute this is sufficiently accurate for a good model. For the purists amongst us if this is not considered sufficiently accurate, a general correction to the entry speed of say 5 % can be made if it is considered that 5% slip at the test torque is likely, then make a more accurate correction in real time.

3.2 The tyre model
A slip model and a resistance model are required.

3.2.1 Calculation of Slip.
There are several methods of modelling a tyres slip in real time. By far the most suitable in this case because of its easy automation is that presented in (3).

This uses an empirical formula, which fits experimental data well. The following inputs to calculate slip are required.

a) Six constants which are derived from tyre test data for any particular tyre type. These can be entered into the model before the test is run.

b) The load on the tyre in kN. Easily calculated in real time from the static load on the axle plus any weight transferred to or from the axle by the overturning moment of the vehicle from a knowledge of vehicle mass, height of the centre of gravity, and the deceleration.

c) The vehicle camber angle which is usually constant for different wheel loads but not necessarily

d) The vehicle speed.

These values can all be updated with ease every 10 milliseconds resulting in a slip value being derived. The slip speed times the torque gives the power absorbed by slip. The motor then mimics this part of the work done in helping braking.

3.2.2 Work done in rolling resistance.
There is a marked increase in rolling resistance at higher braking forces. (5). This is measured on special tyre test rigs usually with two wheels both turning one common drum. Power is used to turn one wheel and power is measured at the other tyre with a controlled load such as a motor. This results in the loss graph of Fig. 3. It will be seen clearly that the tyre resistance increases as the load is increased. This is not as great as the slip loss but is still significant. This graph uses units of torque reduction to separate out this effect from speed reduction due to slip. In both Fig. 2 and Fig. 3 losses occur simultaneously when braking. To model this in software a simple spread sheet table with resistance in N force given in the table at various speeds and decelerations is required. A table as low as 10 by 10 is quite adequate, as linear interpolation is also a quick real time process. This torque is added to the torque required from the motor for other purposes to give a total torque demand.

3.2.3 Operation of the model
Once the energy which would have been dissipated in slip and resistance by the tyre if the test brake had been on a vehicle has been calculated these can be simply added together and converted to an equivalent torque for the speed at the time. This is then added to the actual measured torque in the model to correct both the 30 millisecond fast current loop and the slower outer loop.

4 CONCLUSIONS
The aim of the paper is to present a case for increasing the accuracy of dynamometer testing, eight sources of error (1.d through 1.k) have been identified. Today we routinely correct for six of them on single ended dynamometers. A method has been presented for correcting for tyre losses which have been shown to be the largest error. To accomplish this, data on the tyre, usually from the manufacture, is required. This information is often provided reluctantly and to date two downright refusals have been received. However in future the brake manufacturer will need this information in order to produce the most realistic test. No longer is it possible for each discipline to work in isolation. The brake and friction industry must take the lead in considering the interaction of these elements.

It is already the case that the friction industry take the brunt of the acoustic testing for NVH for the complete corner on its shoulders even though it is often the design of other components and systems which affects the noise. It seems likely that the inclusion of the tyre in the equation will now also be necessary. Significantly this can be done without physically attaching the tyre to the test rig which is a great saving by allowing the use of the existing dynamometers. All that is required to bring this about is the addition to the average dynamometer of a more accurate encoder, at negligible cost, and of course some modern software. Proving all this as an adequate and complete model will probably be the subject of many papers in the future.

It is clear that the brake designer will in future not be able to undertake brake design in isolation from an understanding of all the other forces involved when braking and real time modelling of software on a dynamometer will become an essential tool in achieving these improvements.

References

1. G. A. Harper. Brakes and friction materials. Published by MEP ISBN 1 86058 127 7
2. H. Smales. Friction materials – Black art or science. I Mech E 1994 Chairman's Address 10/8/94.
3. E. Bakker, H. B. Pacejka and L. Lindner. A new tire model with an application in vehicle dynamics studies. SAE Autotechnologies Conference and Exposition 01-23-1989 Monaco No. 890087
4. Data extracted from Fig. 5 of (3) above for an 8,000 kN tyre load
5. A. R. Williams. The influence of tyre and road surface design on tyre rolling resistance. Technical Paper Series, Institute of Petroleum, Number 81/003, April 1981.

Holographic interferometry used to investigate noise from a drum brake mounted on a half vehicle test rig

J D FIELDHOUSE, C TALBOT, C BEVERIDGE, and W STEEL
School of Engineering, University of Huddersfield, UK

ABSTRACT

The paper is a continuation of earlier work (1) and so reviews the characteristics of a drum brake when generating noise on a ¼ vehicle test rig and compares the results to the same drum brake mounted on a ½ vehicle test rig. Holographic interferometry is used to provide whole body visual information of the modes of vibration of the component parts. Although the principal frequency discussed is 850Hz there is consideration and comment provided for frequencies of 1400 Hz & 4600 Hz. With the accumulated information it was possible to predict other possible unstable frequencies and although these were not observed within this series of test the frequencies have been observed on earlier work.

The suspension system plays an important role at the lower frequencies but less so at the higher frequencies. It is suggested that the mode of vibration of the backplate, and its associated interface mounting with the suspension, may provide an answer as to why this is so. Additionally it is suggested that there is an influencing relationship between the trailing arm, the spring pan and the main cross beam, the spring pan exhibiting a complex mode. A series of time related holograms allows the phase relationship of the spring pan structure to be established at 76°.

A noise "fix" is suggested for the low frequency noise and additional suggestions are made to reduce the influence of the suspension system through basic design changes.

1 INTRODUCTION

Low frequency drum brake noise may be experienced when a vehicle is allowed to "just move off" whilst the brake is applied - such as brake controlled roll down an incline.

Dynamometer testing has recreated this noise but effective "noise fixes" on the dynamometer have not proved successful when applied to the vehicle. It is believed that the difference between dynamometer testing and vehicle testing is the influence of suspension members.

Within the braking fraternity there tends to be much discussion regarding the influence of the suspension system on the propensity of a brake to generate noise and when its existence should be acknowledged in any analysis or testing. It is clear that the opinions will vary depending on whether a disc or drum brake is being considered and indeed on the value of the frequency being generated. Much of the reason for such a diversity of opinion results from the test procedures carried out. If a noise is generated on a vehicle but not on the dynamometer then the investigator will be inclined to attribute the difference to the suspension system. Before such generalised sweeping statements are made there should be further considerations given to principal feature differences between disc and drum brake systems.

If a disc brake is considered first it may be argued that the knuckle assembly separates the brake parts from the suspension system and since this is a fairly substantial brake component element it may isolate disc brake instability from the suspension. Indeed this may be the case at the higher squeak frequencies, greater than 4000Hz, where even the calliper may not be relevant since its natural frequencies are too low. In such high frequency cases it is generally accepted that the instability arises from the friction pair comprising the disc and pads. This may be correct at the higher frequencies but when the frequencies considered are of squeal value, 2000Hz, then the influence of the calliper may become significant and in particular the influence of the calliper carrier bridge. Certainly with a disc 3 diameter mode of vibration the calliper body is seen to deform in a symmetrical torsional manner with relatively high amplitudes of excitation (2). At even lower frequencies, the 600Hz "moan" region, the calliper body is known to deform considerably as whole body motion (3). In such cases it would be reasonable to assume that this deformation would tend be transmitted to the suspension system through the knuckle and as a consequence the suspension could then become part of system instability.

With drum brakes the influence of an unstable brake on the suspension is more direct and more complex than with disc brakes. The major brake component parts which are known to contribute to noise generation are the drum, brake shoes and the backplate. The vibration transmission path would be direct from the backplate through its common mounting with the bearing housing to the suspension beam bracket, the latter being of pressed steel form.

It has been shown [1] that at low frequencies, around 1000Hz, the backplate exhibits high amplitudes of vibration with a 2 diametral mode of vibration whereas the drum appears to exhibit little excitation. This backplate mode increases to a 4 diameter mode of vibration at 2500Hz with a very reduced amplitude. This increased mode order and amplitude of excitation is a significant feature in the transfer of excitation forces and displacements. At 5250Hz the backplate mode becomes less clear whereas the drum mode becomes significant - exhibiting a 3 diameter mode of vibration. At this highest recorded frequency the backplate generally exhibits a 4 diameter mode combined with a circumferential mode, this often tending towards a mode which is almost random in nature. Since the backplate mode at these higher frequencies comprises a large number of low amplitude antinodes, spaced radially and circumferentially about the backplate, it is felt that its influence on the suspension would be generally minimal.

At the lowest frequencies, 1000Hz and less, the influence of the backplate on the suspension must be a serious consideration. The work considered here is concerned with a drum brake noise frequency in the region of around 800 Hz, and so the suspension system is included.

2 REVIEW OF ¼ RIG AND RESULTS

2.1 Test Rig
The fundamental details between the ¼ and ½ test rig are the same so the rig discussion at this point will embrace both rigs – the difference with the ½ vehicle test rig being outlined later.

The test rig accommodates a ¼ vehicle suspension and brake as shown generally in Figure 1. It will be noted that the beam was mounted rigid at its centre but this rigid mount was subsequently replaced with a rotationally free mount without any significant change in noise propensity or generated frequency. The second suspension mount comprised a small bracket and pin which supported the elastomer bush assembly within the trailing arm section of the beam. The drum and wheel centre mount was driven from a variable speed drive through the vehicle's shortened driveshaft.

Figure 1: General view of test rig showing mounting of 1/4 vehicle rear suspension and brake.

Figure 2: General view of drum mounting with the support cranked to recreate wheel position under drum. Vertical guides for the sliding section and the curved foot on the base to allow the whole head to move with the swinging suspension arm.

Because the holographic technique requires visual access to the drum it was not possible to mount the assembly on its wheel and tyre. To replicate the wheel and tyre as close as possible it was necessary to support the drum with a stiffness equal to that of the tyre yet still allow it to rotate freely on its own bearings. It was also necessary to allow it to swing in a manner which would accommodate the geometry of the suspension arm. To achieve this a stub shaft was mechanically located and fixed concentric to the drum centre – Figure 2. The stub shaft was then supported in self aligning rolling element bearings which were housed in a support designed to replicate the drum, wheel and tyre mounting as near as possible to that in the vehicle. The mounting comprised a sliding section and a base section.

The sliding section was cranked to support the drum geometrically as would the tyre and wheel yet has vertical movement provided by linear guides in the support base. The guides are aligned with the effective tyre support axis and bear on two stacks of disc springs which equate to the tyre stiffness. The support base is curved with a radius equal to that of the tyre so as to allow the housing to behave dynamically during trials as if it were mounted on the wheel. The suspension beam is then loaded against the tyre springs using the vehicle main spring which is mechanically compressed an appropriate amount equal to the vehicle weight.

Figure 3: General front view showing mirrors positioned to provide additional information of the drum side, backplate, spring pan and main beam.

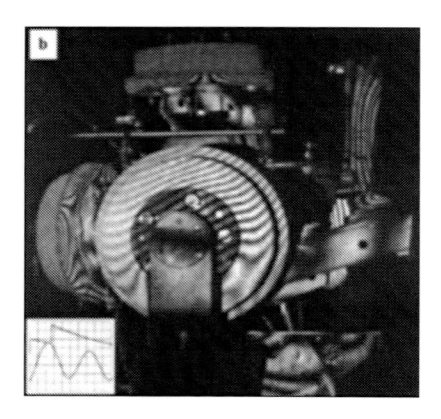

Figure 4: Typical image of a drum brake generating 960Hz on a ¼ vehicle test rig. One from a series of 9.

To maximize the visual information available large mirrors were positioned around the assembly to provide views of the drum side & backplate (left hand mirror and top mirror), spring pan (lower mirror) and the main beam (right hand mirror). Figure 3 shows the positioning of these mirrors and serves to familiarise the reader with the individual brake parts when considering the holograms.

2.2 Overview of Results – ¼ Vehicle Test Rig:

The results from this investigation are presented in detail in earlier work (4) but basically considers an excitation at 960Hz.

A series of 9 time related interferograms were taken to establish the information presented, a typical image shown in Figure 4. It is clear that all the component parts contribute to the dynamic instability of the system as they all exhibit displacement fringes.

Two features were considered in some detail, the backplate and the main cross-beam. The spring pan had deflection but was not analysed or referenced to other features although comment was made regarding its probable deformation.

2.2.1 Backplate

The backplate exhibited a 3 diameter mode of vibration and by observation and measurement of a single node it was possible to determine its direction and rate of movement about the backplate. Figure 5 shows the general movement where it is seen that it is not linear but progresses at a dual rate as shown by the two slopes. This movement is different to that observed on disc brakes where the disc mode movement is smooth, almost sinusoidal. The principal difference between the two systems is that the backplate is constrained in its movement by its connection to the trailing arm and subsequently the main beam. The mode movement is therefore influenced by the stiffness of the connected components. That is not the case with a disc.

The high rate of movement within the central section was measured at 2.22µs/deg and the lower rate at 5.56µs/deg giving an average rate at 3.9µs/deg. The rate for 960 Hz would be 4.3µs/deg if it were moving at a rate equal to twice the frequency of vibration divided by the mode order of 3 in this instance.

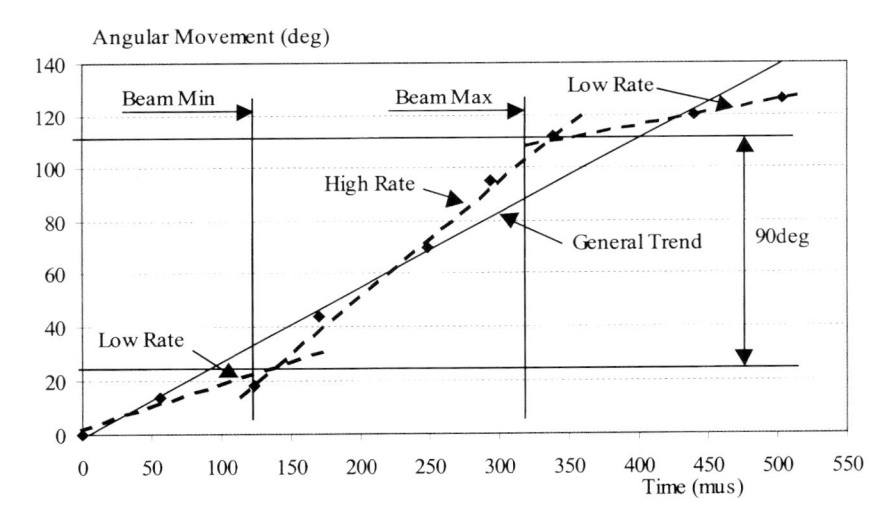

Figure 5: Graph showing rate of movement of node position about backplate - clockwise as viewed in Figure 5.

It is significant to note that the high rate of movement of the backplate node begins approximately as the main beam is at minimum amplitude and continues until the main beam is at maximum amplitude. This takes about ¼ of the cycle. Conversely the low rate of backplate mode movement takes place as the main beam reduces in amplitude.

2.2.2 Main Cross Beam

A typical main beam deformation is shown in Figure 6. By enlarging the image it was possible to count the number of fringes at each stage and plot the number (amplitude) against time, this being shown in Figure 7.

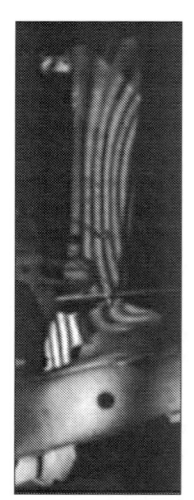

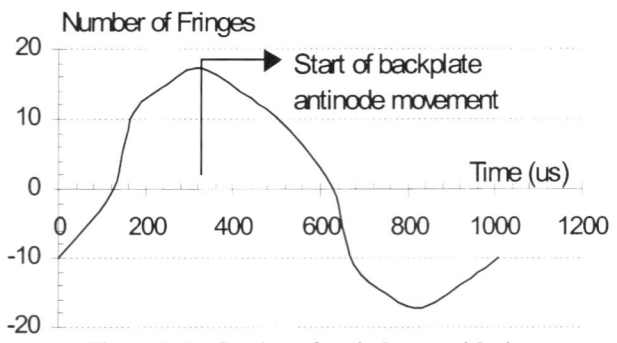

Figure 7: Deflection of main beam with time.

Figure 6: Typical image of main cross beam during vibration, the number of fringes indicating amplitude.

It was significant to note that when analysing the beam deflection it was observed that the backplate antinode movement, as seen in the side mirrors, starts as the main beam reaches maximum amplitude. This is the same point in the cycle that the observed backplate node starts to move at its low rate. In essence the observed node moves at a high rate as the side antinode increases in size and moves at a low rate as the antinode rotates about the backplate. It was felt there was a firm relationship between the backplate characteristics and the main beam deflection.

2.2.3 *Drum:*

Although there was no real drum mode easily observed there was a distinct fringe disturbance as indicated by the image in Figure 8, the reference hologram at zero time. Comparing this to the image at half the period of excitation, Figure 9, shows the disturbance to have moved 180°.

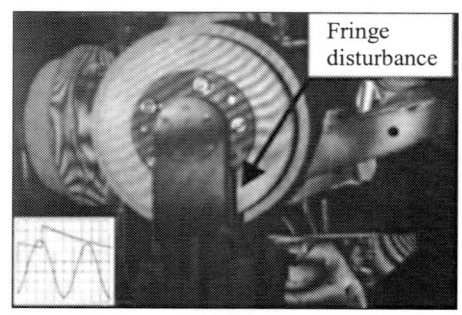

Figure 9: Image at time zero, 0µs.

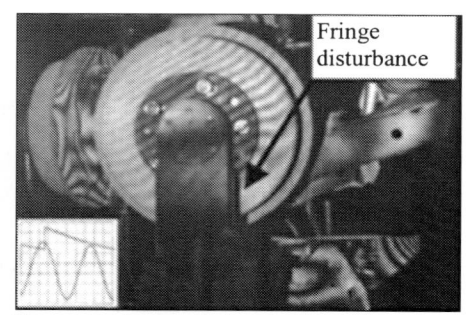

Figure 10: Image 504 µs later in the cycle.

The full series of 9 images showed this disturbance to move progressively about the disc and in the direction of drum rotation. It was stated that this was a possible mode rotation at the noise frequency and in the direction of drum rotation.

3 HALF VEHICLE TEST RIG

The basic rig design for the wheel under investigation remains the same as for the ¼ vehicle test rig but a full axle is now included as shown in Figures 11. The specially designed "mechanical tyre" has been retained for the brake drum under investigation whereas the other wheel includes the vehicle tyre and centre. This "tyred" wheel is allowed to run on bearing mounted rollers as shown but may also be held stationary using a hand brake. Drive to the rear wheels is from a variable speed motor through the vehicle differential and wheel drive axles. The trailing arms of the cross beam are supported in their elastomeric bushes and the whole beam is loaded against the suspension springs as with the ¼ vehicle rig.

Although the problem of rolling resistance is noticeable at very low braking forces, such that the tyred wheel does not rotate, this resistance becomes less significant as the braking force increases. At fairly low nominal braking forces the tyred wheel rotates at the same speed as the "mechanical" tyre.

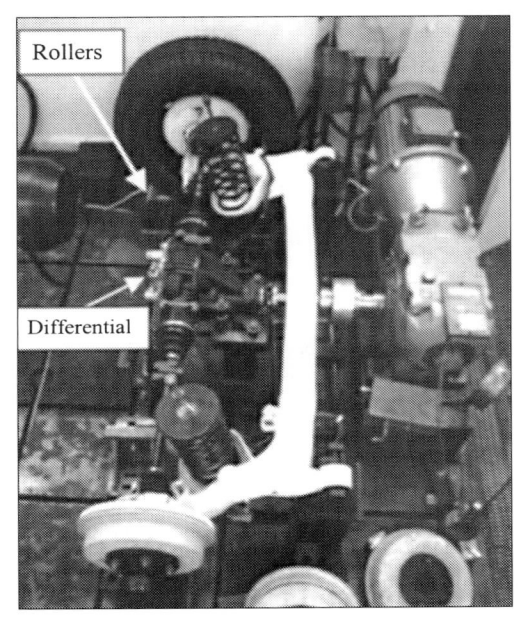

Figure 11: The half vehicle test rig including differential, half-shafts, main cross beam and suspension springs. Note "tyred" wheel on rollers.

4 RESULTS AND DISCUSSION

4.1 System Information – (as supplied by vehicle manufacturer)
Natural frequency of suspension 750 Hz, 820Hz, 1050Hz.
Natural frequency of drum 700Hz
Drum brake noise frequency 850Hz

4.2 Test Results, Holographic Recordings and Discussion.
Noise was generated at 705 Hz, 850Hz, 1400Hz and 4600Hz, the predominant noise being 850Hz. Although noise was detected at 705Hz the noise amplitude was generally very small, resulting in low amplitudes of vibration, with conditions of pressure and temperature being

very similar to those required for 850Hz. The holographic recordings are presented from high to low frequency. The images were recorded using a digital camera.

4.2.1 4600 Hz

General - This frequency is outside the realms of low frequency noise but the results are included because they make a contribution towards the discussion regarding the inclusion suspension parts on test rigs.

Figure 12 shows the full system generating noise at 4600Hz. Although the drum body does not appear to be exhibiting a clear mode of vibration the rim of the drum is seen to be vibrating with a 4 diameter mode. The views through the auxiliary mirrors show the backplate rim to exhibit antinodes at the same position as the drum rim antinodes. This would infer a possible coupling between the two components such that both are caused to vibrate in a similar manner – a 4 diameter mode in this case.

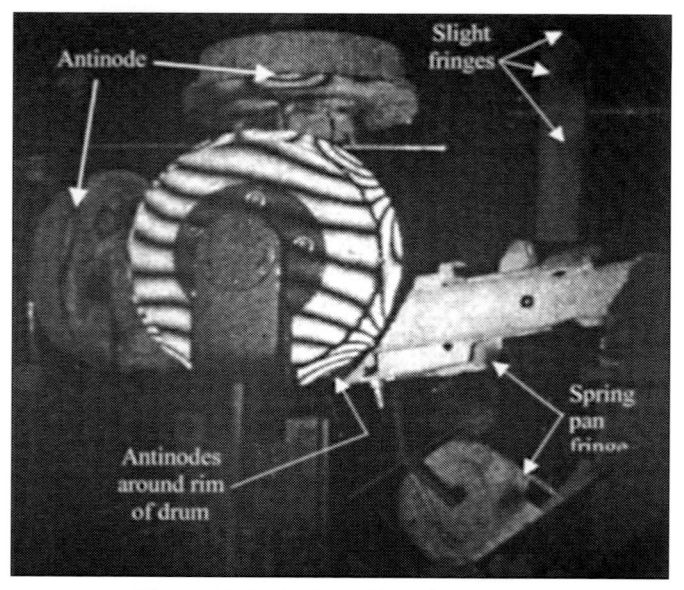

**Figure 12: Full view of interferogram.
Noise frequency 4600 Hz, 93°C, 2.7MPa (400psi)**

A detailed analysis of the top mirror as seen in Figure 13 shows the fringes on the edge of the drum rim to be angled and that they change slope. This change of slope is similar to that of a brake disc rim where it is suspected the slope change is partly due to in-plane motion and mode rotational movement. The slope also indicates the direction of drum rim antinode out-of-plane displacement.

Both the spring-pan and the main beam exhibit only a slight indication of displacement. This observation is in keeping with other recordings at this frequency and it is suggested that these parts may not be so important when analysing the "system" at these higher frequencies, particularly energy.

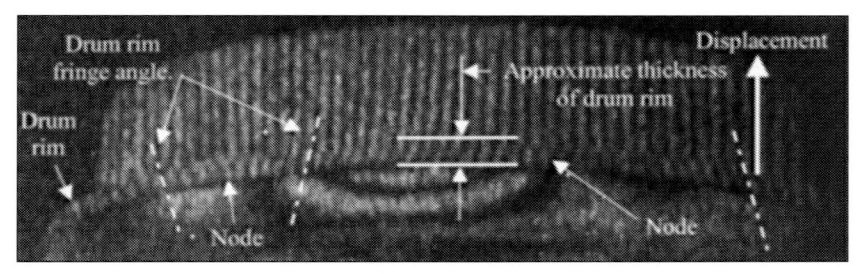

Figure 13: View through top mirror - rim of backplate and drum.
Note change in fringe angle around rim of drum – similar to disc.
Noise frequency 4600 Hz, 93°C, 2.7MPa (400psi)

4.2.2 1400 Hz
The system generating noise at 1400 Hz is shown in Figure 14. It is immediately clear from the hologram that the backplate, spring-pan and cross-beam all are significant components regarding the generation of noise. Even the trailing arm appears to be flexing as there is an indication of a fringe on the body.

The main beam has a series of antinodes along its length as shown in Figure 15. Although the full beam is not shown it must have a total 6 antinodes across its length, the number being significant. With 6 antinodes it means each trailing arm is in antiphase giving a node at the center of the beam – the beam mode being in torsion. The image also shows a slight view of the lower part of the beam where to mode appears to be similar to the upper section.

The spring-pan as shown in Figure 16 is in bending and the pattern of fringes clearly shows the stiffening effect of the side ribs. Spring-pan bending would tend to occur if the trailing arm was purely in torsion, the spring resisting the torsion by imposing a bending in its support pan. If the trailing arm was in bending it is felt the resulting spring force would tend to cause the pan to experience a combined bending and twisting motion, the pan twisting motion (moment) reacting against the trailing arm moment. The spring pan bending would result from the spring force acting some distance away from the neutral axis from the trailing arm. It is felt the trailing arm is in combined bending/torsion during this recording as indicated by the fringe on the trailing arm. It will be noted that there are visible fringes on part of the trailing arm which would indicate an element of torsion.

The backplate is enlarged in Figure 17 and shows a 2 diameter mode order. A large 6 fringe antinode is seen to the side, a smaller 2 fringe antinode is positioned at the top but there is no obvious antinode in the lower section. This asymmetrical flexure is important and may lead to understanding how and when the suspension system should be included in any analysis. Although the side antinode is positioned centrally at the shoe its form is very similar to the 960Hz results as seen on the ¼ vehicle tests where it would rotate in the direction of wheel rotation at a rate determined by the frequency and mode order. It is felt that it is this backplate mode movement that causes the flexure of the trailing arm, which in turn causes flexure of the spring pan and the cross-beam.

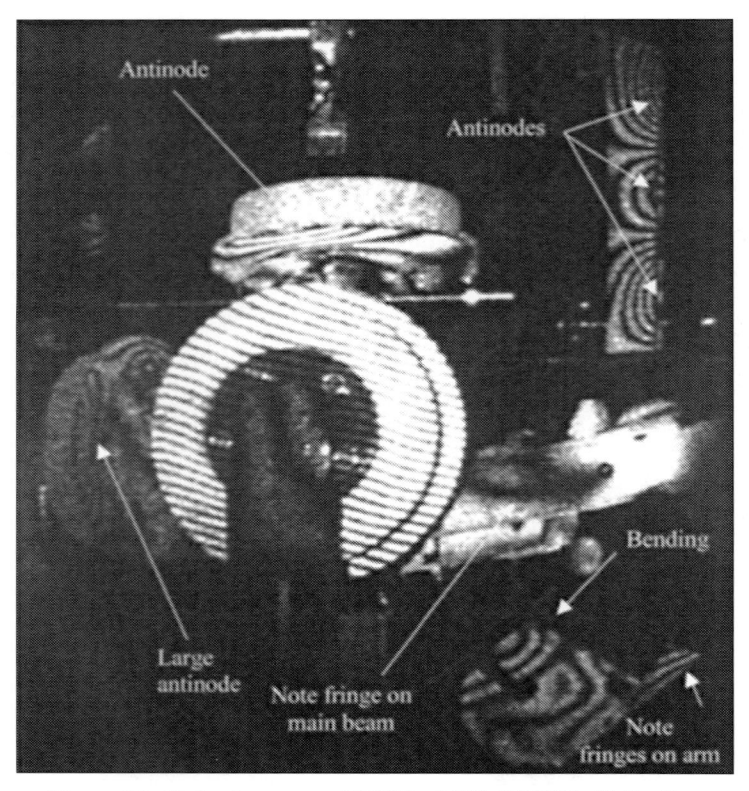

Figure 14: Noise frequency 1400 Hz, 26°C, 0.95MPa (140psi)

The significant differences between 4600Hz and 1400Hz are the mode order of the backplate and the displacement of the suspension parts. With the low mode order (2 diameter for the 1400Hz) the antinode occupies 90° of the backplate and it is evident that this mode is asymmetrical. As this is a travelling mode it will cause the backplate/trailing arm rectangular interface connection to experience a "swashplate" type of movement. If this is the case the trailing arm would first be in bending as the antinode (positioned at the side) increases in size. As the antinode moves about the backplate the trailing arm will experience a change from bending to torsion as the antinode "rolls" from the side to the top. In the case of 4600Hz the antinode is of less amplitude and occupies only 45° of the backplate. As a consequence the swash effect is less and the trailing arm deformation is less.

The trailing arm as seen in Figure 14 has a circular fringe that would indicate localised out-of-plane flexure of the sidewall (buckling). This would be typical of a hollow structure under some mode of bending/torsion and would support the proposal that the main beam is in torsion with an even number of antinodes across its length, and the trailing arm acting as the moment arm. This should be in keeping with the suspension (system) natural frequency but the given values indicate 1050Hz. Possible there is a natural frequency above 1050Hz that corresponds to the 1400Hz. The supplied natural frequencies are given in Table 1 following.

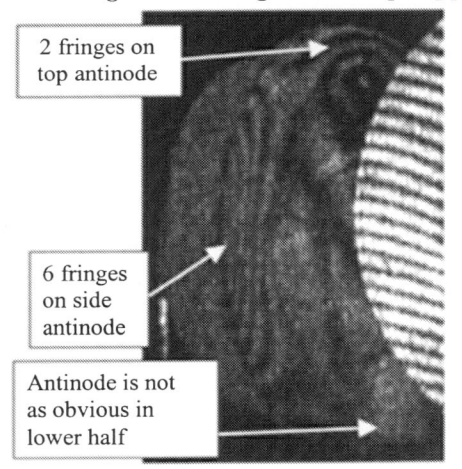

Figure 15: Enlarged view of main beam

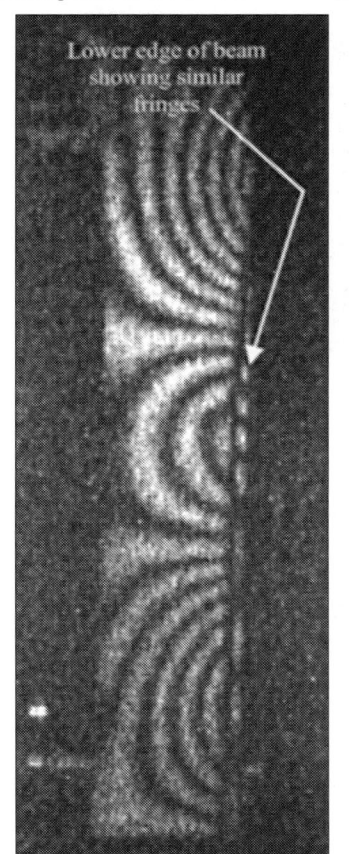

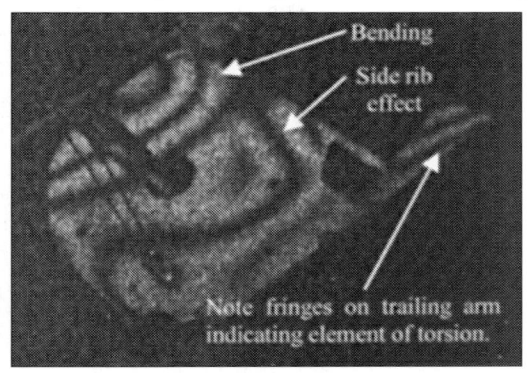

Figure 16: Enlarged view of spring pan.

Figure 17: Enlarged view of side mirror showing large antinode positioned at the shoe.

Table 1: System frequency against some mode order (integer)

Mode order (Integer)	Frequency	Status
1	750	Given and observed
2	820	Given and observed
3	1050	Given but not observed
4	1400	Observed
5	1940	Calculated
6	2541	Calculated
7	3330	Calculated
8	4600	Observed

The additional observed frequencies were included and a generalised exponential curve imposed to match these values as shown in Figure 18. It will be appreciated that it is correct to plot the graph in this manner (frequency against a linear value) because the frequency would normally relate to some integer mode order such as the number of antinodes, a diametral mode and so forth. The curve equation may then be used to calculate other possible unstable frequencies and these are also included in Table 1.

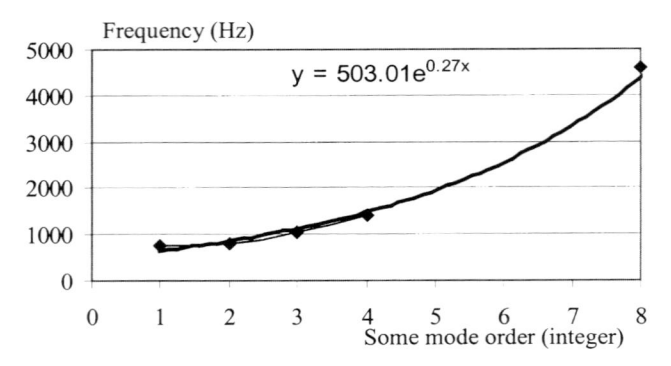

Figure 18: System frequency against some mode order (integer).

An exponential trend curve is placed over the observed to give a trend given by the equation:

$$Frequency = 503 e^{0.27x}$$

where x is an integer.

4.2.3 Suggested design modifications for higher frequency mode orders.

- Make the backplate interface mounting circular and select an odd number of fixings to inhibit cyclical "swashplate" type of movement.
- Include rolled vertical rim on backplate rim to inhibit radial vibration.
- Redesign trailing arm to become insensitive to the type of cyclical input imposed by the backplate.
- Increase axial depth of drum rim to resist axial out-of-plane vibration as seen in Figure 12.

4.2.4 Time Related Series of Hologram - 850 Hz

Two series of 7 holograms were taken of the system generating a noise at 850Hz. Although one set contained information worthy of discussion their quality did not allow for adequate reproduction within this paper. The second set contained 7 images over 749 μs which means the series spans over half a cycle. Six of these are presented as full images in Figure 21 when each hologram has a time delay attributed to it as indicated in Figure 21. All 7 are presented when the spring pan is discussed, Figure 22. The pulse spacing was set at 588μs (half period of excitation).

Points to note when observing the images are the fringe on the main beam, trailing arm, on the spring pan, fringes on side of spring pan, on the backplate edge (as seen in the top mirror), on backplate (as seen in the side mirror) and the fringes on the drum. These features are generally shown in Figure 20.

The laser trigger signal was obtained from an accelerometer mounted on the backplate with an initial delay of 0µs for the first hologram.

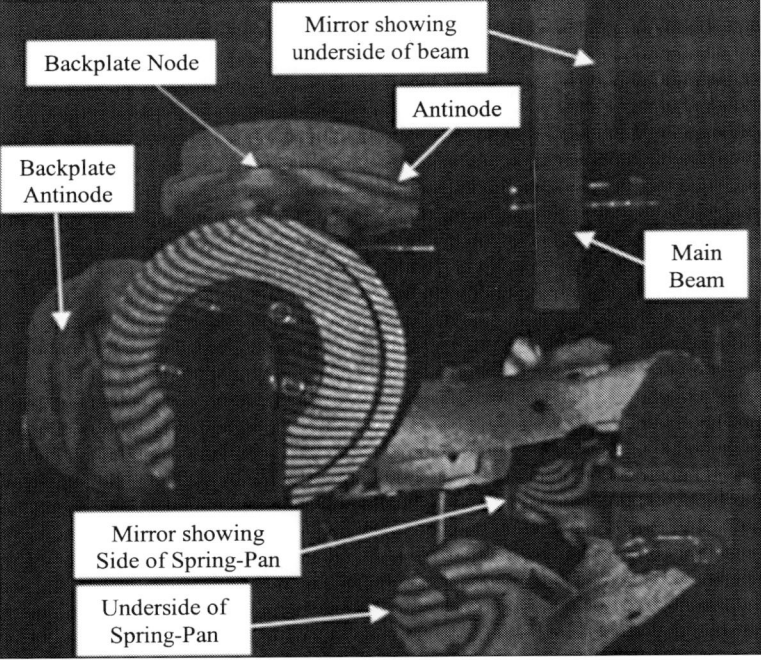

Figure 20 – Typical Image of Drum Brake Generating Noise at 850 Hz.

The main beam is seen to hold 4 low amplitude antinodes across its length indicating torsion. The antinodes are not at zero at the same time indicating that they are not quite in anti-phase. The phase appears to be around 300µs – image 2 to image 4 giving approximately 90° phase relationship. To establish other phase relationships it may be noted that the inner centre antinode is at zero amplitude as the spring pan wall is close to zero.

The backplate may be viewed in the side and top mirrors. At first it would appear that the mode is diametral with a large antinode fixed to the left hand side. The absence of any other clearly defined antinodes could indicate a 2 diameter mode order. Reference to the top mirror would support this as there are 2 antinodes either side of the centre as shown in the top mirrors in the first two images. There is no apparent diametral mode movement. Possibly the arbitrary "mode order integer" referred to in Table 1 could be linked to the number of antinodes on the backplate. If this is correct then the 5 and 7 integers may be neglected leaving 6 at 2541Hz as a possible noise problem. This is supported by earlier work [1] when a noise at 2500Hz was detected on a similar drum brake.

The trailing arm indicates a single fringe that migrates along the beam as it is subjected to a varying bending/torsional load, the type of load predicted by the "swashplate" effect.

Delay 0 μs - Image 1

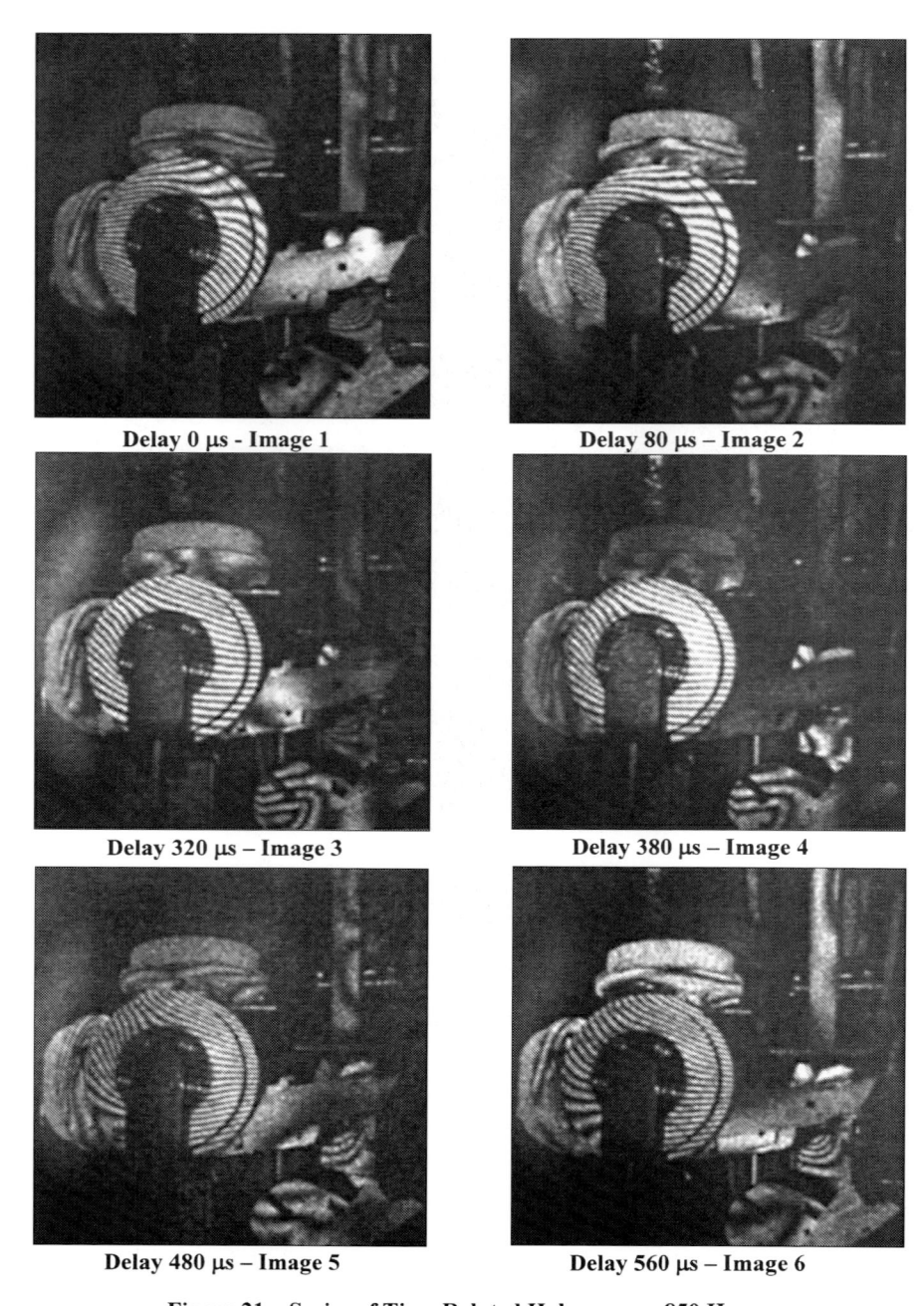

Delay 80 μs – Image 2

Delay 320 μs – Image 3

Delay 380 μs – Image 4

Delay 480 μs – Image 5

Delay 560 μs – Image 6

Figure 21 – Series of Time Related Holograms – 850 Hz.

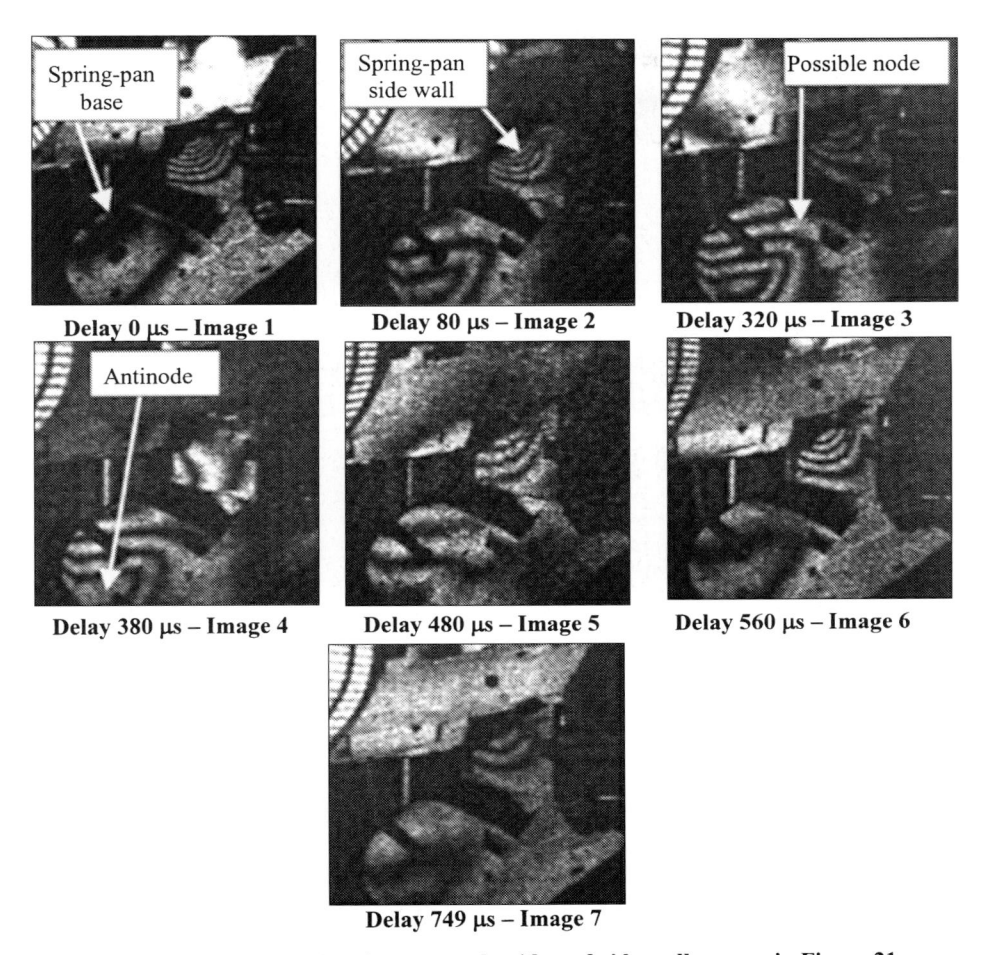

| Delay 0 µs – Image 1 | Delay 80 µs – Image 2 | Delay 320 µs – Image 3 |
| Delay 380 µs – Image 4 | Delay 480 µs – Image 5 | Delay 560 µs – Image 6 |

Delay 749 µs – Image 7

Figure 22 – Enlarged view of spring pan underside and side wall as seen in Figure 21.

The spring pan is the most active member as indicated by the high amplitudes of vibration (number of fringes) on its base and on the side wall. An enlarged view of this area is shown in Figure 22.

The mode of vibration of the side wall is bending with the node fixing being the rear wall and base – this giving the curved fringes. The pan base mode is more complex with a possible nodal line as indicated in image 3. Above the nodal line is bending but below is more circumferential in nature with an antinode at the centre of the "circular" fringes. The additional series of 7 holograms indicated twisting or rotation of the pan about a line normal to the trailing arm, similar to image 5, throughout the cycle. The reason for the difference cannot be explained at this stage but it is suspected that a higher amplitude could be the reason, the initial 7 series having a higher amplitude with 7 fringes. Trials are continuing to expand on the data available to date.

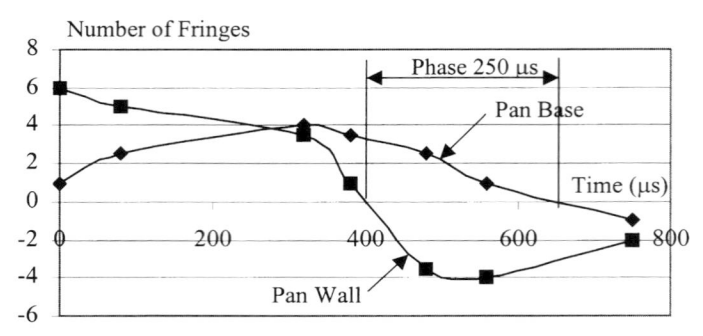

Figure 23 – Number of fringes on pan base and pan side wall over time.

If the number of fringes over time are plotted for both the side wall and base it is possible to determine the approximate phase relationship of the two sections. This is shown in Figure 23 where the phase is approximately 250 µs over half a period of 588µs giving a phase relationship about 76°, the wall leading the base.

4.3 Noise fix – 850Hz Frequency

It has been suggested that the main elements of vibration are the spring pan and trailing arm. It is significant to note that if a mass is added to the spring pan as shown in Figure 24 the noise will be severely suppressed. It is significant to note that the mass is positioned at the maximum deflection of the side wall which has greatest effect on the bending mode frequency of the this section.

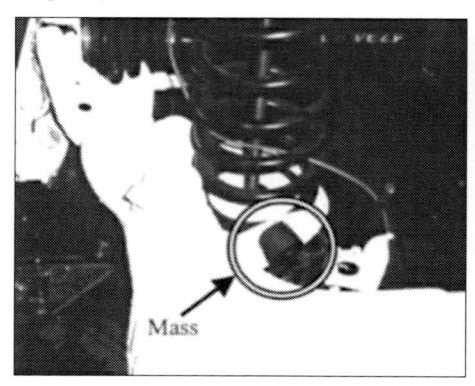

Figure 24 – Mass added as indicated to spring pan of driven wheel suppresses 850Hz frequency.

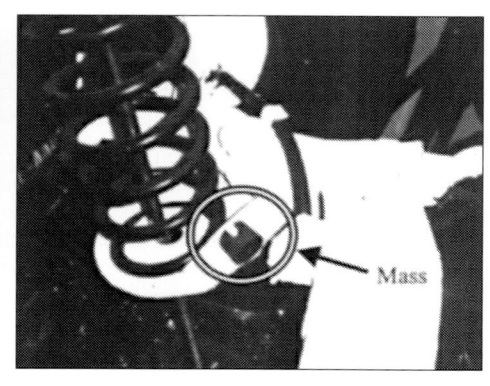

Figure 25 – Position of mass added to "static" tyred wheel that causes noise on driven wheel to increase

It must also be noted that if noise is occurring on the driven wheel and mass is added at a similar position to the static tyred wheel, as shown in Figure 25, then the noise increases.

If a mass is then added to the driven wheel as shown in Figure 24 the noise reduces once more.

This observation indicates that noise may be suppressed using added mass to "shift" the possibility of frequency coupling. The position of the mass will effect both the torsion and bending mode of the spring pan.

Consideration of energy input: The observation of increased noise on the opposing wheel may be significant if consideration is given to a constant energy input during noise. If there is a fixed energy input resulting from a given temperature/pressure/friction level during a brake application then the entire system will become excited to a specific level resulting in noise. If part of the system ceases to be excited (because there is a frequency shift as result of the added mass) then the energy is distributed between the system still being excited leading to a higher noise level from that structure. Adding the additional mass to the remaining system results in a completely stable system. Because the "tyred" wheel is not providing energy input then adding mass to the "mechanical" wheel is sufficient in this case to inhibit noise. If the "tyred" wheel was also being driven (providing energy input) then it is expected that mass would need to be added to both wheels in such a case – as with normal driving.

Conclusions

The investigation confirms the findings and observation outlined in the ¼ vehicle report in that the mode of vibration of the backplate acts as a vibration conduit between the brake and the suspension system. It has been proposed that the backplate mode tends to cause the trailing arm to bend and twist and that this in turn causes deformation and excitation of the spring pan structure and the main cross beam. These components appear to be most significant at around 1400Hz and prove to be less so at the higher and also the lower frequencies of 850Hz and below, with the exception of the spring pan.

For the lower frequencies it is felt the moving backplate mode is central to system excitation. If this is accepted it is necessary to eliminate this motion from the suspension system. Possible ways are as follows:

- Make the backplate interface mounting circular and select an odd number of fixings to minimise cyclical "swashplate" type of movement.
- Review the interface mounting to oppose the transmission of the "swash" effect.
- Increase the radial stiffness of the backplate. This may be achieved by rolling a secondary stiffening rib around the backplate rim.
- Increase the stiffness of the drum to inhibit the very slight "bell mouthing" that appears to rotate about the drum at the noise frequency. This causes the backplate mode to be excited at twice the frequency divided by the mode order, possibly because of the 2 shoes in series.
- To reduce "bell-mouthing" the drum stiffness could be increased by increasing the depth (diameter) of the drum rim and to minimise out-of-plane vibration, as seen at 4600Hz, the width of the rim (distance along drum) needs to be increased.

Additional design considerations may take the form of the following:

- Design the trailing arm to become insensitive to the type of cyclical input imposed by the backplate.
- Design the spring pan to resist the combined torsion / bending deflections observed.
- De-tuning the suspension system by the use of discrete masses.
- Eliminate excessive excitation of the cross-beam by some form of stiffening or damping. Possibly roll or press a vertical stiffening rib along the length of any structure to resist the formation of antinodes as seen at 1400Hz. Damping by painting the beam with some damping medium such as Phenolic Resin Paint.

Work needs to continue through an of introduction the design modifications and additional detailed studies of the ½ vehicle test rig. To do this effectively it is necessary to induce a higher amplitude of excitation within the system.

ACKNOWLEDGEMENTS

The authors would like to thank the Toyota Motor Corporation for their permission to release this information and for their invaluable assistance over the duration of the research project.

REFERENCES

1. Fieldhouse J.D. and Rennison M. "Drum Brake Noise - A Theoretical and Visual Approach" - "Advances in Automotive Braking Technology - Design Analysis and Materials Developments". Chapter 2 pp 25-46 Published October 1996 Mechanical Engineering Publications Limited (MEP) Publishers to the IMechE ISBN 1 86058 039 4.
2. Nishiwaki M. "Generlised Theory of Brake Squeal" Autotech, Birmingham, 1991, IMechE Publication Ref C427/11/001.
3. Fieldhouse J.D. "An Analysis of Disc Brake Noise using Holographic Interferometry". Thesis - The University of Huddersfield, 1993.
4. Fieldhouse J.D. and Beveridge C. "Low Frequency Drum Brake Noise Using a ¼ Vehicle Test Rig". 18th SAE International Congress and Exposition, March 6-9, 2000. SAE Paper number 2001-01-0448.

BIBLIOGRAPHY

1. Nathan. M, Oliver M, O'Reilly and Panayiotis Papadopoulos "Automotive Disc Brake Squeal: A Review" Journal on Noise and Vibration 2002 (To be Published).
2. Ouyang H, Mottershead J.E, "Unstable Travelling Waves in Friction-Induced Vibration of Discs. Journal of Sound and Vibration, 248 (4): 768-779, 2001.

Modelling and simulation of the vibration and squeal of a car disc brake

H OUYANG, Q CAO, and J E MOTTERSHEAD
Department of Engineering, University of Liverpool, UK
T TREYDE
TRW Automotive, Koblenz, Germany
M P CARTMELL
Department of Mechanical Engineering, University of Glasgow, UK

SYNOPSIS

This paper reports on the recent developments in the modelling and simulation of vibration and squeal of disc brake systems by the authors. The stationary components of a disc brake are modelled using many thousands of finite elements. The disc is modelled as a thin plate whereby its analytical modes can be obtained. These two disparate parts are brought together at the disc/pads contact interface in such a way that the friction-induced vibration is treated as a moving load problem incorporating a distinct squeal mechanism. Predicted unstable frequencies are seen to be close to experimental squeal frequencies.

NOMENCLATURE

a, b inner and outer radii of the brake disc.
C   proportional damping matrix of the stationary components.
c viscous damping of the disc.
D flexural rigidity of the disc.
D^* damping of the disc.
E Young's modulus of the disc.
f   nodal force vector for the reduced finite element model of the stationary components.
$\mathbf{f}_o$ nodal force vector of **f** less $\mathbf{f}_p$.
$\mathbf{f}_p$ nodal force vector for the contact nodes on the disc/pads interface.
h thickness of the disc.
i $\sqrt{-1}$.
i, m indices for subscript.
j number of the contact nodes at the disc/pads interface.
K   stiffness matrix in the finite element model of the stationary components.
k, l number of nodal circles and number of nodal diameters in a mode of the disc.

M   mass matrix of the stationary components.
n_i normal (in the z direction) force on the disc from the ith contact node of the pads due to initial static braking pressure.
n   force vector consisting of all n_i at the disc/pads interface.
P   force vector consisting of all P_i
P_i an equivalent force on the disc from the ith contact node of the pads due to the friction couple there.
p   force vector consisting of all p_i.
p_i nodal total normal force on the disc from the ith node of the pads at the disc/pads interface during vibration.
q_{kl} modal co-ordinate for k nodal circles and l nodal diameters for the disc.
$\mathbf{q}_d, \mathbf{q}_p$ modal co-ordinate vectors for the disc and for the stationary components.
R_{kl} mode shape function of the disc in the r direction corresponding to q_{kl}.
r radial co-ordinate of the cylindrical co-ordinate system.
t time.
u,v,w displacements in the r, θ and z directions respectively.
X   mass-normalised modal vector matrix for the stationary components.
x   nodal displacement vector corresponding to **f**.
$\mathbf{x}_o$ nodal displacement vector of **x** less $\mathbf{x}_p$.
$\mathbf{x}_p$ nodal displacement vector of contact nodes at the disc/pads interface.
$\mathbf{x}_w$ displacement vector formed by the w-elements in $\mathbf{x}_p$.
z axial co-ordinate of the cylindrical co-ordinate system.
β_c, β_d damping factors of the stationary components and the disc.
$\delta(\cdot)$ Dirac delta function.
δ_{kl} Kronecker delta.
θ circumferential co-ordinate of the cylindrical co-ordinate system.
μ_i kinetic dry fiction coefficient at the ith node of the pads.
v Poisson's ratio of the disc material.
ρ mass-density of the disc.
ψ_{kl} mode shape function for the transverse vibration of the disc corresponding to q_{kl}.
Ω constant rotating speed of the disc in radians per second.
ω undamped natural frequencies of the stationary components.
ω_{kl} undamped natural frequency corresponding to q_{kl}.

1 INTRODUCTION

Car disc brakes often generate unwanted vibration and noise. The disc brake squeal as a high-frequency noise is very annoying and difficult to fix. An understanding of the mechanisms for squeal generation is crucial in designing quiet brakes and for treating noisy brakes. Accurate modelling of disc brake systems and simulation of their dynamic behaviour is necessary in solving the squeal problem. There have been many investigations into the disc brake squeal problem since the late 1950's. Theoretical models of disc brake systems have become more and more refined and sophisticated.

To ensure that the established model indeed simulates the friction-induced vibration and squeal of a disc brake, three aspects of modelling are thought to be essential. First, a plausible squeal mechanism has to be used. Yang and Gibson (1) reviewed a number of brake squeal mechanisms. North (2, 3) considered an in-plane couple produced by the different friction forces on either side of the disc due to transverse vibration of the disc against the pads. Hulten (4) extended this idea to his drum brake model and Hulten and Flint (5) recently applied the same idea to their simple disc brake model. The authors believe that such an in-plane couple to be a plausible squeal mechanism and therefore have refined the idea in their disc brake model to be presented in this paper.

Secondly, the frictional contact at the disc/pads interface has to be modelled appropriately. Since the disc slides past the pads with friction, the pressure distribution at the contact interface is not uniform and this in turn affects the local contact stiffness and possible the local friction coefficients. Tirovic and Day (6) determined such pressure distribution through nonlinear contact analysis. The information is useful in the subsequent eigenvalue analysis using the finite element method (7, 8). This approach is also adopted in the present paper.

The third issue is the adequate description of the dynamics of the whole disc brake. Only the finite element method is capable of characterising the dynamic behaviour of the stationary components of a disc brake to acceptable accuracy. It is also believed that the vibration and squeal of a disc brake has to be treated as a proper moving load problem (9).

In the present study, the stationary components (the pads, calliper, carrier and slide pins) are described by many thousands of finite elements. The disc is modelled as an annular thin plate and the analytical solution of its mode shapes is obtained. This separate treatment of the stationary components and the rotating disc, and consequently the numerical-analytical combined approach, were first taken by the authors in a previous investigation (10). The whole disc brake system is treated as a moving load problem through the moving contact at the disc/pads interface with friction. The effect of the friction is considered similarly as in the cases of North (2) and Hulten (5) but in a more sophisticated way.

A nonlinear complex eigenvalue formulation is presented for the stability analysis of the equation of motion of the whole disc brake. A realistic disc brake is analysed with different operating conditions and the predicted unstable frequencies are seen to be close to the experimentally established squeal frequencies.

2 FINITE ELEMENT MODEL OF THE STATIONARY COMPONENTS

A car disc brake system contains a rotating disc and certain stationary components comprising the pads, carrier, calliper and mounting pins along which the calliper can slide in a floating calliper brake. Because of its cyclic symmetry, the disc is approximated as a flat plate and can be solved by an analytical method. A vented disc may also be represented by a flat plate of equivalent Young's modulus and mass density. The stationary components, on the other hand, are complex shapes so they are modelled by the finite element method.

When assembled, the contact interfaces between any two of the stationary components must be modelled in a suitable manner. A thin layer of solid elements, or a number of 3-D spring elements, are present at these contact interfaces. The Young's moduli, or the spring constants, are very high in the direction normal to the contact plane but very low in the contact plane.

The contacts at the leading edge and the trailing edge are different. The Young's moduli of the two layers of elements at the disc/pads interface are assumed to depend on the local pressures. The local friction coefficients can vary with the pressure as well. The finite element model of the stationary components is shown in Figure 1.

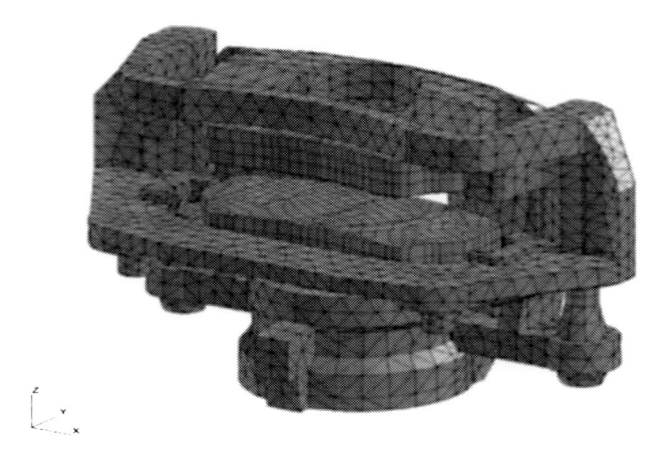

Figure 1. Finite element model for the stationary components (floating calliper)

This model has about 99,000 degrees-of-freedom and therefore necessarily incurs a very heavy computing load in the subsequent parametric study where the influence of a number of parameters on the dynamic instability is investigated. To increase the computational efficiency, a two-level condensation technique is used. At the component level, each stationary component is treated as a substructure and a stiffness matrix is generated and stored corresponding to a reduced set of (retained) degrees-of-freedom using a dynamic reduction technique. The global stiffness matrix is formed from the stored component stiffness matrices after further reduction from the sets of retained nodes of all the stationary components. In this way, if the material, or the configuration, of a stationary component is altered, the super-element stiffness matrix of only this component has to be re-generated and so the global stiffness matrix is re-assembled from it and the stiffness matrices of the other stored components. As a result much time is saved. This process is not elaborated upon here.

When the disc brake system is conceptually divided into the stationary components and the rotating disc (10), the disc/pads interface becomes a free boundary with unknown forces and displacements. The equation of motion of the reduced finite element model of the stationary components is

$$\mathbf{M}\ddot{\mathbf{x}} + \mathbf{C}\dot{\mathbf{x}} + \mathbf{K}\mathbf{x} = \mathbf{f} \tag{1}$$

The nodal force vector and the nodal displacement vector are subsequently divided into two parts. The first part corresponds to the nodes at the disc/pads interface while the second part is for all the other nodes. That is,

$$\mathbf{f}^T = [\mathbf{f}_p^T \, \mathbf{f}_o^T], \quad \mathbf{x}^T = [\mathbf{x}_p^T \, \mathbf{x}_o^T] \tag{2}$$

where

$$\mathbf{f}_p^T = [0, \mu_1(p_1 - n_1), p_1 - n_1, 0, \mu_2(p_2 - n_2), p_2 - n_2, \ldots, 0, \mu_j(p_j - n_j), p_j - n_j],$$
$$\mathbf{f}_o^T = 0, \qquad \mathbf{x}_p^T = [u_1, v_1, w_1, u_2, v_2, w_2, \ldots, u_j, v_j, w_j]$$
(3)

Note that the friction force is defined locally and the friction coefficient can vary from node to node. The first equation in equation (3) allows the nodal friction coefficients to be a function of the local forces n_i due to the initial braking pressure and the apparent relative velocity between the disc and the pads.

Using the modal matrix $\mathbf{X}$ and equation (3), equation (1) can be transformed to

$$\ddot{\mathbf{q}}_p + 2\mathrm{diag}[\beta_c \omega]\dot{\mathbf{q}}_p + \mathrm{diag}[\omega^2]\mathbf{q}_p = \mathbf{X}^T \mathbf{f} = \mathbf{X}_p^T \mathbf{f}_p = \mathbf{X}_\mu^T(\mathbf{p} - \mathbf{n})$$
(4)

where

$$\mathbf{X}_p = \begin{bmatrix} X_{1u}^{(1)} & X_{1u}^{(2)} & \cdots \\ X_{1v}^{(1)} & X_{1v}^{(2)} & \cdots \\ X_{1w}^{(1)} & X_{1w}^{(2)} & \cdots \\ X_{2u}^{(1)} & X_{2u}^{(2)} & \cdots \\ X_{2v}^{(1)} & X_{2v}^{(2)} & \cdots \\ X_{2w}^{(1)} & X_{2w}^{(2)} & \cdots \\ \vdots & \vdots & \cdots \\ \vdots & \vdots & \cdots \end{bmatrix}, \quad \mathbf{X}_\mu = \begin{bmatrix} X_{1w}^{(1)} + \mu X_{1v}^{(1)} & X_{1w}^{(2)} + \mu X_{1v}^{(2)} & \cdots \\ X_{2w}^{(1)} + \mu X_{2v}^{(1)} & X_{2w}^{(2)} + \mu X_{2v}^{(2)} & \cdots \\ \vdots & \vdots & \cdots \\ \vdots & \vdots & \cdots \end{bmatrix}$$
(5)

and $X_{su}^{(r)}, X_{sv}^{(r)}$ and $X_{sw}^{(r)}$ ($r = 1, 2, \ldots, n;\ s = 1, 2, \ldots, j$) are the u, v and w components of the rth eigenvector of the sth contacting node at the disc/pads interface. The left-hand side of equation (4) implies that proportional viscous damping is used. Its right-hand side assumes that the vibration is superimposed on an initial stress problem due to the initial static braking pressure. The same is true in the analysis of the disc in Section 3.

Suppose one solution of equation (4) is in the form of

$$\mathbf{q}_p = \exp(\lambda t)\mathbf{c}_p$$
(6)

where λ is the characteristic exponent and $\mathbf{c}_p$ is a constant vector to be determined from the initial conditions. Mathematical manipulation leads to

$$\mathbf{x}_w = \mathbf{X}_{wp}\mathbf{q}_p = \mathbf{X}_{wp}\,\mathrm{diag}(\lambda^2 + 2\beta_c \omega \lambda + \omega^2)^{-1}\mathbf{X}_\mu^T(\mathbf{p} - \mathbf{n})$$
(7)

where $\mathbf{X}_{wp}$ is part of $\mathbf{X}_p$ comprising only the w-components.

This expression relates the *w*-displacement vector for the nodes of the pads at the disc/pads interface to the normal force vector at the same nodes, and fully reflects the dynamics of the stationary components. It should be noted that $\mathbf{X}_{wp}$ and $\mathbf{X}_\mu$ are both non-square matrices in general and any number of modes may be retained in equation (7).

3 ANALYTIC MODEL OF THE DISC

The disc under investigation has five mounting holes in the top-hat section and many vents, both of which are evenly distributed in the circumferential direction. To facilitate the treatment of moving loads acting onto the disc from the stationary components and to increase computational efficiency, the disc is represented by a classical annular, thin plate. The mode shapes of the plate, as a substitute of the disc, can be obtained analytically. A number of natural frequencies and modes of the disc are identified by modal tests. By adjusting the Young's modulus, the mass density and/or the inner radius of the plate, numerical frequencies of the plate can be brought to be close to the experimental frequencies of the disc to acceptable level of accuracy.

Since the interest here is in the dynamics of the system in the low speed range, gyroscopic and centrifugal effects are very small and can safely be omitted. A cylindrical co-ordinate system fixed to the centre of the plate is used to describe its transverse vibration. When the pads are discretised into finite elements as described in the foregoing section, the pressure from the pads on the disc can now be represented by a series of concentrated forces. Any two normal forces acting on the same (r,θ) position, but on the top and bottom surfaces of the disc, vary with time and are usually unequal as the disc vibrates in the transverse direction. As a result, the corresponding two friction forces on either side of the disc at the same (r,θ) position are usually unequal as well. So they produce a couple acting on the disc.

Ouyang and Mottershead (11) studied the vibration of a disc excited by two rotating mass-spring-damper systems with friction acting on either side of the disc and thus generating a friction-induced couple. That formulation is extended here to cover the rotational frictional contact between the pads and the disc. Note that the single pair of point-wise forces due to the friction couple described in (11) now becomes a vector $\mathbf{P}$ consisting of many such pairs of point-wise forces in the present study, accounting for all such couples acting on the contact nodes of the disc/pads interface.

The equation of transverse motion of the disc plate becomes

$$\rho h \frac{\partial^2 w}{\partial t^2} + c \frac{\partial w}{\partial t} + D\nabla^4 w = -\frac{1}{r}\mathbf{d}^T(r,\theta,t)(\mathbf{p}(t)-\mathbf{n}) + \frac{1}{r}[\mathbf{d}^T(r,\theta,t) - \mathbf{d}^T(r,\theta+\Delta\theta,t)]\mathbf{P}(t) \qquad (8)$$

where

$$\mathbf{d}^T = [\delta(r-r_1)\delta(\theta-\theta_1-\Omega t), \delta(r-r_2)\delta(\theta-\theta_2-\Omega t), ..., \delta(r-r_j)\delta(\theta-\theta_j-\Omega t)] \quad (9)$$

The elements of $\mathbf{P}$, by direct analogy to the pair of point-wise forces in (11), are

$$P_i = \frac{\mu_i(p_i - n_i)h}{2r_i \Delta\theta} \tag{10}$$

The solution of equation (8) can be represented by a modal expansion as

$$w(r,\theta,t) = \sum_{k=0}^{\infty} \sum_{l=-\infty}^{\infty} \psi_{kl}(r,\theta) q_{kl}(t) \tag{11}$$

where

$$\psi_{kl}(r,\theta) = \frac{R_{kl}(r)}{\sqrt{\rho h b^2}} \exp(il\theta) \quad (k = 0, 1, 2,...; l = 0, \pm 1, \pm 2,...) \tag{12}$$

The modal functions satisfy the ortho-normality conditions,

$$\int_a^b \rho h \overline{\psi}_{kl} \psi_{mn} r \mathrm{d}r \mathrm{d}\theta = \delta_{km}\delta_{ln}, \quad \int_a^b D\overline{\psi}_{kl}\nabla^4\psi_{mn} r \mathrm{d}r \mathrm{d}\theta = \omega_{mn}^2 \delta_{km}\delta_{ln}. \tag{13}$$

where the overbar denotes complex conjugation.

Following the procedure presented in (10) and making use of equation (13), equation (8) can be transformed into

$$\ddot{q}_{kl} + 2\beta_c \omega_{kl} \dot{q}_{kl} + \omega_{kl}^2 q_{kl} = -\frac{1}{\sqrt{\rho h b^2}} \sum_i^j R_{kl}(r_i) \exp[-il(\theta_i + \Omega t)](1 - \frac{\mu_i h}{2r_i} il)(p_i - n_i) \tag{14}$$

4 DYNAMICS OF THE WHOLE SYSTEM

The whole system will vibrate at the same frequencies when the stationary components and the rotating disc are coupled together through the frictional contact interface. When compared with equation (6), equation (14) suggests that its solution is in the form of

$$q_{kl}(t) = \exp[(\lambda - il\Omega)t]c_{kl} \tag{15}$$

where c_{kl} is a constant. Substituting equation (15) into (14) and after further manipulation of the resultant equation one obtains

$$\mathbf{q}_d = -\mathrm{diag}[\exp(-il\Omega t)]\mathrm{diag}[(\lambda - il\Omega)^2 + 2\beta_d\omega(\lambda - il\Omega) + \omega_u^2]^{-1}\mathbf{S}'^H(\mathbf{p} - \mathbf{n}) \tag{16}$$

From the contact condition at the disc/pads interface of

$$\mathbf{x}_w = \{w(r_1,\theta_1 + \Omega t, t), w(r_1,\theta_1 + \Omega t, t), ..., w(r_j,\theta_j + \Omega t, t)\}^T \tag{17}$$

it follows that

$$\mathbf{x}_w = -\mathbf{S}\mathrm{diag}[(\lambda-il\Omega)^2 + 2\beta_d\omega_{kl}(\lambda-il\Omega)+\omega_u^2]^{-1}\mathbf{S}'^{H}(\mathbf{p}-\mathbf{n}) \tag{18}$$

where the elements of matrices $\mathbf{S}$ and $\mathbf{S}'$ are

$$\mathbf{S}(i,m) = \frac{R_{km}(r_i)}{\sqrt{\rho h b^2}}\exp(im\theta_i) \quad \mathbf{S}'(i,m) = \frac{R_{km}(r_i)}{\sqrt{\rho h b^2}}\exp(im\theta_i)(1+\frac{\mu_i h}{2r_i}im) \tag{19}$$

Considering equations (7) and (18) together, a new equation is established as

$$\{\mathbf{S}\mathrm{diag}[(\lambda-il\Omega)^2 + 2\beta_d\omega_{kl}(\lambda-il\Omega)+\omega_u^2]^{-1}\mathbf{S}'^{H} + \\ \mathbf{X}_{wp}\mathrm{diag}(\lambda^2 + 2\beta_c\omega\lambda+\omega^2)^{-1}\mathbf{X}_\mu^T\}(\mathbf{p}-\mathbf{n}) = 0 \tag{20}$$

This represents a highly nonlinear eigenvalue problem which is intractable by the use of existing algorithms and can only be solved by searching the complex space of the eigenvalues. The search method described in Reference (9) is used. The real part of an eigenvalue represents (positive or negative) damping and the imaginary part the natural frequency. If the real part is positive, then the vibration involving the frequency of this λ is unstable, indicating squeal.

5 NUMERICAL SIMULATION

To get accurate numerical results for the natural frequencies and mode shapes of the disc, an in-house program in a symbolic software environment is developed and used. For the disc, $a = 0.045\mathrm{m}$, $b = 0.133\mathrm{m}$, $h = 0.012\mathrm{m}$, $v = 0.211$, $E = 1.2\times 10^5 \mathrm{MPa}$, $\rho = 7200 \mathrm{kg \cdot m^{-3}}$. Thirty-five natural frequencies of the disc are calculated by an analytical method, of which the first thirteen distinct values are given in Table 1. Thirty-five disc modes are involved in the subsequent computation of the eigenvalues of the disc brake.

Table 1. Natural frequencies in Hz of the disc

k,l	0,0	0,±1	0,±2	0,±3	0,±4	0,±5	0,±6	1,0	1,±1	1,±2	0,±7	1,±3	0,±8
ω_{kl}	1087	1090	1279	1957	3130	4704	6622	7028	7295	8115	8859	9532	11404

The Young's modulus of the pads depends on the piston line pressure and the temperature. This dependency is worked out indirectly by comparing the finite element results with the experimental results of the pads. The Young's modulus is between 5.4 GPa and 10.8 GPa.

A general-purpose software package is used to compute the natural frequencies and mode shapes of the stationary components. A hundred frequencies (up to 17324 Hz) and modes of the reduced finite element model of the stationary components are used.

The system eigenvalues when the brake is applied without the disc rotation are given in Table 2. These results suggest that the finite element model and the values of system parameters are suitable.

Table 2. Frequencies in Hz of the brake system without disc rotation

Numerical	1182	1432	1931	2067	2312	2931	3398
Experimental	1207	1485	1774	2293	2448	2849	3034
Error	-2.01%	-3.55%	8.89%	-9.87%	-5.55%	2.86%	12.00%

The predicted eigenvalues of the disc brake at different rotating speeds of the disc are presented in Table 3 for $\beta_c = 10^{-3}$, $\beta_d = 0$ and a constant friction coefficient of $\mu_i = 0.7$.

Table 3. The predicted eigenvalues in Hz

Ω (rad·s^{-1})	Eigenvalues of the disc brake system			
0	-1.91±1399i	-2.55±1705i	-1.27±2384i	-7.01±4071i
5	-1.11±1399i	-3.66±1705i	2.07±2385i	-1.59±4071i
10	-0.32±1399i	-4.62±1705i	3.18±2385i	3.66±4071i
15	0.64±1399i	-5.57±1705i	8.60±2386i	9.08±4072i
20	1.43±1399i	-6.53±1705i	11.62±2386i	14.33±4072i

The predicted unstable frequencies are 1399Hz, 2384-2386Hz and 4071-4072Hz. The first one is only marginally unstable while the last two are more strongly unstable. Experiments on disc brake squeal have been carried out at different piston line pressures and/or rotating speeds using a number of non-contact displacement transducers. Those squeal frequencies are listed in Table 4 that are close to the predicted unstable frequencies.

Table 4. Experimentally-established squeal frequencies

Pressure (MPa)	0.2	0.27	0.27	1.39	1.36	0.2	0.2	0.27	0.3	0.27
Speed (rad/s)	6.5	6.5	6.5	6.6	6.7	6.5	6.5	6.5	6.5	6.5
Frequency (Hz)	1390	1416	2354	2398	2406	4184	4186	4220	4238	4244

The experiments repeatedly find a squeal frequency around 4220Hz. The squeal frequency at about 2400Hz also appears several times, though less frequently. The squeal frequencies of 1390Hz and 1416Hz appear occasionally. It can be seen that there is a good agreement between the predicted squeal frequencies and experimental values.

CONCLUSIONS

This paper presents a method for modelling the friction-induced vibration and predicting the squeal frequencies of a car disc brake. The disc brake is treated as a moving load problem consisting of two parts, the rotating disc and the stationary components, which are dealt with by classical analysis and the finite element method respectively. A plausible squeal mechanism for inducing dynamic instability is incorporated into the model, which shows very good agreement with squeal-test results from an experimental rig.

ACKNOWLEDGEMENT

The authors are grateful for the support by the Engineering and Physical Sciences Research Council (GR/L91061), TRW Automotive and TMD Friction UK. The experimental results are obtained by S James and analysed by him and D Brookfield.

REFERENCES

1   **Yang, S.** and **Gibson, R. F.** (1997) Brake vibration and noise: review, comments and proposals. *Int. J. Mater. & Product Tech.*, **12**(4-6), pp.496-513.

2   **North, N. R.** (1972) *Disc brake squeal — a theoretical model.* M.I.R.A. Research Report, No.1972/5.

3   **North, N. R.** (1976) Disc Brake Squeal. *Proc. IMechE*, C38/76, pp.169-176.

4   **Hulten, J. O.** (1995) Some drum brake squeal mechanisms. *Proc. of the 1995 Noise and Vibration Conf.*, Traverse City, Michigan, Vol.1, pp.377-388.

5   **Hulten, J. O.** and **Flint, J.** (1999) An assumed modes approach to disc brake squeal analysis. Society of Automotive Engineers, paper SAE1999-01-1335.

6   **Tirovic, M.** and **Day, A. J.** (1991) Disc brake interface pressure distribution. *Proc. IMechE, J. Auto. Eng.*, **205**, pp.137-146.

7   **Lee, Y. S., Brooks, P. C., Barton, D.** and **C., Crolla, D. A.** (1998) A study of disc brake squeal propensity using parametric finite element model. *IMechE Conf. Trans., European Conf. On Noise and Vibration*, 12-13 May, pp.191-201.

8   **Ripin, Z. B. M.** (1995) *Analysis of Disc Brake Squeal Using the Finite Element Method.* PhD Thesis, Dept. Mech. Eng., University of Leeds.

9   **H Ouyang, J E Mottershead** and **W Li.** Vibration and Dynamic Instability of Moving Load Systems. *The 25th International Conference on Noise and Vibration Engineering* (ISBN 90-73802-75-X), Leuven, Belgium. 13-15 September, 2000, 1355-1360

10 **Ouyang, H., Mottershead, J. E., Brookfield, D. J., James, S.** and **Cartmell, M. P.** (2000) A methodology for the determination of dynamic instabilities in a car disc brake. *Int. J. Vehicle Design*, **23**, pp.241-262.

11 **Ouyang, H.** and **Mottershead, J. E.** Dynamic instability of an elastic disc under the action of a rotating friction couple. *ASME J. Appl. Mech.*, submitted in August 2001.

Brake system noise and vibration – a review

P IOANNIDIS, P C BROOKS, and **D C BARTON**
School of Mechanical Engineering, University of Leeds, UK
M NISHIWAKI
Toyota Motor Corporation, Shizuoka, Japan

SYNOPSIS

This paper contains a critical review of research in the general area of brake noise and vibration that has taken place since 1991 when Crolla and Lang [1] presented the state of the art. The main thrust of the present review focuses on the growth of knowledge in the area of brake squeal that derives from both experimental and theoretical research. It distinctively concentrates on the subcategories that comprise the general brake noise field for both disc and drum brakes, such as low and high frequency squeal. Attention, however, is also given to recent investigations on brake judder and low frequency noise. Traditional analytical methods such as complex eigenvalue and non-linear transient analysis using lumped parameter models as well as advanced finite element models, are discussed, in conjunction with new experimental techniques. A condensed description of the theories behind noise that formed the basis of all recent investigations is also provided. The paper also summarises other current issues of concern to the brake engineer and concludes with a look forward to potential problems that await solution.

1. INTRODUCTION

Significant advances in the refinement of the automotive chassis and drivetrain continue to lower the noise and vibration harshness levels experienced by vehicle occupants and the general public. As a result, awareness of the noise and vibration problems that are associated with the operation of a foundation brake is becoming more acute. The continuously growing expectations of vehicle occupants with respect to a comfortable ride has resulted in high warranty costs for the manufacturing companies. A re-evaluation of customers' requirements puts comfort high on the list of a vehicle's major design considerations, in order to provide a competitive and attractive product to the public. The non-repeatable nature of noise has proved to be a difficult physical problem to "capture" within the confines of a simple mathematical model. Although some early noteworthy pieces of work provided a good understanding of the problem, the utilisation of integrated numerical analysis methods developed in recent years, in conjunction with advances in computing hardware and software capability has contributed to the appearance of the first predictive tools of brake noise propensity. The braking industry continues to be active in its endeavours to understand and control the brake noise and vibration phenomena. It is, therefore, essential that this research receives a periodic review to help identify recent milestones and prevent the re-invention of the wheel.

2. HISTORY OF RESEARCH ON BRAKE NOISE

2.1 Low and High Frequency Brake Squeal

Studies on the brake squeal problem date from 1935 [2]. These early theories, normally supported by trial and error experimental techniques, attributed squeal generation to a higher static than dynamic coefficient of friction between the linings and the rotor. The phenomenon was believed to induce self-sustaining oscillations in the brake at low speeds. Mills [3] and later Sinclair [4] stated that these continuous audible oscillations are produced as a result of the negative gradient characteristics that the dynamic coefficient of friction possess against the relative sliding velocity in the contact interface. Consequently, energy is fed into the system, which drives the brake instability. The above theories are also acknowledged as "stick-slip" and "negative damping" respectively. The latter, nonetheless, is still considered to be the basis of experimentation and analysis of low frequency noise in brake systems.

Spurr [5] proved that self-exciting vibration could occur with a constant coefficient of friction due to a sprag-slip effect of the shoe against the rotor. This new theory, although based on a cantilever-on moving surface- mechanism, proposed that brake squeal depends upon the magnitude of interface coefficient of friction and that the normal and frictional forces that act at the interface can be affected by the stiffness of the supporting structure.

Earles et al. [6] extended the sprag-slip theory to produce sophisticated pin-on-disc models with a greater number of DOFs, representing the disc and both pads of a disc assembly. In general, these models predicted regions of unstable motion and the outcome demonstrated that squeal was not only dependent on the coefficient of friction but also on the magnitudes of the physical parameters of the system. The type of instability was geometrically induced and became known as "kinematic instability theory".

A very important advance in the analysis of brake noise was North's [7] proposal of a binary flutter model for a disc brake. The binary flutter theory is based upon the dynamic characteristics of the rotor. It was shown that instabilities in the system arise due to changes in the direction of both normal and frictional forces. It was evident that energy could be diverted into the vibrating system due to the coupling between rotational degrees of freedom and radial and tangential force components. The eight-degree of freedom model produced not only took into account the geometrical characteristics of the brake components but also incorporated friction coupling, damping and stiffness values of the interactive components. An instability predictive design tool was produced for the first time, which could be accurately correlated to very specific, real brake situations. The mechanism was found to be similar to aircraft wing flutter.

Based on North's binary flutter model, Millner [8] introduced two degree of freedom models for each of the pad and caliper, which were coupled by a kinematic constraint. The operation of a long-standing "fix" of an offset contact between the pad and the actuating piston was explained. North's and Millner's approaches have proved to be quite beneficial in terms of their design usefulness because they are based upon models of physical brake components. However, they consist of a small number of degrees of freedom, which presents an obstacle to their application in the analysis of higher order squeal frequencies.

A breakthrough in the experimental methods employed by NVH researchers came in 1977 when Felske et.al [9] introduced the holographic interferometry technique in the field. The use of the method allowed the visualisation of the modes of vibration and confirmed the existence of diametral modes of the disc during noise generation. Corresponding computational advances also assisted the development of more advanced analytical and numerical methods in the form of finite element analysis. Although several studies had been presented using the finite element method, none was more important than the one presented by Liles [10] in 1989, which is discussed in section 3 below.

2.2 Low Frequency Brake NVH

Low frequency noise is reputed to occur within the range of 100 to 500 Hz. This general noisy area is comprised of several other subcategories. However, clear definitions and the allocation of names of such noises have yet to be standardised. Terms such as chatter, crunch, groan, humming and moan, are commonly encountered within the literature. In general, rigid body oscillations are associated with each of these phenomena [11]. This means that little or no deformation of the brake components actually occurs during the generation of the associated noise. This paper concentrates on the most widely reported noises in the low frequency region such as groan, hum and moan. Brake judder is also included in this survey, since it represents the low frequency vibration domain of the general brake NVH field.

Groan or creep groan is defined as a semi-resonant vibration with frequency of around 100 Hz [11]. Groan vibrations occur at suspension and brake assembly natural frequencies, often involving rotational motions of the caliper and suspension arms. The exciting mechanism is based on the stick-slip phenomenon and occurs at very low speeds, less than 2 km/h [12]. The energy is transmitted through the brake assembly and couples with other chassis components. Since the generation of groan is dependent upon the interface friction (μ) – velocity (v) characteristics, reducing the gradient $d\mu/dv$ seems a possible cure to the problem. In practise, however, such changes are difficult to achieve in isolation from other performance parameters. Other empirical fixes have been suggested such as increasing the damping in the system. For example, increasing the damping of the rotational mode of the suspension by changing the rubber bushes is suggested as a possible noise cure [11].

Hum and moan vibrations are audible and generated in the range of 100 to 400 Hz [13]. Humming is considered to be the lowest of the geometrically induced instabilities and occurs during light brake dragging at motorway speeds with rotor thermal distortion as the cause [14]. Moan is associated with rigid body oscillations and occurs during a brake application at speeds between 2 to 20 km/h due mostly to negative damping effects in the system [12]. Therefore the parameters that affect the generation of these noises are the interface friction, the system's damping and the geometrical characteristics of the brake components, which can be interpreted as the mass and stiffness distribution [15]. The adoption of a systematic approach in changing these parameters is required in order to be able to minimise the noise without affecting the performance characteristics of the brake assembly.

Brake judder is a vibration usually felt rather than heard by the driver on the steering wheel. Judder is classified in the literature based on the operational temperature of the brake system during its generation. It is also a forced vibration, the frequency of which is a multiple of the instantaneous vehicle speed. For temperatures below 100°C the judder is called cold and is caused by geometrical irregularities (D.T.V.=Disc thickness variations) due to machining and mounting of the components, uneven wear and uneven friction film generation [16]. Proposed solutions to the problem include alternative friction materials, increased clearance between stator and rotor and increased piston-seal retraction within the brake caliper.

On the other hand when the temperature threshold, usually around 100°C, is exceeded then the judder transmitted is considered to be hot and is caused by thermal deformations of the rotor, uneven thermal expansion and phase transformation of disc/drum material [17]. Thermal judder is a by-product of Thermo-Elastic Instabilities (T.E.I) active on the surface of the brake rotor, which consequently initiate non-uniform contact and heat generation during the brake application. T.E.I. on the rubbing surface generate hot bands around the rubbing path, which may in turn lead to the development of so-called hot-spots [18].

3. RECENT APPROACHES TO DISC BRAKE NVH

3.1 Introduction

Simplistic mathematical representations of a physical disc brake system have proved to be inadequate in providing a design tool that could predict and eliminate noise, due to the limited number of degrees of freedom involved in the associated theoretical models. Computational advances in the 1990's and the consequent development of advanced finite element analysis software allowed the generation of models with a vast number of degrees of freedom. The integrated analytical techniques developed thereafter have minimised costs related to trial and error experimental techniques and provided a better understanding of the brake systems' physical behaviour and subsequently noise generation. The success of this method is evident by the number of proposed finite element noise-predictive tools that have been developed by the automotive industry.

3.2 Low and High Frequency Squeal

3.2.1 Theoretical Approaches

A 3-DOF theoretical model generated by Watany et al. [19] suggested that the natural frequencies of the brake components have great influence on squeal propensity. The pressure-independent model predicted the natural frequencies of the disc brake system and included no structural damping or friction coupling between the components. Individual experimental analysis of the brake parts was performed to determine the modal stiffness and consequently calculate the effective coupling stiffness between the brake pad and disc incorporated in the mathematical model. It was shown through the theoretical eigenvalue analysis and the experimental testing that the geometric characteristics of the pad seemed to have a major effect on squeal propensity. Although non-linearities need to be added in the above analysis in order to improve accuracy, simple mathematical models can provide guidance in terms of what is to be expected from high fidelity FE models.

In 1999, El-Butch and Ibrahim [20] produced a 7-DOF theoretical model of a disc brake assembly in order to study the influence of both the geometrical characteristics and the contact parameters on brake squeal generation. The Langragian technique was used to derive the governing set of differential equations and the state space formulation to create the first order equivalents. The model included structural damping and friction coupling while the fluid pressure was assumed to vary linearly with time. The time domain obtained results showed that the point of application of the actuation load was responsible for causing instabilities in the system. The positioning of the actuating piston towards the trailing edge of the pad was associated with a stabilised response of the system. Finally, it was also shown that the contact rotary stiffness of the rotor was another parameter that could be used to control the stability of the system.

Similarly Ahmed et al. [21] used the above state space formulation to solve a frequency domain problem corresponding to an 8-DOF theoretical model based on the models of North and Millner [7, 8]. By modifying geometric parameters the model predicted that squeal propensity decreases as the distance between the clamping bolts of the caliper increases. Increase in lining thickness was also shown to decrease the likelihood of squeal; however, the lining width indicated the opposite effect.

In an attempt to investigate high frequency squeal Flint [22] created a mathematical model in which the rotor and the lining were represented as flexible continuous elements as opposed to the discrete elements that correspond to the pistons and the caliper. The Euler-Bernoulli beam theory was employed to obtain the equations of motion of the system. The analysis showed that instabilities that arise in the system are attributed to complex modes of vibration, which can be interpreted as travelling waves. This phenomenon was also experimentally verified by Fieldhouse and Beveridge [23]. Additionally, it was shown that as the system damping increases the number of instabilities decrease. Other important parameters found to affect the stability of the system were the coefficient of friction and the geometric characteristics of the pad.

Tzou et al. [24] investigated the in-plane vibration modes of discs. It was observed that a system's tendency to squeal is attributed to the coupling between in-plane forces and transverse or out of plane motion. By using a solid annular cylinder to represent the disc and the application of the Ritz method to discretize the equations of elastodynamics for both the free and clamped state, it was shown that this coupling is credited to the finite thickness of the disc. The analysis of the coupling showed that certain geometric characteristics are capable of reducing the out of plane motion and consequently reduce noise.

McDaniel et al. [25] examined the dominant paths of the power flow between the interactive brake components in order to determine important design guidelines that could eliminate squeal. Unstable modes in the system were determined through a mathematical model, which incorporated modal and inertial properties of the brake parts as well as the transverse and in-plane displacement ratio of each mode. The study concluded that coupling between in-plane and transverse displacements of the rotor is the primary cause of the generation of dynamic instabilities and consequently brake squeal. It is suggested that the utilisation of power flow maps to characterise a single unstable mode would be an efficient means of analysing instabilities in high fidelity finite element models.

3.2.2 Finite element approaches

The capabilities of high fidelity finite element models, with a vast number of degrees of freedom, have enabled the accurate representation of a brake system. Several types of analyses have been performed on disc brake systems through FEM, in an attempt to understand the problem of noise and develop a predictive design tool. The most commonly used type of analysis is the complex eigenvalue analysis.

Complex eigenvalues usually result from the frictional coupling of brake components due to the off-diagonal terms that arise in the stiffness matrix of the system causing it to be unsymmetrical. Complex eigenvalues with positive real parts are identified as unstable modes, which always appear in complex conjugate pairs. Such unstable modes are more prone to squeal. Traditionally, pads and rotors have been connected by so-called friction springs. In recent years, researchers have suggested an alternative method associated with the direct connection of stator and rotor and the elimination of these "imaginary" springs. Other studies have combined the above method with investigations related to pressure distribution and non-uniform contact between the brake components. The major disadvantage of performing a complex eigenvalue analysis through finite element analysis is that it is time consuming and consequently requires considerable computing power.

The substructure or mode synthesis method utilised by various studies reduces the size of a finite element model, in terms of number of degrees of freedom, while retaining the essential dynamic characteristics of the system. Additionally, individual component characteristics can be altered since the system is analysed by its components, which are coupled to get the system response. Therefore, the overall system's mass, damping and stiffness matrices include modal characteristics of each component (substructure), which can be altered, and through a new analysis the system's response can be re-evaluated. Several researchers have also investigated the influence of geometric characteristics on generation of instabilities by performing a sensitivity analysis. Therefore modifications of pre-specified design variables such as disc thickness, lining width etc. when altered showed changes in the number of instabilities. The procedure, however, has proved to be time consuming. Optimisation procedures can therefore be introduced in order to determine which design variables, for example added masses, can minimise the real parts of the obtained eigenvalues for a specific frequency range.

Finally, transient solution dynamic analyses have been performed that predict the presence of an instability but do not indicate which mode is responsible. In comparison with complex eigenvalue analysis, transient analysis is more time consuming, especially for high frequency problems, but no assumptions are needed to pre-define the contact condition and motion between the pads and the rotor during vibration. The outcome of individual studies is described in the following paragraphs that correspond to the above analytical methods using the finite element approach.

Liles [10] presented the first paper associated with a complex eigenvalue analysis of a finite element disc brake assembly. Through the utilisation of the substructure method flexible finite element models were produced instead of simplistic lumped parameter models to capture the dynamic behaviour of the system. The pioneering characteristic of this study is the inclusion of the friction coupling terms between the pad and the disc. Stiff springs were used to interconnect adjacent nodes of the FE models, resulting in frictional forces which were functions of differences in nodal displacement. The sensitivity analysis that was performed demonstrated that higher coefficients of friction could increase the occurrence of squeal whilst conversely increased structural damping, higher lining stiffness and shorter linings could have the opposite effect. Vehicle testing results confirmed the accuracy of the analysis, which subsequently caused the industry to establish the method as a standard noise predictive tool.

Kido et al. [26] concentrated on low frequency disc brake squeal and adopted the above method to generate reduced finite element models and determine unstable modes of vibration. The analysis concentrated on the relationship between the ratio of the rotor/caliper eigenvalue and squeal occurrence. Previously design guidance was obtained by performing a complex eigenvalue analysis on a 3-degree of freedom mathematical model. The analysis emphasised the importance of the eigenvalue ratio of the rotor and caliper in noise generation. Additional it suggested that the natural frequency of a dynamic absorber should be higher than the coupled natural frequency of the system. FE analysis and dynamometer testing showed, by altering the modal mass of the rotor and using alternative designs respectively, that the higher the eigenvalue ratio the higher the tendency for the system to squeal. The use of steel dynamic absorbers confirmed the above suggestions. Matsushima et al. [27, 28] looked at two types of disc brake designs with floating and opposed type caliper. The finite element method concentrated on parametric studies but at a higher frequency range than Kido [26]. The studies concluded that squeal is generated within a specified region of the applied hydraulic pressure and that this range in which squeal is produced expands as the coefficient of friction is increased. The effect of the natural frequencies of different brake components on squeal generation is also identified. Experimental verification was obtained through the use of a laser Doppler vibrometer.

Friction springs were also incorporated in the finite element modelling of the disc brake assembly by Kung et al. [29]. Observations indicated that low frequency squeal is a result of the modal coupling between the rotor, the caliper and the mounting bracket. A complex eigenvalue analysis was performed for the determination of the unstable modes of vibration. It was shown that the system possessed an unstable mode at 2.5 kHz The calculation of modal participation factors of each of the component modes showed that the rotor contributed 23% of the total system motion and that 53% of the total rotor motion was associated with the third diametral mode at the unstable frequency. The process was repeated for the rest of the components. Hence, the effect that each component exerts on the system and also the effect that its mode has for the associated component could be quantified. Finally it was shown how pairs of real modes with similar frequency could become coupled and form a conjugate pair when alterations in the coefficient of friction and the Young's modulus value of the rotor occur.

Dihua and Dongying [30] studied a squealing disc brake and produced a closed loop-coupling model with 128 DOFs using the finite element method. The assembly included the mounting bracket. Through the substructure synthesis method the vibrational behaviour of each component was determined and boundary conditions and frictional coupling were introduced. They performed a complex eigenvalue analysis and predicted the instabilities. The system proved to be more unstable than the corresponding experimental results, due to the coupling model's lack of structural damping. The substructure synthesis method allowed the alteration of the modal frequencies of any given substructure or component while keeping the associated modes shape constant. By performing an analysis for the changed coupled model, it was determined whether or not the system's response had been stabilised. It was finally observed that the instabilities generated were attributed mostly to the rotor modes in conjunction with the 7th mode of the mounting bracket.

The studies mentioned above are based upon the assumption of permanent contact between the rotor and the pads. Tirovic and Day [31] studied the effect that several geometric characteristics of their disc brake assembly (including a caliper) and the coefficient of friction have on pressure distribution. The introduction of friction was found to cause a significant change in the interface pressure distribution. Additionally, the quasi-static analysis showed that the contact is reduced at the trailing edge during braking and the maximum pressure takes place towards the leading end of the pad.

Pressure distribution studies and complex eigenvalue analysis have been combined by Lee et al. [32] and Nack [33]. A non-linear contact analysis was performed in both cases and the interfacial contact behaviour of a disc brake assembly was investigated. Next, linearisation of the same system allowed extraction of the complex eigenvalues and vectors. Lee suggested that the key to the linearisation process lies in replacing only the non-linear elements, which are closed (in-contact) as a result of the applied force with equivalent linear interface elements. Nack's results indicated that squeal can be attributed to the strength of coupling between neighbouring rotor modes. The coupling mechanism was attributed to the frictional interaction at the rotor-pad interface. Stabilisation of the system and consequently the decoupling of these modes, proved to be a difficult task due to the high modal density of the system. The parametric studies that Lee performed suggested that squeal propensity is reduced if the interface contact between the pad and disc is engineered to be uniform. The propensity of the system to squeal in this study is represented by a single number as against a set of eigenvalues. The number is derived from the standard deviation of all positive instability measures of those eigenvalues within a pre-determined frequency range from the mean value of zero. Finally it was observed that the instability standard deviation is shown to be directly related to the magnitude of rubbing interface friction coefficient but independent of the brake hydraulic pressure. Other parameters investigated were backplate thickness, slots in the lining material and pad abutment conditions.

Blaschke et al. [34] proposed the direct connection of the rotor with the pads, therefore eliminating the "imaginary springs" which could result in errors when a complex eigenvalue analysis takes place within a high frequency region. The new method incorporates important factors such as stick-slip, the effect of braking pressure and rotational velocity of the rotor and temperature in the modelling of the brake system. However, it was assumed that the mating surfaces remain in contact and that the friction forces do not change direction during vibration. The coupling terms that need to be added to the system's matrices were calculated after the equilibrium of the system was determined through a non-linear contact analysis. A simple theoretical model is utilised which consists of the pads and the rotor. The governing equations of motion were determined including terms that simulate the actual behaviour of the system. The method was applied to both disc and drum brakes. However, due to lack of important data such as μ-v characteristics of the lining, it was only possible to assess the stability of the system based upon the stiffness coupling due to friction. Dynamometer and vehicle testing have verified the analytical procedure. Yuan [35] presents a more analytical finite element formulation of the contact problem with kinetic friction and brake pressure effects.

In a paper published by Shi et al. [36] suggestions were made as to how to transform the current complex eigenvalue analysis to an improved design tool for squeal prediction. They introduced forty-six 1-gram concentrated masses, evenly distributed on the back of both brake pads. By varying these design parameters through a finite element optimisation procedure they tried to minimise the real parts of the eigenvalues within a specified frequency range. The outcome showed almost no unstable modes of vibration while the total mass of the pads was increased. Further suggestions were made on how to improve the accuracy of this method.

In 1994, Nagy et al. [37] introduced a new method to predict disc brake squeal occurrence based on the stability of the system. Since the dynamic behaviour of the rotor and caliper remain linear in the physical system, they performed an eigenvalue analysis on the two components in order to determine the modal behaviour of each component. By using the results as constraint equations on the piston and pads components, a non-linear transient analysis was performed. The model also included the non-linearities of friction when the surfaces are in contact. The outcome showed that the stability of the system is not sensitive to relative speeds but mainly affected by frictional coupling between the pads and the rotor.

This work was continued by Hu et al. [38] by relating the non-linear transient analysis method with the contribution of important brake design parameters to the brake squeal generation. Six parameters were selected including the finger length on the caliper, slots on the friction material, chamfers on the friction material, rotor thickness and friction material thickness. Consequently they generated various finite element models corresponding to the design parameters through an automated procedure. In order to predict the relative influence of each parameter among the others and to investigate the interactions between two different factors related to brake squeal, sixteen runs were performed. The design of experiment (DOE) approach found that not only does frictional coupling affect squeal generation but so also do the geometrical characteristics of the components with the more influence exerted by the rotor thickness and the slots on the brake pad.

Mahajan et al. [39] lists the advantages and disadvantages of the three most commonly analytical methods used. These are the non-linear transient, normal mode and complex eigenvalue analyses. After performing a series of analyses on a disc brake using the FEM they compared the methods based on the following criteria: methodology followed, modelling, computation time, technical features, technical limitations, prediction capabilities and typical design variables.

Based on several studies reporting the influence of the caliper's mounting bracket on the generation of squeal, Baba et al. [40] used the finite element method to define an optimised design for the bracket that leads to the reduction of low frequency brake squeal. A direct relation between the vibration characteristics of the bracket and the generation of squeal was established. An important factor determined for the redesign is the amplitude difference of the outer and inner sides of the bracket, which are subjected to vibration due to the circumferential motion of the disc (in-plane vibration). They concluded that the smaller the difference the less the tendency to squeal. The outcome was experimentally correlated.

Mottershead and Chan [41] analysed the instability of brakes based on a multi-degree of freedom finite element treatment of follower forces with the assumption of simple Coulomb friction. The research established a frictional follower force model to calculate the equivalent stiffness matrix induced by the non-conservative frictional interface force. The results show that a disc is prone to flutter at doublet mode frequencies even when the pressure load is small. Since the doublet modes possess the property that the two modes with identical amplitudes are out of phase with each other by 90°, they are uniquely susceptible to squeal. The two complex conjugate modes have been observed to travel in opposite direction around the disc.

Chung et al. [42] addresses the problems associated with complex eigenvalue analysis including geometric tolerances, difficulty in finding the responsible unstable modes, time and computing power. A modal domain formulation is proposed in which all the parameters that comprise the governing dynamic equations of motion can be obtained from the normal mode solution of the non-friction system. Hence the manufacturing variability can be easily incorporated in the formulation and determination of the complex eigenvalues can be obtained at a faster rate. Also, the strength of the modal coupling was proposed as a new method for quantifying a system's instability. The strength of the modal coupling is associated with the rate at which two modes converge. Formulation and definitions of this stability criterion are also included in the paper. The method was applied to a brake system and results were comparable to those determined through dynamometer testing.

3.2.3 Experimental Approaches

Matsui et al. [43] found that the pad lining was undergoing shear friction vibration at the contact interface due to a decreasing µ-sliding velocity ratio. It was also observed that squeal occurred when the brake assembly demonstrated resonance vibrations. In their experiments, the caliper, rotor and outer pad vibrated in a coupled manner at a resonant frequency identical to the squealing frequency. It was suggested that the friction vibration is amplified when the system's coupled resonance mode is unstable, therefore leading to squeal generation. Finally, it was concluded that increasing the stiffness of the caliper and optimising the geometry of the pad pressure surface were potential noise cures.

Ichiba and Nagasawa [44] experimentally confirmed that the excitation energy, which causes squeal, is generated due to fluctuations of the friction forces in the contact interface. Through the use of accelerometers it was possible to measure six natural squeal modes and record the vibrations associated with the rotor, the backplate and the lining in both the transverse and in-plane direction. It was determined that the friction-velocity characteristic of the lining is not as important as the ratio of change in the frictional force in the generation of squeal.

Ishihara et al. [45] confirmed experimentally that when the coefficient of friction is constant, low-frequency brake squeal was generated due to geometric dynamic instabilities. After determining the squeal region of the given disc brake assembly, random oscillation waves were applied to the caliper by an electromagnetic shaker in the normal and circumferential direction of the rotor. Accelerometers attached on the caliper were used to measure response while the rotor was rotating at 30 rpm on a dynamometer and pressure was applied. It was shown that as the friction increased the caliper had coupled vibrations between the normal and in-plate direction of the rotor. It was also suggested that the caliper's diagonal deformation and the linear stiffness of the lining material had a great effect on the generation of squeal.

Fieldhouse and Newcomb [46] showed through holographic interferometry that the modes of vibration of the disc rotate relative to the disc at a specific rate dependant on the generated frequency and the disc mode order. The squealing frequency was found to be near the individual natural frequencies of the components, a fact that drew the conclusion that noise propensity might be influenced by the coupling of these natural frequencies when they are close together. It was also confirmed that the disc mode is always diametral and the instability in the system is probably driven by the pad vibration. In a further investigation [47] on disc brake noise, the effect of piston-pad contact position was studied by inserting a wire in various positions between the piston and pad with the resulting squeal favouring a leading edge position up to a certain distance beyond which the effective sound pressure level was reduced. It was also added that if the brake system is arranged such that it operates within the region of instability, frictional changes due to temperature and pressure have no further influence on squeal generation.

Fieldhouse [48] also proposed a method to predict the noise frequency of a disc brake, derived from experimental testing using the holographic interferometry technique. This was possible from knowledge of the vibration characteristics of the rotor in the free state and the geometrical characteristics of the disc and the pad. He demonstrated that the preferred frequencies of excitation of any disc brake system might be directly related to the free mode frequency of the disc.

The bending behaviour of the rotor, as the primary cause for squeal generation, has been studied extensively. Matsuzaki and Izumihara [49] attributed the squeal generation to the in-plane vibration of the disc. Experimental testing showed that squeal was evident at frequencies of 8.5kHz, 12.8 kHz and 17.7 kHz that were associated with the 2nd, 4th and 6th circumferential mode order respectively. Only even mode orders were recorded. Sound intensity analysis was used not only to visualise the acoustic field but also estimate vibration modes. The method confirmed that longitudinal vibration of the rotor itself is responsible for the squeal generation. The use of radial slots cut into the rubbing surface of the rotor was the proposed method for the reduction of brake squeal.

Kai et al. [50] came to the same conclusions and presented a paper that concentrates on high frequency squeal generated due to the disc rubbing surface vibration in the circumferential direction. It was suggested that interrupting this circumferential vibration of the rotor by introducing cut-outs and slots on the rubbing surface eliminated brake squeal; however, other problems arose related to strength and thermal deformations. Hence they performed parametric studies using the finite element method in order to minimise the in-plane vibrations. It was shown that increasing the rigidity of the disc's hat region reduced the amount of in-plane vibrations, therefore decreasing squeal.

Eriksson et al. [51] investigated the influence of air humidity and pad humidity on the coefficient of friction and consequently the generation of squeal. In general the study showed that air humidity has a limited influence on the coefficient of friction and hence on squeal.

In a separate piece of work Nishizawa et.al [52] developed a system to counteract the vibration of the rotor thereby eliminating squeal. The concept of the Electronic Control Cancelling Noise (ECCN) is based upon the generation of a vibration whose phase is an inverse of the rotor vibration. The system consists of four piezoelectric elements and one electronic control unit. Its operation was successful on both the dynamometer and the actual vehicle for squeal generated within the region of 2-4 kHz.

Finally, Abendroth [53] summarises all the advanced experimental tools and techniques used by NVH engineers including dynamometers, holography, Electronic Speckle Pattern Interferometry (ESPI), laser Doppler scanning systems and thermal imaging systems. Additional information can also be found in Edwards et al. [54] and McDaniel et al. [55].

3.3 Low Frequency Noise

A multibody system simulation has been presented by Riesland et al. [56] regarding the investigation of the low frequency noise called moan. A fully non-linear dynamic model of a disc brake is generated which incorporated friction coupling between the components and allowed the study of altered design parameters and their influence on noise generation. Moan is associated with rigid and sometimes flexible body motions of brake and chassis components. Vehicle testing revealed that moan would occur when both new and worn lining materials were used. The time histories obtained from laboratory testing showed that this specific noise can be described as a series of stick-slip pulses. The 92 degree of freedom model used for the simulations showed that no noise generation occurred until the axle was integrated with the brake assembly. The simulation results correlated well with experimental results while the parametric studies showed the great effect that the axle tube properties, in conjunction with pad contact stiffness, have on moan generation.

Similar behaviour was observed by Bettela and Sharp [57] through experimental investigations on an actual vehicle. However, creep groan was the main subject of their study. Results from various tests on the vehicle were recorded by single axis and triaxial accelerometers and sound pressure level meters. Several tests such as forward and reverse motion, braking of the vehicle on an incline and of a jacked car with and without wheels contributed to the derivation of the following conclusions. The creep groan phenomenon can be characterised by two phases in which initially the phenomenon is friction driven followed by a period of some system natural frequency excitation. It was also concluded that the noise generated is principally associated with the suspension arms with a possible additional effect from the large parts of the brake assembly such as the rotor and caliper. Contributors to the creep groan noise were found to be the tyre torsional stiffness and wheel inertia. Finally, it was demonstrated through measurement on the chassis and body that the boundary for the creep groan noise investigation is the suspension system rather than the whole car.

Dunlap et al. [58] performed an in-vehicle experimental testing in order to investigate the phenomenon of disc brake groan using the operational deflection shape (ODS) technique. Cabin noise was recorded while accelerometers were recording the response of brake and suspension components. The ODS technique can give the deflection shape by measuring vibration at pre-specified points of the structure and consequently performing a Fourier analysis. The noise recorded had a frequency of 240 Hz and was associated with rotational rigid body motion of the caliper housing, caliper bracket and knuckle. Attempts to alter the characteristics of these components in order to minimise groan proved unsuccessful. Noise was still generated but at a shifted frequency. The researchers concentrated on changes of the lining material in an attempt to reduce the forcing function. Finally, lining material constituents were identified, that were associated with groaning and non-groaning effects. These included abrasives and filler constituents.

3.4 Brake Judder Phenomena

Brake judder is a forced vibration of a complex nature due to its high sensitivity to friction changes, relative velocities, temperature gradients, wear and phase transformations. This low frequency vibration is the result of cyclic fluctuations in the brake torque output, otherwise known as brake torque variations (BTV). Brake torque variation is mainly attributed to a non-uniform circumferential disc rotor thickness. Various phenomena such as, the geometrical irregularities of the friction surface, uneven wear and heating, uneven friction film generation and external forces can lead to disc thickness variations (DTV). There are two dominant paths that this low frequency vibration follows during brake application. The first one is from the suspension to the steering system and car body and the vibration can be felt on the driver's hands or on the floor. The second path is through the brake hydraulic system, while the vibration can be felt on the brake pedal. In the latter case, brake torque variation is translated to brake pressure variations and consequently oscillations within the hydraulic system [59]. For the former case, researchers have determined that the suspension system amplifies the magnitude of the vibration generated in the braking system due to coincidental eigenfrequencies of suspensions components, which leads to resonance. Vehicle desensitisation to vibration transfer has been the proposed solution for the elimination of judder in many studies [60,61]. This can be achieved by shifting the natural frequencies of the suspension components above the frequency range of the first wheel order [62].

Other studies [63], have suggested that the elimination of the mechanical defect known as "run-out" of the disc brake can consequently quench possible brake pressure variations, which would otherwise lead to judder. The minimisation of "runout" can be achieved by tightening the manufacturing tolerances of the disc itself or by a careful check of its vertical alignment during assembling. The running clearance between the pad and rotor is another proposed method for reduction of DTV. De Vries and Wagner [64] showed that positive retraction is more achievable with a pad with low compressibility,. However, off brake contact will take place when higher normal forces are applied on the disc. The automotive industry has also concentrated on increasing the effectiveness of the caliper piston seal rollback.

Additionally, researchers have concentrated on the generation of judder due to thermal effects. It is known that DTV can take place under elevated operating temperatures. DTV is produced due to thermal distortions of the rotor (warping and coning), excessive pad material deposits on the rubbing surface leading to uneven glazing of the rotor and focal hot spots. The high-localised temperatures due to the latter cause phase transformation of the iron to martensite, which result in a permanent non-uniformity in the friction surface [1,12]. Many studies [65] have utilised the finite element method in the investigation of the thermo elastic instabilities and have suggested a number of possible cures, which mostly involve the thermal properties of the lining and the geometric characteristics of the rotor.

4. RECENT APPROACHES TO DRUM BRAKE NVH

4.1 Introduction

The high warranty costs associated with disc brake noise have limited the amount of investment and consequently research on drum brake noise. However, cities and towns are still greatly suffering from the noise emitted by heavy-duty commercial vehicles such as buses and lorries, the majority of which still use radial or drum brakes. Although radial brakes incorporate shoes instead of pads and a drum instead of a disc, the principle decelerating mechanism is identical to axial (disc) brakes. Therefore the noise emitted from any of these brake types are not to be considered as independent problems despite the differences of noise frequencies or structural shape. Noise eliminating and predicting methodologies previously adopted on disc brakes have recently been employed on drum brake assemblies.

4.2 Low and High Frequency Squeal

4.2.1 Theoretical Approaches

In 1993 Nishiwaki [66] showed that all NVH problems related to low frequency noise and squeal are essentially instability problems, which can be analysed by performing a complex eigenvalue analysis. In the generalised theory developed, the kinetic and potential energy terms of each component were calculated for both drum and disc brakes, while the kinetic energy per cycle was determined. The Lagrangian approach was utilised to determine the governing equations of motion. It was shown that dynamic instabilities arising in any brake system are associated with the increase of kinetic energy per cycle. The instabilities are associated with unstable modes of vibration, which are derived from a complex eigenvalue analysis.

Hulten [67] presented a theoretical method using a two-degree of freedom mathematical model to investigate the instabilities that appear within a brake system. The complex eigenvalue analysis that was performed showed that unstable modes of vibration exist regardless of the constant coefficient of friction assumed in the theoretical model. Two theoretical analyses performed on both disc and drum brakes, assuming a very stiff rotor, showed instabilities with a non-synchronous motion for both the shoe and the pad respectively. A general conclusion was that the shoes and pads are the motors that drive squeal vibration, which propagates to other components within the brake assembly.

In a further study by Hulten [68] on drum brake squeal it was shown that there are four mechanisms during squeal generation. Principally these mechanisms cause instability due to their association with the non-conservative frictional forces at the interface between the drum and the shoes and the generation of bending moments about the neutral axis of either the drum or the shoe. These mechanisms create four waves that are superposed and move in different directions. Three of them move in the same direction as the rotor while the other moves in the opposite direction. If waves with opposite directions are superposed the solution is forced towards a stable standing wave, which could reduce the squeal noise. Hence a new method is required to eliminate the instabilities associated with these waves and consequently minimise the squeal.

In a separate piece of work, Hulten [69] investigates the influence of the four squeal excitation theories described above and the lining induced instability theory described in [67] on squeal generation. Using a mathematical model of a shoe and a drum, he showed that the follower forces theory, associated with the binary flutter model, and the kinematic constraint theory have from little to no effect on squeal generation. However, μ-v slope and lining induced instability theory seem to create instabilities in the system with the latter generating more unstable modes. The stick-slip theory was ignored in this study.

4.2.2 Finite Element Approaches

Day and Kim [70] performed modal analyses of an S-cam drum brake assembly using the finite element method. The analysis was performed in conjunction with investigations of the pressure distribution at the friction interface. The latter analysis determined the coupling stiffness between the lining and the rotor. The calculated value was incorporated within a finite element drum brake assembly but did not take into account the frictional coupling. The extraction of the eigenvalues predicted paired modes of the same order in the squeal frequency region. Based on Lang's experiments [71], a decoupling process of these modes was suggested to be an effective method in reducing squeal propensity. It was also suggested that the form of the contact pressure distribution between the brake lining and the drum at the friction interface contributes to the squeal propensity of the brake. Partial contact conditions cause a brake to have a greater squeal propensity than full contact and that crown contact is less prone to squeal than other partial contact conditions.

In 1999 Hamabe et al. [72] applied the complex eigenvalue method to a drum brake FE assembly incorporating a constant coefficient of friction. A two-degree of freedom mathematical model was used to explain the formation of a complex eigenvalue with a positive real part through the coupling of two eigenvalues associated with natural modes, a phenomenon that indicates self excited vibration. For this specific study, the finite element method indicated that self excited vibrations could occur even with a constant coefficient of friction due to the coupling of two eigenvalues of the drum. However, similar instabilities were obtained in the original drum-shoe assembly due to the introduction of the backplate. Decoupling of the appropriate modes was suggested as a means for the elimination of squeal. The results were verified experimentally through the use of holographic interferometry.

4.2.3 Experimental Approaches

Lang et al. [71]. used doublet mode decoupling by added mass to achieve stability on a drum brake and studied the effect of added stiffness when a practical mass was added to the drum. Their study showed the cyclic decoupling effect when the mass rotated with the drum producing cyclic squeal. Added stiffness along the drum periphery was found to be effective in separating the doublet modes with the arc length having great influence on the separation; a longer arc length was more effective for separation of the lower modes.

Hulten et al. [73], verified that both the shoes and the drum exhibit a combination of a wave motion and a synchronous motion by using the ODS technique. The experimental method allowed the measurement of the vibration characteristics of the drum assembly on the actual vehicle. The ODS technique can give the deflection shape by measuring vibration at pre-specified points of the structure and consequently performing a Fourier analysis. It was noted that the Doppler effect must be taken into account for an accurate recording of results. The outcome illustrated that the vibration is a combination of a wave motion and a synchronous motion and it is exhibited by both the shoes and the drum. It was found that these waves move in the direction of the drum's rotation. This behaviour is only associated with the binary flutter theory and the lining induced instabilities which incorporate a constant coefficient of friction.

Fieldhouse [74,75] replicated a one-quarter vehicle of a rear wheel drive vehicle, including the suspension system on a dynamometer to investigate the cause of low frequency noise. Through the use of mirrors, which were strategically placed around the drum brake assembly, he was able to observe the modes of vibration of each component such as the backplate and the drum through holographic interferometry. The results showed that the mode of vibration of the backplate is not stationery with respect to the backplate but has a tendency to move in the direction of the drum rotation. It was also shown that the rate of rotation of the backplate mode tends to be the frequency divided by half the mode order. However, the influence of the suspension on squeal propensity was not identified.

Fieldhouse and Beveridge presented a further paper in 2001 [23] that compares the disc and drum brake rotor mode movement. Based further on the theories previously established [47,74], it was suggested that both mode movements may be related to two opposed travelling waves of the same frequency but different amplitude and that the wave travels in the direction of the rotor rotation.

5. CONCLUSIONS AND FUTURE TRENDS

Advances in computer technology during the last decade have enabled researchers on brake noise to realise analytical models of great complexity, with the use of FEM as the predominant brake noise prediction tool. In particular, research has concentrated on performing complex eigenvalue analysis to determine the unstable modes of vibration and relate noise generation to certain geometric characteristics of any given brake assembly. This type of analysis, however, was often based on assumptions that did not truly reflect actual brake behaviour and led to inconsistent results. Therefore, the above analysis was combined with non-linear contact analysis, which determines pressure distributions over the contact regions and optimisation procedures in order to automatically generate instability-free brake assemblies based on geometric changes. Other important features associated with noise generation have been combined with the complex eigenvalue analysis such as stick-slip, the effect of braking pressure, the rotational velocity of the rotor and even thermal effects. The simultaneous combination of the above-mentioned factors has proved to be an impossible task due to the high complexity of the models produced and the associated vast computing power and time required. For the purpose of minimising the complexity of the finite element models, researchers have introduced the modal synthesis method, which reduces the size of a FE model in terms of number of degrees of freedom while retaining the essential dynamic characteristics. A transient analysis can also be employed to study the dynamics of the brake system, but each individual analysis is characterised by long run times because of the small time steps required to capture the high frequency components of the system response. This in turn impacts on the practicalities associated with the conduct of a parametric study.

Multibody system tools have been introduced in order to help the investigations on low frequency noise generation, which is mostly associated with rigid body motions. Suspension components have been integrated with the brake parts to determine potential resonant behaviour within the assembly. Advanced experimental tools and techniques have been utilised by NVH engineers including accelerometers, dynamometers, laser Doppler scanning systems, sound intensity measuring devices and thermal imaging systems. Undoubtedly, the holographic interferometry technique has been the basis of experimentation providing a visual description of the brake's vibration behaviour during squeal generation.

The majority of the studies presented in this paper conclude that noise occurs when the brake system demonstrates resonance vibrations. These may involve modal coupling between any combination of brake and sometimes suspension components and it has been shown that this resonant frequency is identical to the squealing frequency. Additionally, recent experimental studies are indicating that substantial in-plane motion occurs during a squeal event in addition to the recognised out-of-plane motion. This behaviour must be quantified and the extent of coupling determined through numerical analysis.

The evolution of new ideas and suggestions on noise elimination has been evident throughout the period covered by this paper. Researchers have concentrated on the development of theoretical and finite element models in order to produce an integrated brake noise predictive tool. Such tools have been presented here but their capabilities are generally limited to very specific operating conditions and only partial predictions can occur. The fugitive nature of brake noise, in conjunction with the specifications required by the automotive industry for a reliable design tool, have put NVH engineers under great pressure. Also, it is evident that due to the confidentiality behind the production of friction materials, there is a lack of public domain information which slows the development process. Growing interest in the use of disc brake installations on commercial vehicles, coupled with impending noise/environmental legislation, is likely to shift the focus of research to this class of vehicle during the foreseeable future.

REFERENCES

1. **Crolla, D. A.** and **Lang, A. M.**, "Brake Noise and Vibration –The State of the Art" (1991), Vehicle Tribology Leeds-Lyon Tribology Series,18, p.165-174
2. **Lamarque, P. V.**, "Brake Squeak", The Experiences of Manufacturers and Operators, Report No. 8500 B,Inst. Auto. Engrs. , Research and Standardisation Committee, 1935
3. **Mills, H.R.**, "Brake Squeak: First Interim Report. Report No. 9000B, Institution of Automobile Engineers, Automobile Research Committee,1938
4. **Sinclair, D.**, Frictional Vibrations, Journal of Applied Mechanics, pp.207-214, 1955
5. **Spurr, R. T.**, "A theory of brake squeal", IMechE, Auto Div Proc. No. 1,1961/62, pp.33-40

6. **Earles, S.W.E., Lee, C.K.**, "Instabilities arising from the friction interaction of a pin-disc system resulting in noise generation", Trans. ASME., Journal of Engineering for Industry, Vol.98, Series B., No.1, pp.81-86
7. **North, M.R.**, " A mechanism of disc brake squeal", 14th FISITA Congress, Paper 1/9, June 1972
8. **Millner, N.**, "An analysis of disc brake squeal", SAE Paper, 780322,pp.1-11,1978
9. **Felske, A., Happe, A.**, "A vibration analysis by double pulse laser holography", SAE Paper No. 770030,1977
10. **Liles, G.D.**, "Analysis of disc brake squeal using finite element methods", SAE Paper No 891150,1989
11. **Lang, A. M., Smales, H.**, "An Approach to the solution of Disc Brake Vibration Problems", IMechE, C37/83,1983
12. **Abdelhamid, M. K., Blaschke, P., Wang, W. A., Yuan, S.**, "An overview of brake noise and vibration problems", I Mach Conference in Florida, p.57-61 (2001)
13. **North, M. R.**, "A survey of published work on vibrations in braking systems", A (8_23), M.I.R.A, 1969, Bulletin, 4, pp.8-12
14. **Kreitlow, W., Schrodter, F., Matthai, H.**, "Vibration and "Hum" of disc brakes under load", SAE Paper, 850079,1985
15. **Gouya, M., Nishiwaki, M.**, "Study on disc brake groan", SAE Paper, 900007,1990
16. **Rhee, S. K., Jacko, M. G., Tsang, P. H.**, "The role of friction film in friction, wear and noise of automotive brakes", Wear, Vol.146., pp.89-97, Oct. 1990
17. **Little, E., Kao, T. K., Ferdani, P., Hodges, T.**, "A dynamometer, investigation of thermal judder", SAE Paper, 982252, 1998
18. **Inoue, H.**, "Analysis of brake judder caused by thermal deformation of brake disc rotors", SAE Paper, 865131
19. **Watany, M., Abouel-Seoud, S., Saad, A., Abdel-Gawad, I.**, "Brake squeal generation", SAE Paper, 1999-01-1735, 1999
20. **El-Butch, A. M., Ibrahim, I. M.**, "Modeling and analysis of geometrically induced vibration in disc brakes considering contact parameters", SAE Paper, 1999-01-0143, 1999
21. **Ahmed, I. L. M., Leung, P. S., Datta, P. K.**, "Theoretical investigations on disc-brake squeal noise", IMechE 2000, C576/027/2000
22. **Flint, J.**, "Modelling of high-frequency disc-brake squeal", International Conference on Automotive Braking –Brakes 2000, IMechE, Leeds, ISBN 1-86058-261-3, pp.39-50
23. **Fieldhouse J., Beveridge C.**, "A comparison of disc and drum brake rotor mode movement, Seoul 2000 FISITA, World Automotive Congress, F2000H241, 2000
24. **Tzou, K.I., Wickert, J. A., and Akay, A., 1998**, "In-plane vibration modes of arbitrary thick discs", Journal of Vibration and Acoustics, Transactions of the ASME, Volume 120, Number 2, pp. 384-391
25. **McDaniel, J. G., Li, X., Moore, J., Chen, S. E.**, "Analysis of instabilities and power flow in brake systems with coupled rotor modes", SAE Paper, 2001-01-1602,2001
26. **Kido, I; Kurahachi, T; Asai, M.** " A study on low-frequency brake squeal noise", Proceedings of SAE, Technical Paper 960993, 1996.
27. **Matsushima, T.; Masumo, H.; Ito, S.; Nishiwaki M.**, "FE analysis of low-frequency disc brake squeal", Proceedings of SAE, Technical Paper 982251,1998
28. **Matsushima, T.; Masumo, H.; Ito, S.; Nishiwaki M.**, "FE analysis of low-frequency disc brake squeal", Proceedings of SAE, Technical Paper, 973020, (1997)

29. **Kung, S., Dunlap, K. B., Ballinger, S.**, "Complex eigenvalue analysis for reducing low frequency brake squeal", SAE Paper, 2000-01-0444, 2000
30. **Dihua, G. and Dongying, J.**, "A study on disc brake squeal using finite element methods", SAE Paper, 980597, 1998
31. **Tirovic, M., Day, A. J.**, "Disc brake interface pressure distributions", Proc. Instn Mech. Engrs Vol. 205, IMechE 1991, p.137-146
32. **Lee, Y.S.; Brooks, P.C; Barton, D.C.; Crolla D.A.**, "An integrated finite element approach to modelling disc brake squeal", Proc. FISITA. No B0404, 1996
33. **Nack, W.**, "Brake squeal analysis by finite elements", SAE Paper, 1999-01-1736, 1999
34. **Blaschke, P., Tan, M., Wang, A.**, "On the analysis of brake squeal propensity using finite element method", SAE Paper, 2000-01-2765, 2000
35. **Yuan, Y.**, "An eigenvalue analysis approach to brake squeal problems", Proceedings of the Dedicated Conference on Automotive braking Systems, 29th ISATA, Florence, 1996
36. **Shi, T. S., Dessouki, O., Warzecha T., Chang, W. K., Jayasundera A.**, "Advances in complex eigenvalue analysis of brake noise", SAE Paper, 2001-01-1603, 2001
37. **Nagy, L. I., Cheng, J., Hu, Y. K.**, "A new method development to predict squeal occurrence", SAE Paper, 942258, 1994
38. **Hu, Y. K., Mahajan, S., Zhang, K.**, "Brake squeal DOE using non-linear transient analysis", SAE Paper, 1999-01-1737, 1999
39. **Mahajan, S., Hu, Y. K., Zhang, K.**, "Vehicle disc brake squeal simulations and experiences", SAE Paper, 1999-01-1738, 1999
40. **Baba, H., Okade, M., Takeuchi, T.**, "Study on reducing low frequency brake squeal- from modal analysis of mounting bracket", SAE Paper, 952697, 1995
41. **Mottershead, J. E. and Chan S.N.**, Brake Squeal- An Analysis of Symmetry and Flutter Instability, DE-Vol. 49, Friction-Induced Vibration, Chatter, Squeal and Chaos, ASME, pp.87-97, 1992
42. **Chung, C. H., Steed, W., Kobayashi, K., Nakata, H.**, "A new analysis method for brake squeal Part 1: Theory for modal domain formulation and stability analysis, SAE Paper, 2001-01-1600
43. **Matsui, H., Murakami, H., Nakanishi, H., Tsuda, Y.**, "Analysis of disc brake squeal", SAE Paper, 920553, 1992
44. **Ichiba, Y., Nagasawa, Y.**, "Experimental study on disc brake squeal", SAE Paper, 930802
45. **Ishihara, N., Nishiwaki, M., Shimizu, H.**, "Experimental analysis of low-frequency brake squeal noise", SAE Paper, 962128, 1996
46. **Fieldhouse, J. D. and Newcomb, T.P.**, "The application of holographic interferometry to the study of disc brake noise", SAE Paper, 930805, 1993
47. **Fieldhouse, J. D. and Newcomb, T.P.**, "An experimental investigation into disc brake noise", IMechE, C444/036/93, 1993
48. **Fieldhouse, J. D.**, "A proposal to predict the noise frequency of a disc brake based on the friction pair interface geometry", SAE Paper, 1999-01-3403, 1999
49. **Matsuzaki, M. and Izumihara, T.**, "Brake noise caused by longitudinal vibration of the disc rotor", SAE Paper, 930804, 1993
50. **Kai, H., Suga, T., Baba, H.**, "Study on reduction of brake squeal caused by in-plate vibration rotor", SAE Paper, 982254, 1998

51. Eriksson, M., Lundqvist, A., Jacobson, S., "A study of the influence of humidity on the friction and squeal generation of automotive brake pads", Proc. Instn Mech. Engrs., Vol. 215, Part D, D05600, p. 329-342, 2001
52. Nishizawa, Y., Saka, H., Nakajima, S., Arakawa, T., "Electronic Control Canceling System for a disc brake noise", SAE Paper, 971037, 1997
53. Abendroth, M., "Advances/Progress in NVH brake test technology", SAE Paper, 982241, 1998
54. Edwards, C., Taylor, N., Williams, D., Dale, M., Buckberry, C., Reeves, M., "The use of high-speed ESPI and near-field sound pressure measurements to study brake-disc modal behaviour", International Conference on Automotive Braking –Brakes 2000, IMechE, Leeds, ISBN 1-86058-261-3, pp.73-84
55. McDaniel, J. G., Moore, J., Chen, S. E., Clarke, C. L., "Acoustic radiation models of brake systems from stationary LDV measurements", Proceedings of IMEC, ASME, Nashvile, Tennessee, 1999
56. Riesland, D., Janevic, J., Malosh, J., Stringham, W., "Rear disc brake moan – experimental investigation and ADAMS simulation", 2nd International Seminar on Automotive Braking, 1998, Leeds, IMechE, pp.67-76
57. Bettela, M., Sharp, R. S., "A measurement method for automotive creep groan induced vibration with data processing and analysis", C576/020/2000, IMechE2000, pp.603-617
58. Dunlap, K. B., Riehle, M. A., Longhouse, R.E., "An investigative overview of automotive disc brake noise", SAE Paper, 1999-01-0142, 1999
59. Engel, H.G., Bachman, Th., Eichorn, U., and Saame, Ch., "Dynamical behaviour of brake-disc geometry as cause of brake judder", Proceedings, EAEC 4th International Conference on Vehicle and Traffic System Technology, Vol.1, pp.465-481, Strasburg, 1993
60. Gassmann, S., Engel, H. G., "Excitation and transfer mechanism of brake judder", SAE Paper, 931880, 1993
61. Stringham, w., Jank, P., Pfeifer, J., and Wang, A., "Brake roughness-Disc brake torque variation, rotor distortion and vehicle response," SAE Transactions, Section 6, pp.1235-1247, SAE Paper 930803,1993
62. Jacobson, H., "The brake judder phenomenon. Classification and problem approach" The 2nd International Seminar on Automotive Braking, IMechE, Leeds, 1998
63. Vikulov, K., Tron, B., Buonfico, P., "Brake vibration and disc thickness variation (DTV)", 2nd International Seminar on Automotive Braking, 1998, Leeds, IMechE, pp.77-84
64. de Vries, A., and Wagner, M., "The brake judder phenomenon", SAE transactions, Section 6, Vol.101, pp.652-660. SAE paper 920554, 1992
65. Kao, T. K., Richmond, J.W., Douarre, A., "Thermo-mechanical instability in braking and brake disc thermal judder: an experimental and finite element study", 2nd International Seminar on Automotive Braking, 1998, Leeds, IMechE, pp.231-263
66. Nishiwaki, M., "Generalized theory of brake noise", Proc. Instn. Mech. Engrs Vol. 207, D02392, p.195-202, 1993
67. Hulten, J., "Brake Squeal- A self-exciting mechanism with constant friction", SAE Paper, 932965, 1993
68. Hulten, J., "Some drum brake squeal mechanisms", SAE Paper, 951280,1995
69. Hulten, J., "Friction phenomena related to drum brake squeal instabilities", ASME Paper, DETC97/VIB-4161, Sept.1997

70. **Day, A J., Kim, S A.** "Noise and vibration analysis of an S-cam drum brake", Proceedings of IMechE, Vol. 210, Paper No: D03894, 1996.
71. **Lang, A. M., Schafer, D. R., Newcomb, T. P., Brooks, P. C.,** "Brake Squeal –The influence of rotor geometry", IMechE, C444/016/93, 1993
72. **Hamabe, T; Yamazaki, I; Yamada, K; Matsui, H; Nakagawa, S; Kawamura, M.** "Study of a method for reducing drum brake squeal", Proceedings of SAE, Technical Paper 1999-01-0144, 1999.
73. **Hulten, J., Flint, J., Nellemose, T.,** "Mode shape of a squealing drum brake", SAE Paper, 972028,1997.
74. **Fieldhouse, J. D., Rennison M.,** "An investigation of low frequency drum brake noise", SAE Paper, 982250,1998
75. **Fieldhouse, J.D.**, "Low frequency drum brake noise investigation using a ¼ vehicle test", SAE Paper No.2000-01-0448, 2000
76. **Nishiwaki, M.**; "Review of Study on Brake Squeal", JSAE Review Vol.11 No.4, pp.48-54, October 1990

Evaluation of disc brake pad pressure distribution by multibody dynamic analysis

W RUMOLD
Robert Bosch GmbH, Schwieberdingen, Germany
R A SWIFT
Bosch Automotive Chassis, South Bend, Indiana, USA

ABSTRACT

Common problems associated with the design and utilisation of disc brakes often relate to the effects of pad loading. Non-uniform pad loading can result in taper wear of the linings, which can result in greatly reduced pad life and potential rotor damage in extreme cases. In order to examine pad loading characteristics, a multibody dynamics model has been utilised to perform an investigation into caliper geometry changes that could be employed to effectively improve pad loading uniformity. This investigation explored modifications to caliper finger position fore and aft (relative to the housing centreline), and piston offsets from the caliper centreline. Pressure distributions were measured during the constant torque region of the brake applies. The utility of the multibody dynamics approach to extract useful design trend and design sensitivity information is described.

1 INTRODUCTION

Non-uniform pad loading can result in uneven lining wear in disc brakes. In pronounced cases, this tapered wear can significantly lower lining life. Low pad lining life also contributes to customer complaints related to brake system maintenance. In order to investigate techniques that can be used to improve pad loading, a multibody dynamics model of a medium-heavy truck disc brake is employed. Multibody simulations have been utilised in many automotive applications (1-6), and analyses of this type have been applied to resolve a number of interesting braking problems (7,8).

A goal of this investigation was to look at minor changes to caliper housing geometry to improve pad loading characteristics, with the desire to make pad loading as uniform as possible. In this study, minor changes are considered as those that can be achieved through minor alterations of caliper casting geometry or simple machining operations to the existing casting. Two different caliper modifications are considered. The first involves the modification of the caliper fingers, shifting their effective location by ± 5mm in the leading or trailing direction (as defined by rotor rotation). The second modification involves shifting the pistons relative to the caliper centreline, again by ± 5 mm. These two modifications influence the pressure differences between the leading and trailing edge of the pad lining.

2 DESCRIPTION OF THE DISC BRAKE MODEL

This study of pad loading involved a complete system model of a twin piston medium truck disc brake. This particular brake design utilises two 73mm pistons with a 380 mm diameter rotor, and is designed for applications up to a gross vehicle weight of 146.8 kN. The model is composed of twelve separate components, representing the anchor, tie bar, pistons, pads, guide pins, guide pin locating bolts, caliper housing, and rotor. The geometry, mass, and inertia of each component are taken from the solid models of production hardware. The geometric information is complete, including dimensional tolerances on all components and features. Joints, contact definitions, bushings, and motions are applied to the dynamics model, representing the physical characteristics of the actual hardware. The individual components of the assembly are attached using 163 contact and joint entities. Figure 1 shows the disc brake model. The ADAMS (9) simulation package is employed.

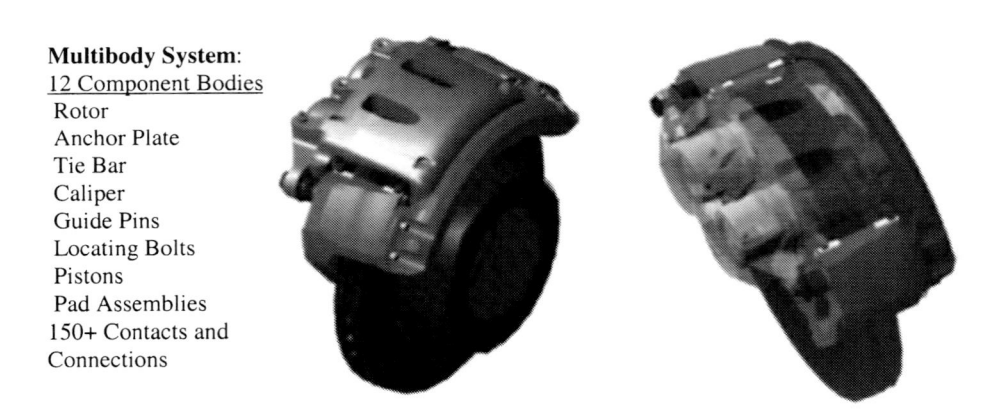

Multibody System:
12 Component Bodies
Rotor
Anchor Plate
Tie Bar
Caliper
Guide Pins
Locating Bolts
Pistons
Pad Assemblies
150+ Contacts and Connections

Figure 1. ADAMS multibody dynamic model of a medium-heavy truck brake

The model can be employed for virtually any type of braking condition that can be expected. For the investigations presented here, braking pressures in the range of 1.38 MPa to 6.90 MPa are considered, with vehicle velocities in the range of 16.1 to 48.3 km/h.

3 PAD LOADING CALCULATIONS

The inner and outer pads each have 12 contact points defined with the rotor. Figure 2 describes these contact point locations for the inner pad. These points are placed in a symmetric fashion around the perimeter of the lining. Note that although Figure 2 indicates a chamfer, the mathematical model has no chamfer present. Thus, contact points 2 and 7 are, in fact, attached to the unchamfered lining.

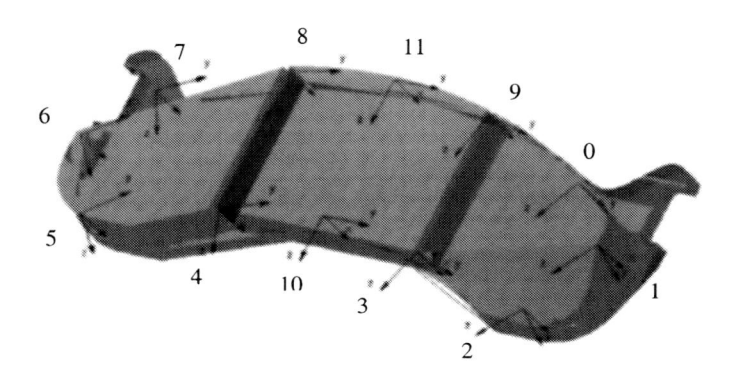

Figure 2. Inboard pad model with contact points indicated

During any given brake application, forces will be generated at these 12 contact points. These forces will have components in all three coordinate directions, as defined by the contact model and the applied coefficient of friction. A standard ADAMS IMPACT formulation is employed, as described in Figure 3. IMPACT uses a sphere to plane contact assumption, and the resultng contact force is a function of four variables (STEP is an intrinsic ADAMS function).

$$Force = \begin{cases} Max\left(0, k(x_1 - x)^e - STEP(x, x_1 - d, c_{max}, x_1, 0)\dot{x}\right) : x < x_1 \\ 0 : x \geq x_1 \end{cases}$$

where
k = contact stiffness
e = exponential constant
d = maximum penetration depth (the depth at which full damping is applied)
c_{max} = maximum velocity dependant damping

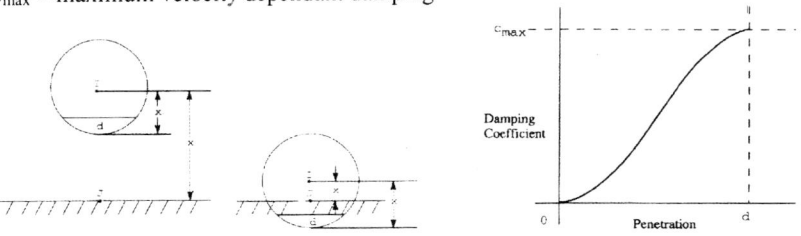

Figure 3. Sphere to plane contact definition employed in the ADAMS IMPACT function

Pressures are approximated by dividing the contact force by the effective portion of the pad area,

$$\sigma_i = F_i / \left(\frac{A_{pad}}{n} \right)$$

where n is the number of contact points. Thus we assume that the force associated with each of the 12 contact points acts over an area equal to one twelfth the total lining area. Figure 4 represents an example time history of the contact forces (normal to the pad surface) from a 1.38 MPa brake application with an initial velocity of 16.1 km/hr. Here, the contact forces reach nearly constant values once the hydraulic pressure has reached its nominal value (at t=1.35 s). A constant torque region of the brake application is effectively entered. The vehicle comes to a stop at t = 5.7 s, and pressure is released. The calculated pressure values for the inner pad are plotted in Figure 5 as a function of the horizontal position. The green line in Figure 5 indicates the solution of an analytical calculation of the pad pressures on the basis of the applied hydraulic pressure and the pad area,

$$\sigma_a = P_{hydr} \, 2 A_{pist} / A_{pad} = 1.379 MPa \left(\frac{m^2}{1(10^6) mm^2} \right) 8371 mm^2 / 13451.5 mm^2 = 0.86 N/mm^2$$

where σ_a is the calculated pressure, p_{hydr} is the hydraulic pressure and A_{pist} and A_{pad} are the piston and pad areas.

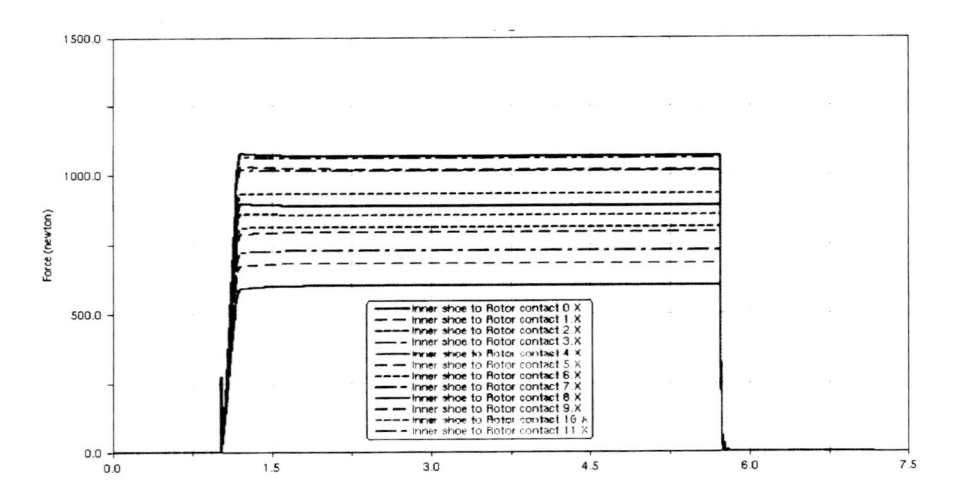

Figure 4: Contact forces versus time (1.379 MPa, 16.1 km/hr)

The results shown in Figure 5 clearly describe both the radial pressure distribution differences as well as the differences between the leading and trailing edge of the lining (leading edge is to the right in the figure).

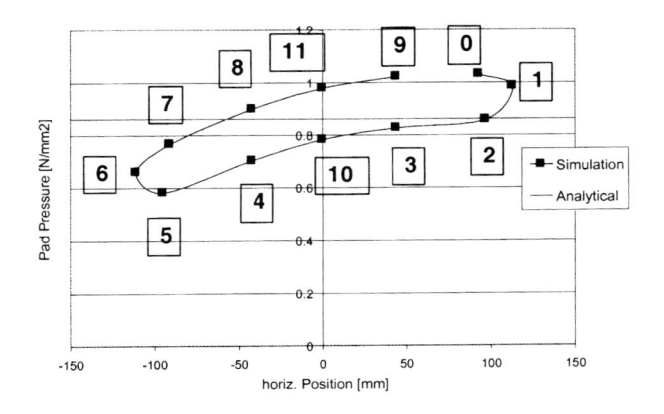

Figure 5. Inner pad contact pressures versus horizontal marker position

4 PAD LOADING DESIGN PARAMETER STUDIES

The multibody dynamics model is now used to investigate the impact of brake design parameter changes on the pad load distribution. The goal of these investigations is to identify simple modifications that have the propensity to improve overall pad loading characteristics. The effects of caliper finger position and piston position on both the inner and outer pad load distributions are considered. In all of the following descriptions, the results reflect pressure characteristics obtained from a 16.1 km/hr, 1.379 MPa brake application. Pressures are calculated during the constant torque region of the brake application, at t= 3.0 seconds.

4.1 Caliper Finger Position Location

Modification of the caliper finger position relative to the caliper centreline will alter the pressure differential between the leading and trailing edge of the lining, due to a shift in the reaction force locations on the shoe steel. All three fingers are offset in unison. Figure 6 describes this modification.

Figure 7 describes the results for the inner pad. The difference between the leading and trailing pressure decreases when the caliper fingers are shifted towards the trailing edge. Moving the fingers toward the leading edge increases the pressure difference. On the outer pad (refer to Figure 8), the same changes occur, although they are more pronounced.

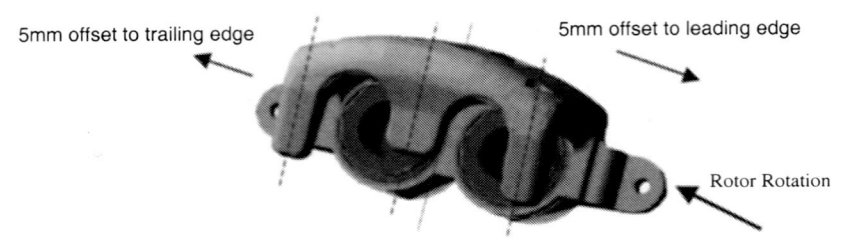

Figure 6: Caliper finger offset description

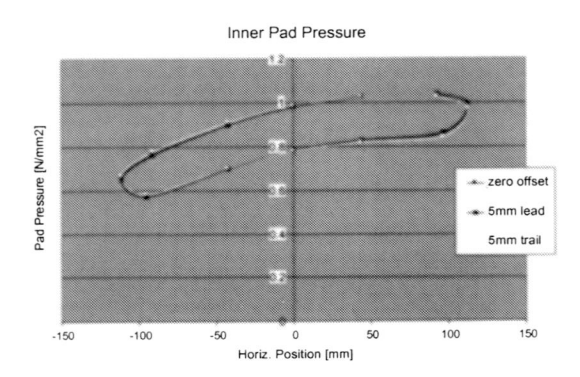

Figure 7: Inner pad pressure distribution as a function of the caliper finger positions

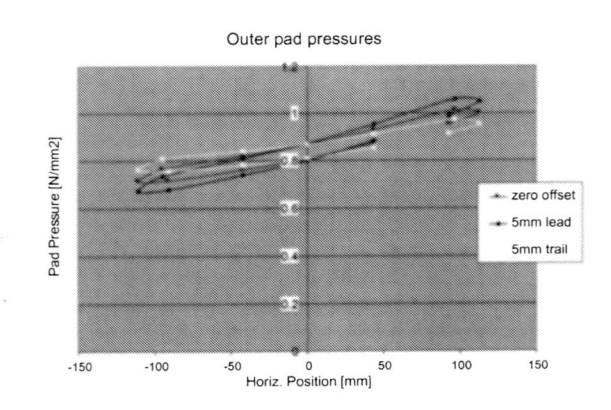

Figure 8: Outer pad pressure distribution as a function of the caliper finger positions

4.2 Piston Location Relative to Caliper Centreline

Another design parameter change that can positively influence pad loading is piston position. Figure 9 describes the geometry involved. Both pistons are moved in unison. Figures 10 and 11 describe the inner and outer pad pressures, respectively. A shift to the trailing edge is beneficial as it decreases the differences between the leading and the trailing pressures, by offsetting the reaction forces on the shoe steel of the inner pad. Both the inner and the outer pad pressures show significant response to this modification.

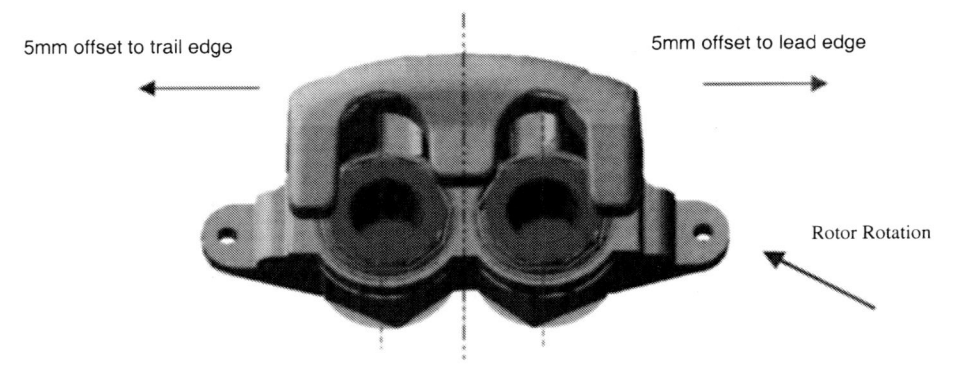

Figure 9: Piston centreline offset description

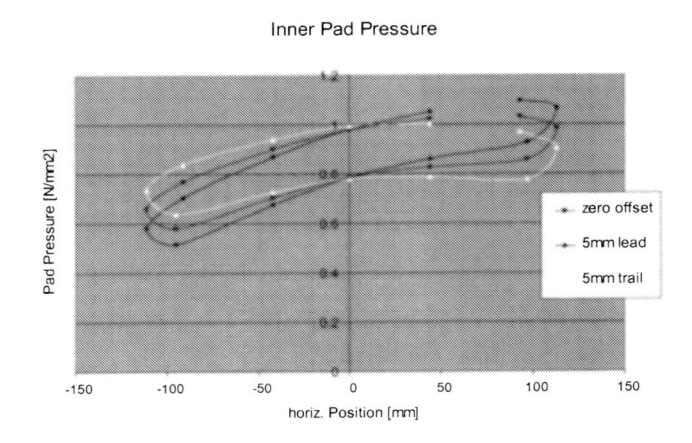

Figure 10: Inner pad pressure distribution as a function of the piston positions

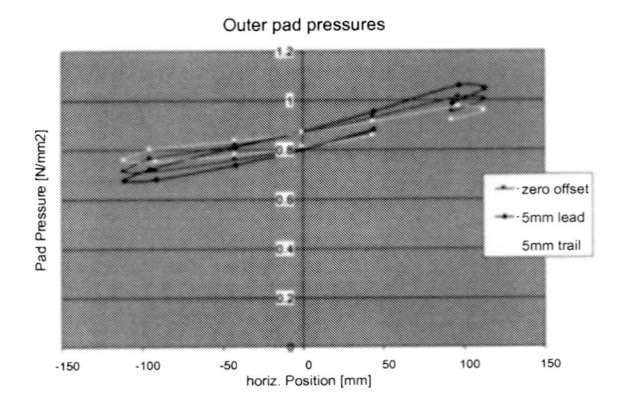

Figure 11: Outer pad pressure distribution as a function of the piston positions

The model behaviour indicates that when the piston positions are changed, the caliper has the tendency to "clock" slightly with respect to its original position, changing both the inner and outer pad pressure distributions. This effect is also present when the caliper finger positions are changed, but the inner pad pressures are only slightly altered, as the piston force on the inner pads is primarily dependent on the hydraulic pressure (which does not change).

5 CONCLUSIONS

An ADAMS multibody dynamics model has been utilised to investigate the behaviour of a 73mm twin piston medium truck brake. Design parameter studies centred on improving pad loading characteristics were performed. In all cases, improvements to the overall inner and outer pad load distributions could be achieved. The next step in this study will be the experimental measurement of pad pressure distributions for model correlation. Pad linings instrumented with strain gages will be the primary source of this pad loading information. Pressure sensitive mylar film sensors will also be employed. Finally, as dictated by the model correlation effort, flexible bodies may be selectively introduced to further refine the pad loading prediction.

6 REFERENCES

1. SAE 2000-01-3439 - "Integration of Virtual Prototyping with Instrumented Testing of Vehicles"; Greg Wheeler, Anand Mantrala, S. Colin Ashmore, Henry C. Hodges Jr. (Nevada Automotive Test Center)

2. SAE 2001-01-2809. "Drivetrain Noise and Vibration Troubleshooting"; Shan Shih, Jacinto Yruma, Phil Kittredge (ArvinMeritor Inc.)

3. SAE 2001-01-1138 - "Stability Control of Combination Vehicle"; Pahngroc Oh, Hao Zhou, Kevin Pavlov (Visteon)

4. SAE 2000-01-0303 - "Utilization of ADAMS to Predict Tracked Vehicle Performance"; Jesper Slattengren (Mechanical Dynamics, Inc.)

5. SAE 980901 - "Design of a Light Weight Suspension Component Using CAE"; Cheon Soo Jang, En Wha Jung, Jae Wook Jeon, Young-Duk Yoo, Choi Bok Lok (Hyundai Motor Co.)

6. SAE 1999-01-3760 - "Application of Computer Aided Engineering in the Design of Heavy-Duty Truck Frames"; Carlos Cosme, Amir Ghasemi, and Jimmy Gandevia (Western Star Trucks, Inc.)

7. SAE 2001-01-3131 - "Multi-body Dynamic Simulation for the Evaluation of Disc Brake Slide Force"; Richard Swift (Bosch)

8. SAE 2000-01-2769 - "Brake Moan Simulation using Flexible Methods in Multibody Dynamics"; Anthony Gugino, John Janevic (Mechanical Dynamics, Inc.), Laszlo Fecske (Bosch Braking Systems)

9. Mechanical Dynamics Inc. – "Using ADAMS Solver, Version 11.0", Ann Arbor, MI, 2000

Investigation of the time development of disc brakes generating noise using laser holography

C TALBOT and **J D FIELDHOUSE**
School of Computing and Mathematics and School of Engineering, University of Huddersfield, UK

ABSTRACT

The technique of double pulsed laser holography to investigate real disc brake systems generating noise has been developed to investigate the time evolution of the vibrations. By using electronic triggering devices, a series of time related holographic images can be obtained, corresponding to different times in the cycle of vibration of the braking system.

By digitising the holographic images it has been possible, using image processing techniques, to develop algorithms that determine the co-ordinates of the hologram fringe lines on the exposed portions of the disc and pad surfaces. From this information three-dimensional representations of the vibrating disc and finger pad surfaces can be constructed. Furthermore, from the series of time related images, a three dimensional animation of the real motion of the disc and pad surfaces can be obtained.

Using least-squares mathematical approximations it is also possible to generate accurate Fourier Series type representations of the vibrating disc and pad surfaces, showing both in-plane and out-of-plane motion. Holograms of the disc rim show how the vibrating motion varies throughout the thickness of the disc. It is believed that this investigation will be important in developing mathematical models of disc brake squeal.

1. INTRODUCTION

Holographic interferometry has been successfully developed over several years to investigate brake systems generating noise. By using a test rig that reproduces the real conditions in which a disc brake operates, driving the noisy brake at a suitable rotational speed and steadily applying pressure, it is possible to obtain a continuous "squeal" that can be investigated with a view to analysis and possible cure. Brake noise is notorious for its unpredictability and it is often necessary to apply a range of pressures, allowing for temperature build-up of the pads and rotor, to obtain continuous noise. Transitory effects may also be observed at a range of different frequencies. However if a persistent squeal is observed in vehicles under road conditions, it is usually possible to produce a continuous squeal at the same frequency under test conditions. Changing the angular velocity of the brake rotor (typically fixed at about 10 r.p.m.) does not appear to effect the generation of brake noise at the higher "squeal" frequencies.

The instability of the disc rotor, pad and other components undergoing continuous vibration at a definite frequency can be recorded using double pulsed laser holography. The dark lines or fringes on the resulting hologram can be used to determine the mode order of the vibrating disc rotor (typically a diametrical mode of order between 3 and 10) as well as giving information about the instantaneous amplitude of vibration of the various components at the time of recording the image. A considerable amount of information can be obtained from this method, especially if suitable mirrors are mounted around the rig enabling side and top views of the brake components to be made.

However the method is limited to the production of a hologram that "freezes" the system at one point in time, it does require considerable experience to interpret and it may not reveal the complex dynamics of the vibrating system. Over the last decade the method has been developed further to enable a series of time related holograms to be recorded over a cycle of oscillation of the noisy brake system [1]. By the use of electronic triggering devices located on the brake pad, the time in the cycle of vibrations at which a zero displacement with a positive velocity gradient occurs can be precisely established. By using this "zero" point the time during the cycle of vibration at which the holograms are taken can thus be accurately determined also. From this series of time related holograms much more information can then be obtained about the motion of the various parts of the brake system, particular the disc and pad.

Using image processing techniques the data obtained from the fringe lines in the holograms can be used to generate a three dimensional surface that represents the actual displacement (typically of the order of multiples of 100 nm) of the brake components at the given instant in time when the hologram is recorded. By combining together (using interpolation) a series of such surfaces obtained from time related holograms of the brake disc, it has been shown to be possible to generate an animation representing the actual vibratory motion of the disc surface and pad surface [7].

A further development has involved fitting the fringe lines for the disc obtained in the holograms to mathematical equations using Fourier analysis. By this means the wave motion on the disc surface can be analysed as it evolves in time. Two results from this investigation appear to be common to all disc brake systems.

Firstly the waveform on the surface of the disc is not stationary but travels round it. It can be analysed into a forward travelling wave that moves in the same direction as the disc's rotation together with a smaller backward travelling wave moving in the opposite direction. The angular velocity of the travelling waves is that of the angular frequency of the brake noise divided by the mode order.

Secondly the vibration of the disc, as well as having an out-of-plane component, appears to have a significant in-plane component. The in-plane motion would appear to vary throughout the thickness of the disk, depending on the perpendicular distance from the mid-plane of the disk.

In this paper these features are demonstrated to occur for a four-piston opposed type disc brake system that is used on an off-the-road vehicle. Since the pad is totally obscured by the calliper for this system it was not possible to consider the pad motion investigated for other systems. However the holograms (seen through a mirror) of the rim of the disc for this case enabled more detailed calculations of in-plane motion to be performed than has previously been possible.

It is believed that the results obtained will assist in developing a deeper understanding of the mechanism by which squeal occurs in a disc brake system, and assist in the development of realistic mathematical models for this phenomena. A variety of analytical and finite element approaches have been made to modelling disc brake noise (see e.g. [2-5]) but arguably do not reproduce all the complexities of the real world system.

2.INVESTIGATIVE SET-UP

A photograph of the test rig is shown in Figure 1. The holograms are recorded by using a 100 mJ pulsed Ruby laser with a pulse duration of 30 ns. Time between the pulses can be varied to give the optimum holograms required being adjustable from 1μs to 999 μs. A piezo-electric crystal mounted on the pad is used to trigger the first laser pulse with an adjustable timer varying from 43 μs to 1000 μs. Using the pad vibrations as a reference, the time taken from the zero point with positive gradient can be determined accurately for each hologram.

Figure 2 shows six out of the eight time-related holograms taken for this system. In each case the holograms are labelled with the time taken in μs from the start of the cycle of vibration. The brake system was kept rotating at a constant angular velocity of 1.05 rads/s (10 r.p.m) in a counter-clockwise direction. Note that the fringe patterns on the disc surface indicate the number of nodes (zero displacement) and anti-nodes (maximum or minimum displacement) on the disk surface. Allowing for the portions that are obstructed it can be seen that there are 8 of each, representing a 4 diametrical mode of vibration. The movement of the fringe pattern in time in a counter-clockwise direction around the disc can be observed. Since the number of fringes will be shown to be proportional to the amplitude of vibration, variation of amplitude over the time of a cycle can also be seen.

Figure 1 Overall view of test rig

| 0 μs | 11.61 μs | 23.22 μs |
| 92.88 μs | 116.10 μs | 150.93 μs |

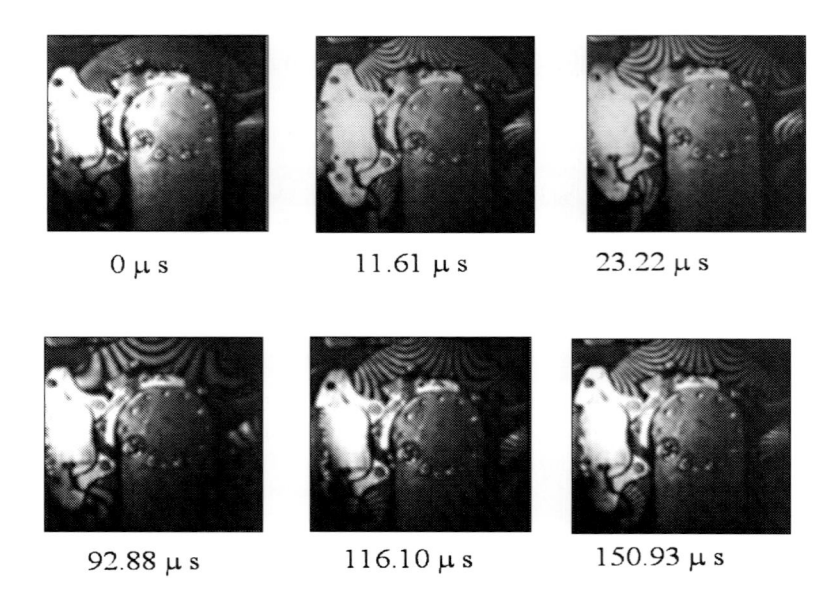

Figure 2 Six of the time-related holograms for the OTR vehicle brake. Frequency 2900 Hz

Mirrors are mounted at the top of the brake disc to show the interference fringes on the disc rim. A close-up through the mirror is shown in Figure 3.

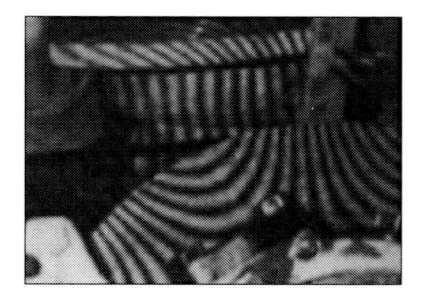

Figure 3 Top view of disc rim seen through mirror

3. HOLOGRAPHIC INTERFEROMETRY

Standard holographic interferometry shows that the intensity of fringe lines at point P on the surface of the object is given by [6]:

$$I(P) = 2I_0(P)(1 + \cos(\Delta\phi(P)))$$

Here $I_0(P)$ is the intensity of the light of both the first and second pulses of light and $\Delta\phi(P)$ is the phase difference between the two light wavefronts. The dark fringes are then given at points P where $\Delta\phi(P)$ is equal to an odd integer multiple of π. Approximating the light wave pulses by rays of light, it can be shown that [6]:

$$\Delta\phi(P) = \frac{2\pi}{\lambda}\delta(P)$$

where λ is the wavelength of the laser light and $\delta(P)$ is the difference in length of the light rays travelling from the illumination point to the observation point. Another calculation then shows that [6]:

$$\delta(P) = \mathbf{d}(P).(\mathbf{b}(P) - \mathbf{s}(P))$$

where $\mathbf{d}(P)$ is the displacement vector at point P on the surface between being illuminated by the first and second light pulse, $\mathbf{s}(P)$ is the unit vector in the direction of the incident light ray towards the surface at P and $\mathbf{b}(P)$ is the unit vector in the direction of the reflected beam away from the surface at P.

For the purpose of this work the light rays can assumed to be parallel, i.e. $\mathbf{s}(P)$ and $\mathbf{b}(P)$ are independent of P. Let Δt be the time between laser pulses, so that the point P on the surface of the disc at time t has a displacement $\Delta\mathbf{w}$ in time Δt, i.e. taking $\mathbf{d}(P) = \Delta\mathbf{w}$ in the above, the basic equation for destructive interference (dark fringes) is then :

$$\Delta\mathbf{w}.(\mathbf{b} - \mathbf{s}) = (N - \tfrac{1}{2})\lambda \tag{1}$$

where N is an integer (positive or negative).

Assume that the x and y axes lie in the brake disc (x horizontal, y vertical) with the z-axis pointing out of the disc (towards the light source). A further simplifying assumption is that the x, y coordinates may also be used in the holographic image, that is to say orthogonal projection rather than perspective projection is used. Then $\mathbf{s}$ and $\mathbf{b}$ will be as shown in the schematic representation in Figure 4.

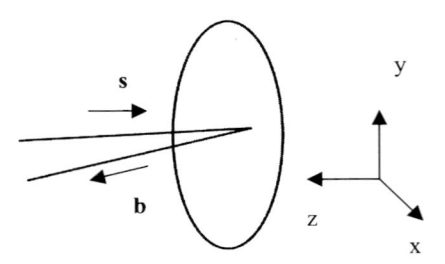

Figure 4 Schematic diagram showing notation used

Let a, b and c be the x, y and z components, respectively, of $\mathbf{b}-\mathbf{s}$. Since the angles in the optical arrangement can be measured, a, b and c can be calculated from the direct cosines of the incident and reflected light directions $\mathbf{s}$ and $\mathbf{b}$.

Taking the origin of the coordinate system to be at the centre of the brake disc it is convenient to work in polar coordinates. If r and θ are the usual polar co-ordinates in the x-y plane, denote the standard unit orthogonal vectors in the r, θ and z directions by $\mathbf{e}_r, \mathbf{e}_\theta$ and $\mathbf{e}_z$, respectively. With respect to this basis:

$$\mathbf{b}-\mathbf{s} = (a\cos\theta + b\sin\theta)\mathbf{e}_r + (-a\sin\theta + b\cos\theta)\mathbf{e}_\theta + c\mathbf{e}_z$$

(In the test rig above, $\mathbf{s}$ and $\mathbf{b}$ are at $4°$ and $10°$, respectively, to the z direction, and lie approximately in the horizontal plane so $a = 0.10$, $b \cong 0$ and $c = 1.98$).

First consider constant velocity rotation with angular velocity Ω in a counter-clockwise direction. In this case,

$$\Delta\mathbf{w} = r\Omega\Delta t \mathbf{e}_\theta$$

Then (1) implies:

$$r\Omega\Delta t(-a\sin\theta + b\cos\theta) = (N - \tfrac{1}{2})\lambda$$

i.e. $\Omega\Delta t(-ay + bx) = (N - \tfrac{1}{2})\lambda$

(for integer N), i.e. a series of parallel lines. Such lines, though obscured by the brake housing in this case, can often be seen in the central region of disc holograms

Consider next the disc undergoing vibration at a fixed frequency. Assume the displacement of the disc surface due to vibration is given by

$$\mathbf{w} = w_r(r,\theta,t)\mathbf{e}_r + w_\theta(r,\theta,t)\mathbf{e}_\theta + w_z(r,\theta,t)\mathbf{e}_z$$

where it is assumed that w_r, w_θ, w_z are sinusoidal in ωt ($\omega = 2\pi \times 2900$). To gain maximum contrast fringe patterns, Δt, the time between laser pulses, is normally taken to be half a period of oscillation ($\frac{1}{2 \times 2900} = 172.4\ \mu s$ in this case). However this choice produced too many fringe lines for a viable analysis, so a smaller choice of $\Delta t = 32\ \mu s$ was made. It can then be assumed that

$$\Delta \mathbf{w} = -p(w_r(r,\theta,t)\mathbf{e}_r + w_\theta(r,\theta,t)\mathbf{e}_\theta + w_z(r,\theta,t)\mathbf{e}_z)$$

where $p = 0.575$, as shown in Appendix 1.

If the constant-speed rotation of the disc is combined with the displacement due to vibration, equation (1) then yields:

$$-pw_r(a\cos\theta + b\sin\theta) - pw_\theta(-a\sin\theta + b\cos\theta) - pcw_z + r\Omega\Delta t(-a\sin\theta + b\cos\theta)$$
$$= (N - \tfrac{1}{2})\lambda \qquad (2)$$

for fringe lines, where N is an integer. Thus for each value of t in the cycle of vibration, and for $N = \ldots -2, -1, 0, 1, 2, 3, \ldots$ equation (2) gives the polar co-ordinate equation of the fringe lines seen on the disc surface in the holograms.

It seems reasonable to neglect the in-plane motion terms in (2) as both incident and reflected light beams producing the hologram were chosen nearly perpendicular to the disc surface, giving a and b much smaller than c. This point is reconsidered in Section 5.

The term for of the rotation of the disc, though relatively small in this case, can be calculated and the resulting equation (2) can be used to calculate w_z along each fringe line. This enables an accurate mathematical representation of w_z, the out-of-plane displacement as a function of r and θ to be made for each hologram.

Note that if rotation is neglected, (2) becomes:

$$-pcw_z = (N - \tfrac{1}{2})\lambda \qquad (3)$$

and the fringe lines will effectively become contour lines for $w_z = const.$ successive lines representing an increase (or decrease) in displacement of $\frac{\lambda}{2pc}$. Given that the wavelength of light $\lambda = 633$ nm for the given set-up, this gives a value of 515.6 nm for the out-of-plane displacement between fringe lines.

4. FRINGE SKELETONISING AND NUMBERING

The series of time-related holograms were converted to a digitised form in order to subject them to further analysis. A number of methods can be used to process the digitised images. To enable easier determination of the fringe lines it is possible to adjust the distribution of greyscale values to improve the contrast and to use filters to remove background noise. Edge detection techniques were used to find the outline of the disc in each of the holograms as well as picking out key features to use as reference points.

Since there are unavoidable small displacements from one holographic image to the next, a crucial question is to find the coordinates (i.e. pixel position) of the disc and pad boundaries in each image to enable the transformation to a standard "real-world" coordinate system to be found in each case. The boundaries of the disc in each image are slightly elliptical, so it is necessary to apply a transformation so that the elliptical boundaries are transformed to concentric circles.

After experimenting with various techniques it was decided that sufficient information about the fringe patterns in each hologram could be obtained by treating the dark fringes as simple curves, a process known as "skeletonisation" in the holography literature [6]. Enough information about the position of these curves could be obtained by drawing three equally spaced concentric circular paths between the inner and outer boundary of the disc (i.e. circular in the transformed "real-world" system but elliptical on each hologram). The angles at which each fringe line intersects the circular path could then be accurately determined for the three circular paths on each holograms.

Next it is possible to consistently number the fringe lines as each path is followed around the hologram of the disc, moving from zero at the nodes (zero displacement) to maximum positive or negative at the antinodes (see Figure 5). Whether to take a positive or negative value for N (i.e. whether the out-of plane displacement is positive or negative) is somewhat arbitrary at this stage. If assumptions are made about the relation between out-of-plane and in-plane motion it is possible to determine the sign of N by a comparison between the fringe lines on the disc and the slopes of the fringe lines seen on the disc rim through the mirror, considered in Section 6.

Neglecting terms due to in-plane motion, equation (3) enables the displacement w_z to be calculated at each given value of r, θ. For each time related hologram there was found to be adequate co-ordinate data to construct a three dimensional representation of part of the disc surface by using standard spline interpolation.

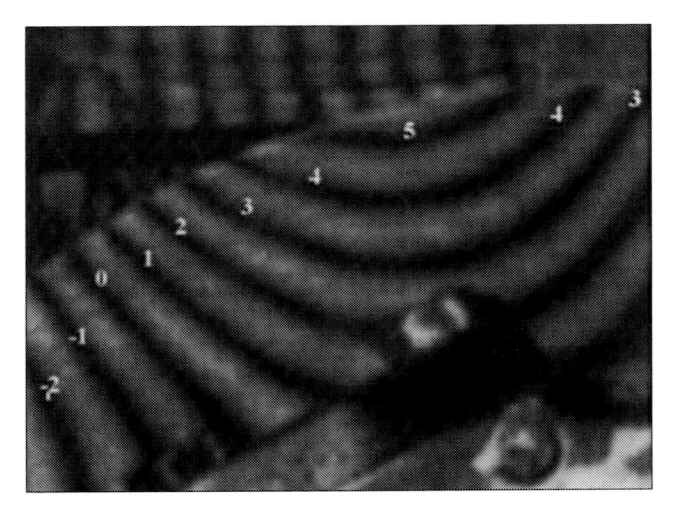

Figure 5 Part of hologram of disc showing a circular path. Angles where the path intersects fringes are measured and fringes are numbered as shown.

It is then possible to bring together the surfaces obtained from the time-related holograms to make a three dimensional animation for part of the disc surface. Using the sinusoidal assumption about the surface motion of the disc, it is possible to determine $A(r,\theta)$ and $B(r,\theta)$ giving a least squares fit to w_z in the form:

$$w_z = A(r,\theta)\cos\omega t + B(r,\theta)\sin\omega t$$

The values of A and B can then used to find w_z at equally spaced values of t throughout a full cycle enabling a three dimensional animation of the disc surface, based on the real data, to be constructed.

5. FOURIER APPROXIMATION TECHNIQUES

To give further information about the real disc and pad motion it was decided to investigate fitting the data with Fourier series type expansions of the form:

$$\frac{a_0}{2} + \sum_{n=1}^{M} a_n \cos n\theta + \sum_{n=1}^{M} b_n \sin n\theta$$

for the displacements w_r, w_θ and w_z along each circular path and in each hologram. There is insufficient data to fully solve the equations for out-of-plane as well as in-plane displacements independently, so w_r and w_θ were at first neglected. Substituting the values of θ for given fringe numbers N into equation (2) enables the Fourier coefficients for w_z to be determined by the least squares method. (It would be better to use approximate integrals to determine the Fourier coefficients but this is not possible due to lack of data for a full circle $\theta = -\pi$ to π).

For each hologram there will then be three expansions, one for each path. These can be combined by making a_n and b_n simple functions of r (e.g. linear or quadratic).

To investigate the change of w_z in time it was decided however to use only the outer path for each hologram but to make a_n and b_n vary sinusoidally in ωt. Thus w_z was approximated by terms of the form:

$$\frac{a_0}{2} + \sum_{n=1}^{M}(A_n \cos n\theta \cos \omega t + B_n \cos n\theta \sin \omega t + C_n \sin n\theta \cos \omega t + D_n \sin n\theta \sin \omega t)$$

and fitted, using equation (2), to the outer path for all the time-related holograms. The restricted number of data points available make it feasible to solve for coefficients for only a few values of n either side a peak at $n = 4$.

Note that terms of the form:

$$a \cos n\theta \cos \omega t + b \cos n\theta \sin \omega t + c \sin n\theta \cos \omega t + d \sin n\theta \sin \omega t$$

can be converted to the equivalent form:

$$R_1 \sin(n\theta - \omega t + \alpha) + R_2 \sin(n\theta + \omega t + \beta)$$

i.e. a superposition of forward and backward travelling waves. This form is also useful in that $R_1 + R_2$ gives the maximum amplitude for each value of n.

For the dominant term ($n = 4$) the values obtained for w_z were:

$$R_1 = 3.60\,\lambda, \ R_2 = 1.42\lambda, \ \alpha = 3.02, \ \beta = 2.69$$

where λ is the wavelength of the laser light. Thus there is a forward travelling wave with a smaller backward travelling wave with amplitude 38.9% of the forward wave. The angular speed of the wave is given by $\frac{\omega}{4}$, i.e. the angular frequency divided by the mode order. Similar travelling waves, also with an angular speed of angular frequency divided by the mode order have now been observed for a number of disc brake systems producing squeal [7].

The amplitudes for $n = 2$ through to $n = 6$ found by the Fourier analysis are given as follows:

n	2	3	4	5	6
Amplitude	1.50λ	1.21λ	5.52λ	0.84λ	0.76λ

The significance of the terms other than $n = 4$ could simply be interpreted by concluding that the forward and backward travelling waves are not perfect sine waves. However the Fourier series terms either side the main mode order terms (i.e the $n = 3$ and $n = 5$ terms) may also be explained by the w_r and w_θ in-plane terms in equation (2). These terms could contribute to n

= 4 terms multiplied by $\sin\theta$ and $\cos\theta$, using standard trigonometrical identities such as $\sin 5\theta = \sin(4\theta + \theta) = \sin 4\theta \cos\theta + \cos 4\theta \sin\theta$, etc. If all the $n = 3$ and $n = 5$ terms are interpreted in this way, the smallness of the components a and b of $\mathbf{b}-\mathbf{s}$ in (2), gives an amplitude as high as 28λ for the forward travelling wave in the angular in-plane displacement w_θ. However even if such high amplitudes are ruled out, the existence of considerable in-plane oscillation can be deduced from the holograms of the disc rim explained in the following section.

6. HOLOGRAMS OF DISC RIM

Consider the images in the holograms of the top view of the disc rim (Figure 6). Because the incident and reflected light rays are turned through $90°$ by the mirror above the disc, equation (2) determining the equation of the fringe lines in the holograms must be suitably modified to give:

$$-2w_r(a\cos\theta + c\sin\theta) - 2w_\theta(-a\sin\theta + c\cos\theta) + 2bw_z + r\Omega\Delta t(-a\sin\theta + c\cos\theta) = (N - \tfrac{1}{2})\lambda \tag{4}$$

Apart from being sinusoidal in t it is convenient to regard the displacements on the rim of the disc to be dependent on co-ordinates x ($= r_1 \cos\theta$, where r_1 is the outer radius of the disc) and z, the co-ordinate perpendicular to the disc, with $z = 0$ interpreted as the mid-plane of the disc. Thus in Figure 6, x will be the horizontal co-ordinate (from left to right) and z the vertical co-ordinate (increasing upwards).

If constant rotation only is considered (4) becomes:

$$KS(\theta) = N - \tfrac{1}{2}$$

where $K = \dfrac{r\Omega\Delta t}{\lambda}$ ($r = r_1$ is constant) and $S(\theta) = -a\sin\theta + c\cos\theta$. As N varies this equation can be solved to give $\theta = \theta_N$, giving nearly equally spaced vertical lines $x = x_N = r_1 \cos(\theta_N)$.

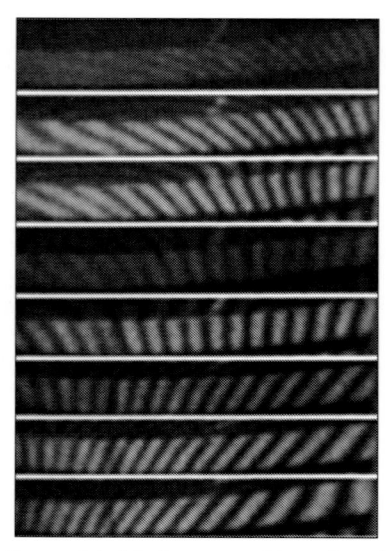

**Figure 6 Eight time related holograms of top view of rim of the disc.
Increasing time in cycle from top to bottom.**

Apart from constant rotation the other main contribution will be from the $-2c\sin\theta w_r + 2c\cos\theta w_\theta$ terms, although all the terms, including that in w_z, were considered in the calculation to explain both the variation in distance between the fringe lines as well as the sloping fringes that are seen in Figure 6 (see Appendix 2). Expanding in powers of z, it may be assumed that for a given value of the time t, $w_\theta = H_0(\theta) + \dfrac{z}{h}H_1(\theta) +$ terms in z^2, with similar expansions for w_r and w_z. Here h is half the thickness of the disc ($=7\times 10^{-3}$ m) so that $w_\theta = H_0 + H_1$ on the disc surface. Given that both the distance between and the slope of the fringe lines vary cyclically with time (see Figure 6) it is likely that w_θ and w_r are approximately sinusoidal in 4θ, as with the out-of-plane vibration.

It is seen that the main contribution to the variation in the distance between the fringe lines comes from the H_0 term (angular in-plane vibration independent of z) whereas the slope in the fringe lines will come mainly from the H_1 term (angular in-plane "shear" vibration with opposing directions of displacement depending on the distance from the $z = 0$ mid-plane). A conservative estimate for the amplitude of H_0 is 3λ and for H_1, 4λ (see Appendix 2)

CONCLUSION

It is believed that the analysis made in this work enables a more detailed understanding to be made of the dynamical behaviour of a disc in a real brake system generating noise. Holograms taken of a variety of disc brake systems suggest that the main features are common to most systems where sustained noise is generated. Accurate information is obtained regarding the out-of-plane vibration and the advance of the waves around the disc surface. Approximate but nevertheless useful estimates of the in-plane motion with significantly large amplitudes that can be expected for such a system are also obtained.

Models of disc brake squeal, both analytic and using finite element techniques, have so far been limited in their reliability and certainly in their ability to predict whether a brake system is likely to generate noise and at which frequency. There is still no agreed analysis of what mechanism generates the self-excited oscillations although many different theories have been put forward over the years. It is believed that the information obtained above can be applied in the development of more satisfactory models and explanations, especially taking into account in-plane motion.

It is clear that a more complete knowledge of in-plane motion could be obtained by taking holograms with the incident and reflected light rays pointing in three independent directions thus enabling a solution of equation (2) to be obtained for all three displacement components. Using mirrors this could be extended to the disc rim and even portions of the rear of the disc. Investigations of this type are now under way.

REFERENCES

[1] J.D.Fieldhouse and T.P.Newcomb, Double Pulsed Holography Used to Investigate Noisy Brakes, *Optics and Lasers in Engineering*, Vol. 6, No. 25, 1996, pp. 455 – 494.
[2] Y.S.Lee, P.C.Brooks, D.C.Barton, D.A.Crolla, A Study of Disc Squeal Propensity Using a Parametric Finite Element Model, Paper No. C521/009/98, *International Conference - Braking of Road Vehicles, I.Mech.E.*, London, , 1998, pp 191 – 201.
[3] M.Nishiwaki, Generalised Theory of Brake Noise, *Proceedings of the Institute of Mechanical Engineers*, Vol 27, *Part D: Journal of Automobile Engineering*, 1993, pp 195-202.
[4] J.Hultén and J.Flint, An Assumed Modes Method Approach to Disc Brake Squeal Analysis, *SAE Brake Colloquium and Engineering Display*, 1999.
[5] J.E.Mottershead, H.Ouyang, M.P.Cartmell and M.I.Friswell, Parametric Resonances in an Annular Disc, with a Rotating System of Distributed Mass and Elasticity; and the Effects of Friction and Damping, *Proceedings of the Royal Society A*, Vol 453, 1997, pp. 1-19.
[6] T.Kreis, *Holographic Interferometry*, Akademie Verlag, 1996.
[7] C.J.Talbot, and J.D.Fieldhouse, Animations of a Disc Brake Generating Noise, *Proceedings of the SAE 2001 Brake Colloquium*, New Orleans, pp 23-29, SAE Paper Number 2001-01-3126.

APPENDIX 1

Consider the disc undergoing vibration at a fixed frequency. Assume the displacement of the disc surface due to vibration is given by

$$w = w_1(r,\theta)\cos\omega t + w_2(r,\theta)\sin\omega t$$

where $\omega = 2\pi \times 2900$. Then using

$$\Delta\cos\omega t = \cos\omega(t+\Delta t) - \cos\omega t = -p\cos(\omega t - \phi)$$

where $p = \sqrt{(1-\cos\omega\Delta t)^2 + \sin^2\omega\Delta t}$, $\phi = \tan^{-1}\left(\dfrac{\sin\omega\Delta t}{1-\cos\omega\Delta t}\right)$ and a similar calculation for $\Delta\sin\omega t$, we have

$$\Delta w = -p(w_1(r,\theta)\cos(\omega t - \phi) + w_2(r,\theta)\sin(\omega t - \phi))$$

Note that Δt remains constant at $32\mu s$ for all the holograms considered, giving $p = 0.575$. It is possible to make a new choice of origin for the time t so that:

$$\Delta w = -p(w_1(r,\theta)\cos\omega t + w_2(r,\theta)\sin\omega t) = -pw$$

APPENDIX 2

For a given value of t, write $w_r = (R_0(\theta) + \dfrac{z}{h}R_1(\theta) +$ terms in $z^2)\lambda$, $w_\theta = (H_0(\theta) + \dfrac{z}{h}H_1(\theta) +$ terms in $z^2)\lambda$, and $w_z = (Z_0(\theta) + \dfrac{z}{h}Z_1(\theta) +$ terms in $z^2)\lambda$.

Substituting into (4) and putting $z = 0$ gives:

$$F_0(\theta) + KS(\theta) = N - \frac{1}{2} \tag{5}$$

where $F_0 = p(S'R_0 - SH_0 + bZ_0)$. Dash denotes differentiation with respect to θ. Suppose that $\theta = \theta_N$ is the solution associated with steady rotation only for each value of N (i.e. $KS(\theta_N) = N - \dfrac{1}{2}$). For the complete top of the disc rim, $\theta = 0$ to π, there should be 54 vertical fringes corresponding to $N = -26$ to 27. The part of the rim visible in the holograms in Figure 6 is restricted to $\theta = 1.75$ rads (right hand side) to $\theta = 2.54$ rads (left) with corresponding N values.

Let $\hat{\theta}_N$ be the solution of (5) for each N. Then a short calculation gives

$$\frac{\theta_{N+1} - \theta_N}{\hat{\theta}_{N+1} - \hat{\theta}_N} \cong 1 + \frac{F_0'(\theta_N)}{KS'(\theta_N)}$$

i.e. the contraction or expansion of the distance between the vertical fringe lines is given by the term $\frac{F_0'}{KS'}$. That this term must reach the value of at least ± 0.5 to give the variation in spacing between the lines seen in the holograms gives minimal estimates for the amplitude of H_0. It can be seen that

$$\frac{F_0'}{KS'} = A_R R_0' + B_R R_0 + A_H H_0' + B_H H_0 + A_Z Z_0'$$

where A_R, B_R, A_H, B_H, A_Z contain S, S', K and p, and so can be calculated for the range of values of θ given. The term $A_Z Z_0'$ is negligible given the small value of b and the known amplitude of the out-of-plane term Z_0. If it is assumed that R_0 and H_0 are sinusoidal in 4θ so that their amplitudes are multiplied by a factor of 4 in R_0' and H_0', a conservative estimate of the amplitude of H_0 (assuming that the H_0 terms contribute to at least half of the total) is 3.

To investigate the slopes of the fringe lines, substitute the above expressions for w_r, w_θ and w_z into (5), differentiate with respect to θ, and put $z = 0$ (i.e. consider the slope along the mid-plane of the disc). Then:

$$\frac{dz}{d\theta} = -\frac{KS'(\theta)}{F_1(\theta)}$$

where $F_1 = \frac{p}{h}(-R_1 S' - H_1 S + bZ_1)$. Since $x = r\cos\theta$, it follows that

$$\frac{dx}{dz} = \frac{r\sin\theta \, F_1}{KS'} = C_R R_1 + C_H H_1 + C_Z Z_1$$

where C_R, C_H and C_Z depend on S, S', K, p, r and $\sin\theta$, so can be calculated for the range of θ under consideration. Again the out-of-plane term $C_Z Z_1$ is negligibly small. The main contribution for larger θ (to the left of the holograms) is C_H so for $\frac{dx}{dz}$ with a value of at least 3, as seen in the first hologram, a calculation shows that the amplitude of H_1 must be at least 4.

Stick–slip vibration in a disc brake system

Q CAO, H OUYANG, J E MOTTERSHEAD, D J BROOKFIELD, and S JAMES
Department of Engineering, University of Liverpool, UK

SYNOPSIS

In this paper, the stick-slip vibration of a three degree-of-freedom model of a disc brake is presented. A new type of stick-slip behaviour is illustrated in a numerical example where the flexibility of the connection of the disc through the wheel and tyre to the road surface are included. It is shown that the natural frequency of the nonlinear stick-slip system can be considerably lower than the natural frequencies of the two underlying linear systems that describe sliding and sticking separately. Computational periodic behaviour is discovered.

1 INTRODUCTION

Stick-slip vibrations has been widely studied and rich dynamical behaviour has been found including bifurcations and chaos, see for example, Popp and Stelter (1) and Galvanetto et al. (2,3). One reason for disc brake groan, judder and squeal may be friction induced stick-slip vibration based on the roughness of the disc surface as discussed by Vadari (4), Jacobsson (5) and Ibrahim (6,7).

Past studies on the stick-slip vibration have mostly been focused on the behaviour of elastic slider on a belt moving at a constant speed v_0 with a constant normal load on the slider with friction (1-3). Recently, some authors began to pay attention to the non-constant situation, non-constant sticking condition or non-constant normal load. Leine et al. (8) presented a violin spring model with constant speed but a non-constant sticking condition. Ouyang et al. (9) studied the friction-induced stick-slip vibration in a disc brake and included the effect of dynamic loading from the slider system on the transverse vibration of the disc.

The brake disc is mounted flexibly through the wheels and tyres to the road surface and therefore undergoes torsional vibration when the brake is applied. The pads vibrate on the surfaces of the disc which itself is vibrating. This introduces a behaviour not previously studied in disc-brake dynamics. A new disc-brake model having several degree-of-freedom becomes necessary.

In this article the torsional oscillations are superimposed on the constant driven-speed of the disc. An adjustable shooting method is introduced to find the stick-slip and the slip-stick transition points within a small tolerance. The FFT is employed to detect the frequencies of the system in stick-slip vibration. Though the results presented are preliminary ones, rich nonlinear behaviour is displayed. Periodic solutions are found, which may possibly lead to quasi-periodic or chaotic solutions and thereby to unwanted noise from the brake. The study indicates that the nonlinear behaviour is sensitive to damping and frication.

2 A THREE-DEGREE OF FREEDOM SYSTEM FOR DISC BRAKE SYSTEM

In this section a nonlinear three degree-of-freedom model is introduced to present complicated disc-brake vibrations, possibly including judder, groan and squeal. The model is illustrated in Figure 1.

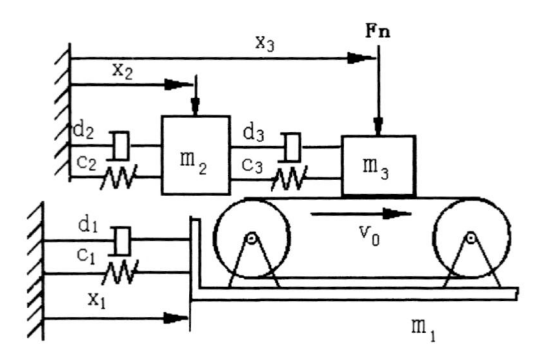

Fig. 1 Three-degree of freedom model for disc brake system

m_1 represents the inertia of the disc and the inertia effects of the shaft, wheel and tyre; v_0 is the constant rotating speed of the disc; and c_1 and d_1 are the torsional stiffness and damping respectively of the shaft, wheel and tyre. m_2 represents the mass of the calliper and its housing and m_3 the mass of the pads. The stiffness and damping of the pads is denoted by c_3 and d_3, whereas c_2 and d_2 are equivalent stiffness and damping parameters for the vehicle structure to which the slider is considered.

The equation of motion for the system given in Figure 1 can be written as

$$\begin{cases} m_1\ddot{x}_1 + c_1 x_1 + d_1 \dot{x}_1 = -F_r(v_r) \\ m_2\ddot{x}_2 + c_2 x_2 + d_2 \dot{x}_2 + c_3(x_2 - x_3) + d_3(\dot{x}_2 - \dot{x}_3) = 0, \\ m_3\ddot{x}_3 + c_3(x_3 - x_2) + d_3(\dot{x}_3 - \dot{x}_2) = F_r(v_r) \end{cases} \quad (1)$$

where

$$F_r(v_r) = \begin{cases} -\mu(v_r)F_n, & v_r > 0 \\ m_3\ddot{x}_3 + c_3(x_3 - x_2) + d_3(\dot{x}_3 - \dot{x}_2), & |m_3\ddot{x}_3 + c_3(x_3 - x_2) + d_3(\dot{x}_3 - \dot{x}_2)| < \mu_0 F_n, v_r = 0, \\ \mu(v_r)F_n, & v_r < 0 \end{cases} \quad (2)$$

μ and μ_0 are the dynamic and static friction coefficients respectively, and

$$v_r = \dot{x}_3 - v_0 - \dot{x}_1 \quad (3)$$

is the relative speed between the disc and pads surfaces and $v_r = 0$ is the sticking condition which is different from the constant belt speed in literature (1).

Considering the sticking condition $v_r = 0$ and differentiating both side of (3) implies that $\ddot{x}_3 = \ddot{x}_1$. This indicates that both the disc, mass 1, and the pads, mass 3, are oscillating at not only the same velocity but also the same acceleration when sticking. The sticking force in (2) now is changed to $F_r = m_3\ddot{x}_1 + c_3(x_3 - x_2) + d_3(\dot{x}_3 - \dot{x}_2)$.

Oscillations of the system will generally involve both sticking and slipping behaviour, but two extreme cases will be considered when system is slipping without sticking and vice-versa. The characteristic equation for the 'slipping system' is given by

$$\left(\omega^2 - \frac{c_1}{m_1}\right)\left(\omega^4 - \omega^2\left[\frac{c_3}{m_3} + \frac{c_2 + c_3}{m_2}\right] + \frac{c_2 c_3}{m_2 m_3}\right) = 0. \quad (4)$$

and the three natural frequencies may be obtained as

$$\omega_{1,2}^2 = \frac{1}{2}\left\{\frac{c_3}{m_3} + \frac{c_2 + c_3}{m_2} \mp \sqrt{\left[\frac{c_3}{m_3} + \frac{c_2 + c_3}{m_2}\right]^2 - \frac{4c_2 c_3}{m_2 m_3}}\right\}, \quad \omega_3^2 = \frac{c_1}{m_1}. \quad (5)$$

The 'sticking system' has two degree-of-freedom and the natural frequencies may be determined from

$$\omega_{1,2}^2 = \frac{1}{2}\left\{\frac{c_1 + c_3}{m_1 + m_3} + \frac{c_2 + c_3}{m_2} \mp \sqrt{\left[\frac{c_1 + c_3}{m_1 + m_3} + \frac{c_2 + c_3}{m_2}\right]^2 - \frac{4(c_1 c_2 + c_1 c_3 + c_2 c_3)}{m_2(m_1 + m_3)}}\right\}. \quad (6)$$

Although these frequencies have an important influence they are not the natural frequencies of the system in Figure 1, which exhibits stick-slip behaviour. New variables (x_1, x_2), (x_3, x_4) and (x_5, x_6) are defined to describe the displacements and the velocities of mass 1, mass 2 and mass 3 in Figure 1 respectively. Equation (1) can be rewritten as

$$\begin{cases} \dot{x}_1 = x_2 \\ \dot{x}_2 = (-c_1 x_1 - d_1 x_1 + F_r)/m_1 \\ \dot{x}_3 = x_4 \\ \dot{x}_4 = (-(c_2 + c_3)x_3 - (d_2 + d_3)x_4 + c_3 x_5 + d_3 x_6)/m_2 \\ \dot{x}_5 = x_6 \\ \dot{x}_6 = (-c_3(x_5 - x_3) - d_3(x_6 - x_4) + F_r(v_r))/m_3 \end{cases} \quad (7)$$

3 NUMERICAL ANALYSIS

In this section numerical integration by Runge-Kutta method is employed to obtain the solutions of equation (7). A shooting procedure is introduced to find the stick-slip and the slip-stick transition points within a given error.

The stick-slip and the slip-stick transition points play an important role in numerical analysis. A point $x(k)$, $x = (x_1, x_2, ..., x_6)^T$, at the k-th integration step is said to be the stick-slip transition point in the numerical sense, if the preceding point $x(k-1)$ is in the sticking regime and the difference between the sticking friction force $F_r(k)$ and the static friction capacity $F_{r0} = \mu_0 F_n$ is no more than the given small error, $\eta = 1.0E - 05$. The step size $h = 0.01$ was taken in the numerical computations. A slip-stick transition point can be defined in a similar way.

Point $x(k)$ is assumed to be the last point in the sticking regime satisfying $|F_r(k) - \mu_0 F_n| \leq \eta$ at time step h and the successive point $x(k+1)$ the friction capacity is exceeded, $|F_r(k+1) - \mu_0 F_n| > \eta$. A new integration time step h_1, which better locates the transition, can be determined by the ratio.

$$\frac{h_1}{h} = \frac{F_{r0} - F_r(k)}{F_r(k+1) - F_r(k)}. \quad (8)$$

Successive steps can be followed until the stick-slip transition point has been found to meet the requirement of the error criterion. This is the idea of the adjustable shooting method.

In the same way, the slip-stick transition can be found by using the difference between the successive steps of the relative speed, v_r, crossing zero.

Figure 1 is considered as a lumped model in the following numerical analysis and the parameters used as non-dimensionalised. Friction is described using the dynamic friction coefficient $\mu = 0.2$, the static friction coefficient $\mu_0 = 0.6$, and the disc rotating speed $v_0 = 0.5$ in the following calculation. The system parameters $m_1 = 1.0$, $c_1 = 4.0$, $d_1 = 0.08$, $m_2 = 3.0$, $c_2 = 1.0$, $d_2 = 0.005$, $m_3 = 1.0$ and $d_3 = 0.04$ are used for the purpose of illustration. The natural frequencies 'slipping system' are found to be $\omega_1 = 0.997$, $\omega_2 = 1.290$ and $\omega_3 = 2.00$. While for the 'sticking system', $\omega_1 = 0.757$ $\omega_2 = 1.608$.

In the phase portrait discussed below the disc is rotating with constant velocity v_0, all three masses have zero initial displacements and velocities, except mass 3 which has an initial velocity equal to the disc speed $v_0 = 0.5$. This assumption represents the initial sticking behaviour of the system. A periodic solution with the normal force $F_n = 7.5$ is discussed first.

The phase portraits, spectra and time series are shown in Figures 2-10 for each of the three masses. The spectra show a dominant peak at around 0.2 rad/sec corresponding to the periodic of about 32 seconds of the periodic motion evident in the time series. This dominant peak represents the fundamental natural frequency of the system, which is considerably lower than any of the frequencies of the 'sticking' or the 'slipping' system. It is clear from the time series data that the disc undergoes more than three oscillations in sticking and two and a half slipping oscillation in one period of 35 seconds. The 'Squiggly' lines across the horizontal in Figures 2 and 8 represents the sticking phase of the motion. The large concentric ellipse in Figure 2 occurs when the pads slip over the surface of the disc. Figure 8 shows that the pads slow in the middle of their excursion when slipping over the disc.

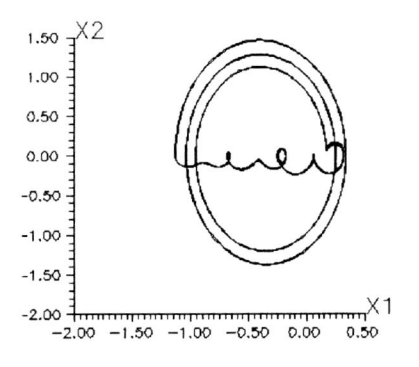

Fig. 2 Disc: Periodic solution

Fig. 3 Disc: Spectral analysis

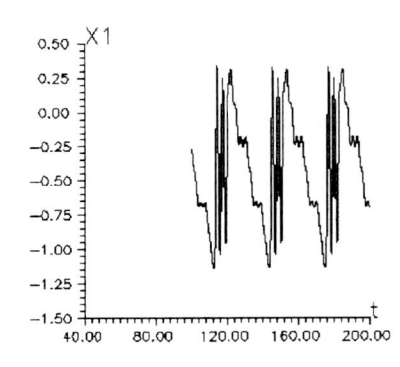

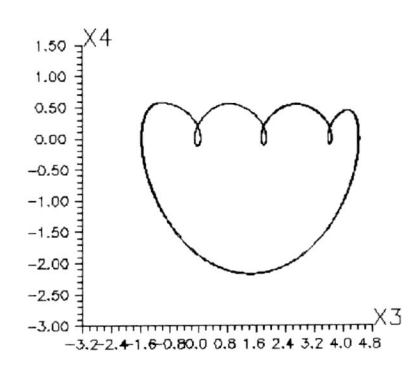

Fig. 4 Time series: Disc displacement

Fig. 5 Calliper: Periodic solution

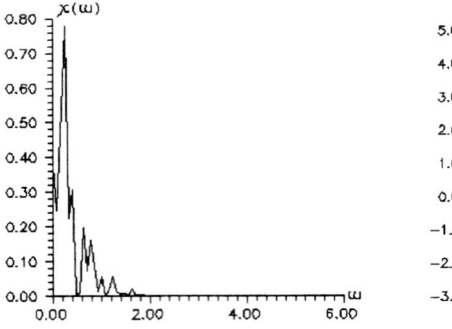

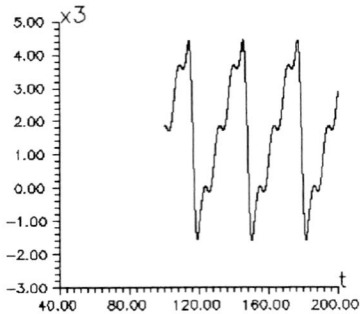

Fig. 6 Calliper: Spectral analysis

Fig. 7 Time series: Calliper displacement

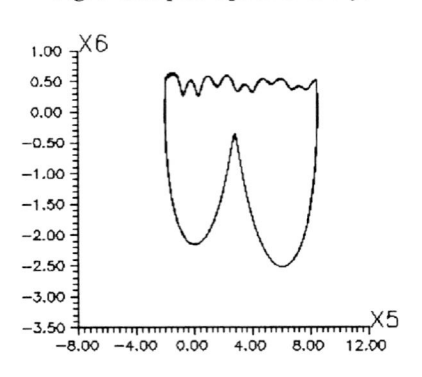

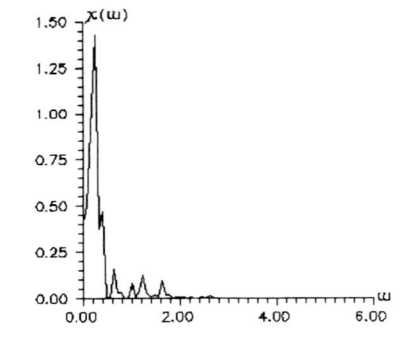

Fig. 8 Pads: Periodic solution

Fig. 9 Pads: Spectra

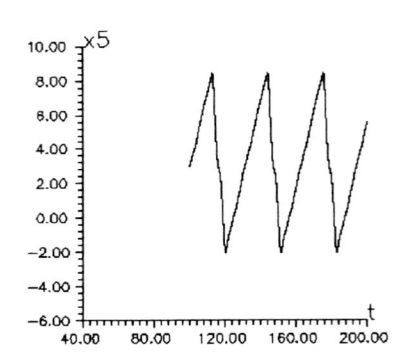

Fig. 10 Time series: Pads displacement

The natural frequency will be reduced further if the normal force (or the friction coefficient) is increased because this will lead to a longer time in the sticking regime, and therefore to more 'sticking' period within the fundamental period of the overall stick-slip system.

In Figures 11-16 phase portraits and the spectra at the normal force $F_n = 5.0$ are shown. The other parameters remain unchanged. The behaviour through still periodic is more complex.

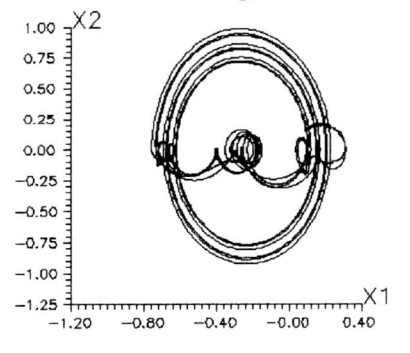

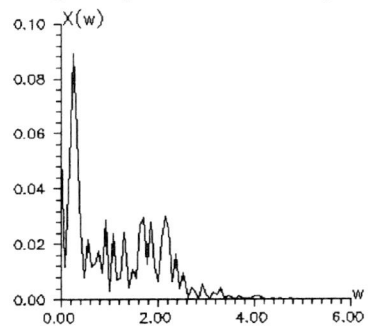

Fig. 11 Phase portrait: Disc:

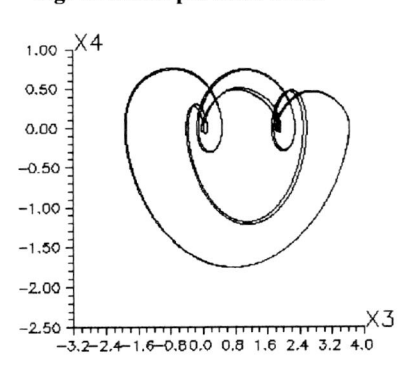

Fig. 12 Spectra: Disc

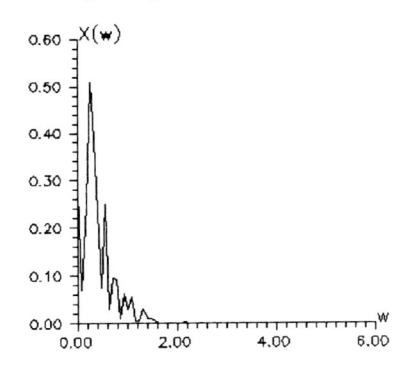

Fig. 13 Portrait: Calliper

Fig. 14 Spectra: Calliper

Fig. 15 Portrait: Pads

Fig. 16 Spectra: Pads

4 CONCLUSION

In this paper, a stick-slip model with three degree-of-freedom is presented for the investigation of nonlinear behaviour in disc brakes. A new type of stick-slip vibration is illustrated where not only does the relative velocity between the pads and the disc disappear to zero but also the angular acceleration of the disc and pads are the same. The natural frequency of the overall 'stick-slip' system can be considerably lower that any of the natural frequencies of the system in 'sticking ' or 'slipping' separately. Complicated periodic behaviour has been discovered in this preliminary analysis.

ACKNOWLEDGEMENTS

The authors are grateful for the support by the Engineering and Physical Sciences Research Council (GR/L91061), TRW Automotive and TMD Friction UK.

REFERENCES

1. **Popp K.** and **Stelter P.**, (1990) Stick-slip vibrations and chaos, Phil. Trans. R. Soc. London, 332,pp.89-105

2. **Galvanetto U.** and **Bishop S. R.**, (1999) Dynamics of a simple damped oscillator undergoing stick-slip vibrations, Mechanica, pp. 637-651

3. **Galvanetto U.**, (1997) Bifurcation and Chaos in a four dimensional mechanical system with dry friction, Journal of Sound and Vibration, Vol.204, No. 4,pp.690-695

4. **Vadari V.** and **Anatrol R.**, (1999) Creep-Groan in Vehicles, 17th Annual Brake Colloquium & Engineering Display, Miami Beach, Florida, USA, October 10-13, pp.247-269

5. **Jacobsson H.**, (1996) Frequency Sweep Approach to Brake Judder, Thesis for the degree of Licenciate of Engineering, Machine and Vehicle Design, Chalmers University of Technology, S-412 Göteborg, Sweden.

6. **Ibrahim R.A.**, (1994) Friction induced vibration, chatter, squeal and chaos Part 1: Mechanics of contact and friction, Applied Mechanics Review, Vol.47, No.7, 209-226

7. **Ibrahim R.A.**, (1994) Friction induced vibration, chatter, squeal and chaos Part 2: Mechanics of contact and friction, Applied Mechanics Review, Vol.47, No.7, pp.227-253

8. **Leine R.I., Campen D.H. Van,** and **Kraker A.DK,** (1998) Stick-slip vibrations induced by alternate friction models, Nonlinear Dynamics, Vol.16, pp. 41-54

9. **Ouyang H., Mottershead J.E., Cartmaell M.P.,** and **Brookfiled D.J.,** (1999) Friction-induced vibration of an elastic slider on a vibrating disk, International Journal of Mechanical Sciences, Vol.41, No.3, pp. 325-336

Brake-squeal measurement using electronic speckle pattern interferometry

R KRUPKA, T WALZ, A ETTEMEYER, and **R EVANS**
Ettemeyer AG, Germany and AG Electro-Optics Limited, Tarpoley, UK

ABSTRACT

The demand for increased braking reliability and safety requires detailed knowledge about the vibration behaviour of the brake disc. This paper will present results from vibration measurement using a technique called Electronic Speckle Pattern Interferometry (ESPI). This non-contact, full-field, three-dimensional technique provides fast behaviour analysis of squealing brake discs.

1. INTRODUCTION

ESPI is a well-established technique, in research as well as industry, for non-contact full field measurement of deformations of various objects under different loading conditions (1).

1.1 The principle of dynamic ESPI measurements

The principle of the pulsed three-dimensional ESPI techniques is as follows. The object under investigation is illuminated with a short light pulse from a pulsed laser (illumination time is several nanoseconds) and is simultaneously observed from 3 different directions with high-resolution cameras.

The use of a double-pulsed laser, allows the acquisition of two images within a very short duration, given by the separation of the two laser pulses. This allows the analysis of the object deformation even during highly dynamic processes. For each of the pulses, a speckle interference image of the illuminated surface is recorded. The difference between the two images reveals information about the deformation change of the sample between the two pulses.

From these images, the quantitative displacements of each object point in the field of view is automatically calculated within a few seconds. Using three cameras simultaneously, the complete 3D-deformation vector is calculated for every point of the inspected surface. A 3D map of the deformation can then be produced using analysis software (Figure 1).

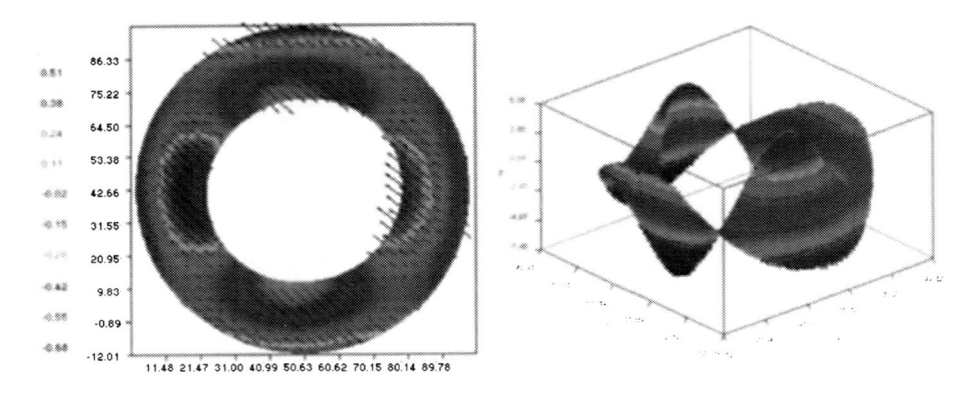

Figure 1: Map of surface deformation

2. BRAKE SQUEAL MEASUREMENT USING ESPI

The measurement of brake squeal vibrations is a difficult problem, because the rotating disc does not allow any physical contact with the sample during measurement. The use of accelerometers is difficult as well as very time consuming. The recent interest of researchers has focused not only on the one-dimensional out-of plane deformation, but also on the determination of the complete, three-dimensional vibration behaviour. The pulsed ESPI non-contact technique, as shown in figure 2, was found to be one of the few methods available capable of making such measurements.

2.1 Brake disc measurement methodology

The component under investigation, a rotating brake disc with engaged calliper, is illuminated by two very short laser pulses with a predetermined time separation. The images at these two laser illuminations are recorded with a special high-resolution camera and analysed (2). The calculated result is a deformation map of each point of the brake disc and calliper, this shows the movement of the individual points of the brake between these two laser pulses. Therefore, this technique literally freezes two moments of the movement of the brake and gives an instantaneous view of the overall deformation.

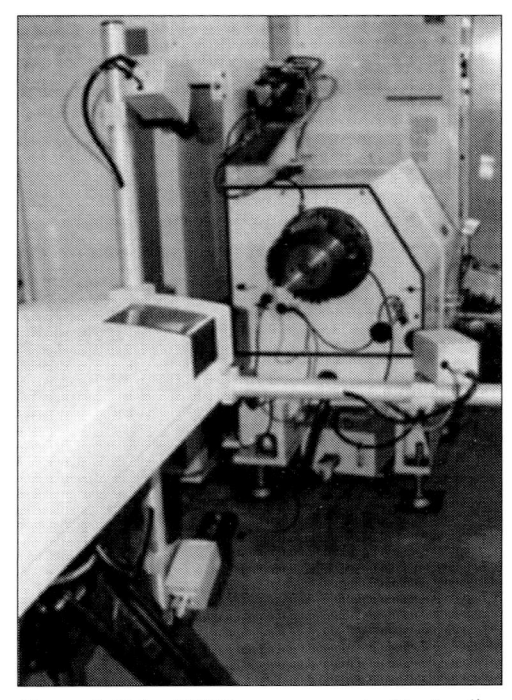

Figure 2: Pulsed ESPI measurement of brake disc.

2.2 Vibration of a brake disc

A brake disc was excited using an electro-dynamic shaker (2). Figure 3 shows the three live camera images acquired in the three different viewing directions of the system. To take into account the three different viewing directions of the sensors, a perspective image correction is applied, before calculating the components for in-plane and out-of plane object deformation (figure 4). The perspective corrected phase images of the three directions are taken for the calculation of the in-plane components Vx and Vy as well as for the out-of plane component Vz of the disc deformation during vibration.

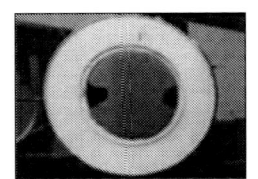

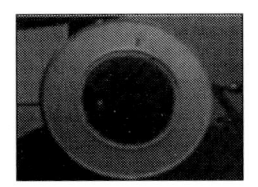

Direction 1   Direction 2   Direction 3
Figure 3: Perspective corrected images of the three viewing directions.

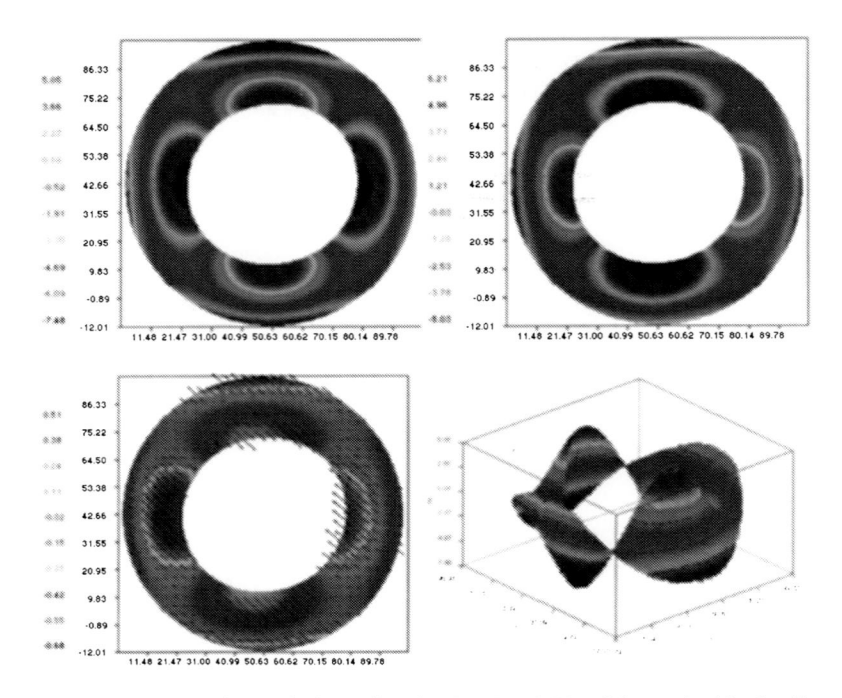

Figure 4: In-plane and out-of-plane vibration Vx, Vy & Vz of the excited brake disc (f=5046Hz).

The combination of the different vibration components in a single graphic allows a better understanding of the complete vibration. An easy separation of in-plane and out-of plane vibration is possible. Figure 5 and 6 show an example of vibration including strong in-plane vibration.

2.3 Rotating disc
In real applications the brake disc is rotating while being measured with the pulsed ESPI system. In order to demonstrate the capabilities of the system under these conditions, a brake disc was measured during rotation and the deformation due to electro-dynamic excitation was calculated. To extract the in-plane and out-of plane components of deformations, the components due to the rotation of the disc have to be compensated (2), as shown in Figures 7 and 8.

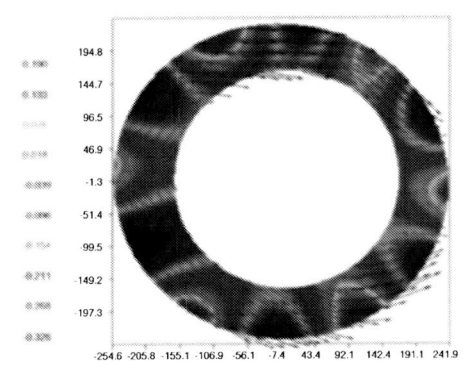

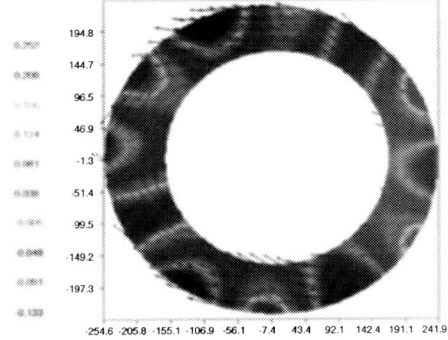

Figure 5: In-plane and out-of plane deformation of a shaker excited brake disc (7417Hz).

Figure 6: In-plane and out-of plane deformation of a shaker excited brake disc (7417Hz). In contrast to figure 5 the instant of acquisition was shifted by 180° with respect to the excitation.

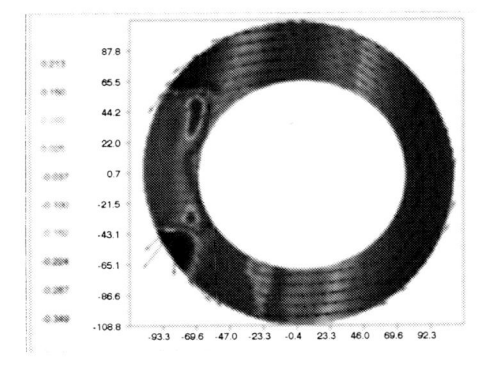

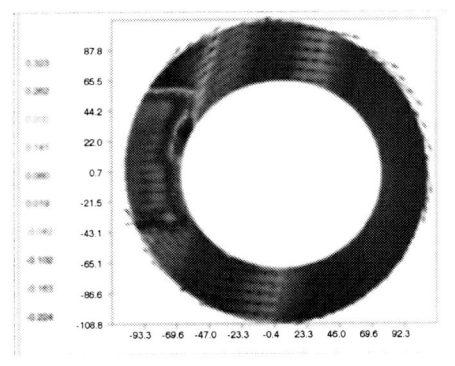

Figure 7: Out-of plane and in-plane deformation of the rotating disc bottom (anti-clockwise rotation).

Figure 8: Out-of plane and in-plane deformation of the rotating disc bottom (clockwise rotation).

2.4 Automatic brake test system

For the application of the three dimensional pulsed ESPI system for serial inspection of brake discs, the system has been further improved for automatic testing. The system automatically recognises the brake squeal during the brake testing (2). From the analysis of the squealing signal, the timing conditions of the system (laser pulse separation and absolute time instants) are automatically computed, the laser is fired, the images are taken for each direction during both laser pulses and the complete three dimensional deformation map is calculated. One complete cycle takes about 3 seconds. The system is then ready for the next event within 10 seconds (the time required for laser recharging). This results in a maximum possible testing cycle of 6 squeals per minute.

During automatic testing of the aforementioned brake disc and calliper sample, it was suggested that the vibration of the calliper was responsible for the squealing.

3. ANALYSIS OF BRAKE CALLIPER

Further measurements of a brake calliper were carried out in the pulse holography laboratory of AlliedSignal, Reinbek/Germany(4).

Figure 9 shows the pulsed ESPI camera, positioned in front of a dynamo brake test bench. The dynamo test bench allows the selection of rotation speed, brake pressure and direction of rotation. It was operated to produce different frequency brake squeal effects. Simultaneously, brake pressure, rotation speed and temperature were recorded.

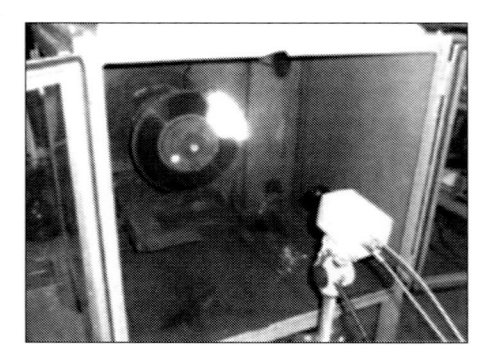

Figure 9: Pulsed ESPI camera observing a rotating brake disc for dynamic vibration analysis.

The microphone of the automatic brake test system was used to observe the brake squeal phenomenon and trigger the laser with respect to the noise signal. Figure 10 shows the complete trigger chain, which enables the exact positioning of the laser pulses to analyse the required event.

The microphone signal is displayed on an oscilloscope and analyzed by a frequency analyzer. The selected frequency can be filtered from the original signal by a band pass filter. This serves as the trigger signal for the pulsed ESPI system. The pulse separation times and the time settings of each laser pulse are defined for each measurement to record the required effect.

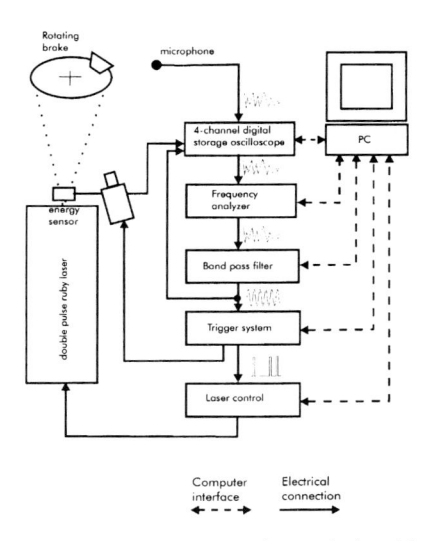

Figure 10: Trigger scheme for brake noise analysis with pulsed ESPI.

Figure 11 shows the camera view of the brake. The fringes on the brake show lines of constant amplitude, the difference between two lines being approximately 0.4 µm. The brake was excited to vibrate at 2100Hz by slightly braking at 18 rpm rotation speed. The laser pulse separation was 100 µsec. At this rotation speed nearly no influence of the rotation can be observed in the measuring result. Even at increased rotation speed up to 40rpm good results are achieved without the requirement of any special optical alignment, use of de-rotator etc, see figure 12. The little distortion of the phase map due to the rotation can be compensated numerically after evaluation. Figure 13 shows the vibration mode at 2100Hz (3).

Instant measurement: In figure 14 the advantage of instant measurement with pulsed ESPI becomes obvious. In case of modulated amplitude or frequency of the brake squeal, the pulsed ESPI system can instantly record the desired event. The actual state is frozen in the camera and no constant operation mode has to be maintained. If the event to be analysed is known, pulsed ESPI is able to record this event within 1msec. In the example an absolute trigger level was defined to record high amplitude or low amplitude vibration modes.

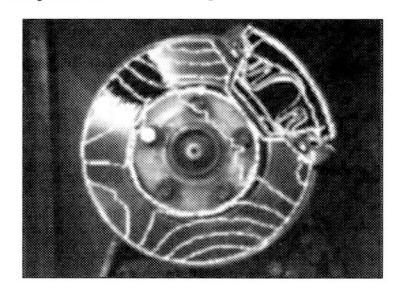

Figure 11: Vibration mode at 2100Hz

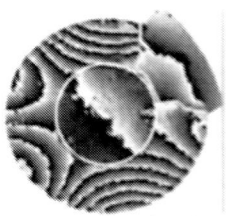

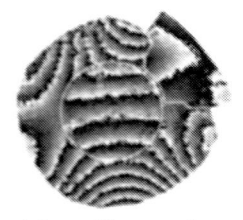

Figure 12: Phase map of brake at 2100Hz: left at 18rpm, right at 40rpm, pulse separation 100μsec

Figure 13: Vibration mode at 2100Hz (18rpm)

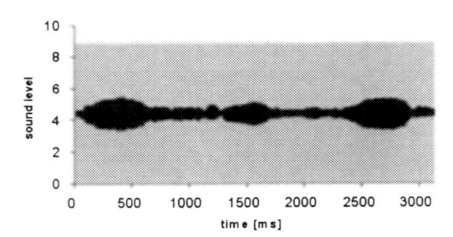

Figure 14: Microphone signal with amplitude modulated signal (frequency 2100Hz)

For brake calliper analysis (Figure 15) the full field instant measuring technique can show the vibration modes of calliper and pad. Figure 16 shows the behaviour of only the disc at different frequencies. The calliper and brake can show the similar vibration behaviour but differences can be seen if the rotation direction is changed. The symmetric vibration mode in counter-clockwise rotation can be shifted to a non-symmetric mode if the rotation is changed to clockwise direction. Figure 17 shows the influence of rotation direction on the pad vibration. At high vibration frequencies the different vibration modes of calliper and pad can also be observed. Figure 18 shows different vibration bending axis for calliper and pad (3).

Figure 19 shows a measurement of the aforementioned brake disc and calliper. It clearly shows that the vibration of the calliper was responsible for the squealing. The vibration was caused by differential movement between the calliper and the disc, which can be seen in the zones highlighted in figure 19.

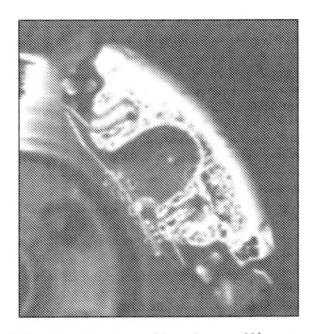

Figure 15: Analysis of brake callipper and pad

3.35kHz, 33rpm   5kHz, 21rpm
Figure 16: Vibration amplitudes at different frequencies

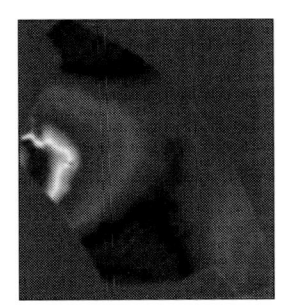

clockwise   counter clockwise
Figure 17: Influence of rotation direction on the pad vibration

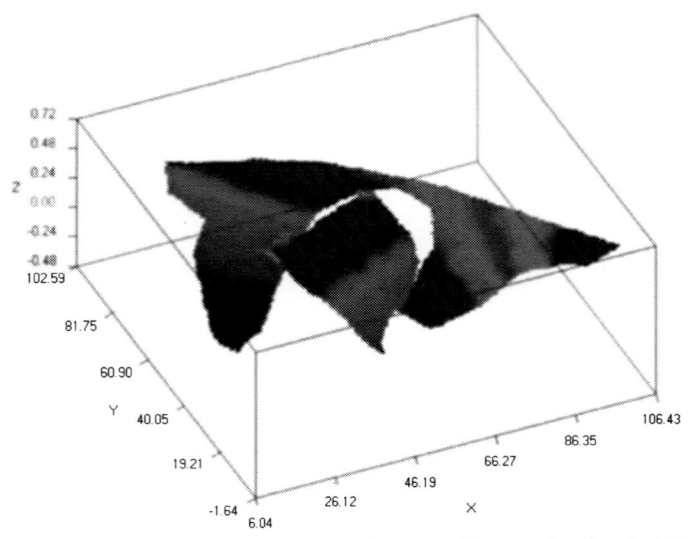

Figure 18: Different vibration modes of brake calliper and pad at 8.1kHz

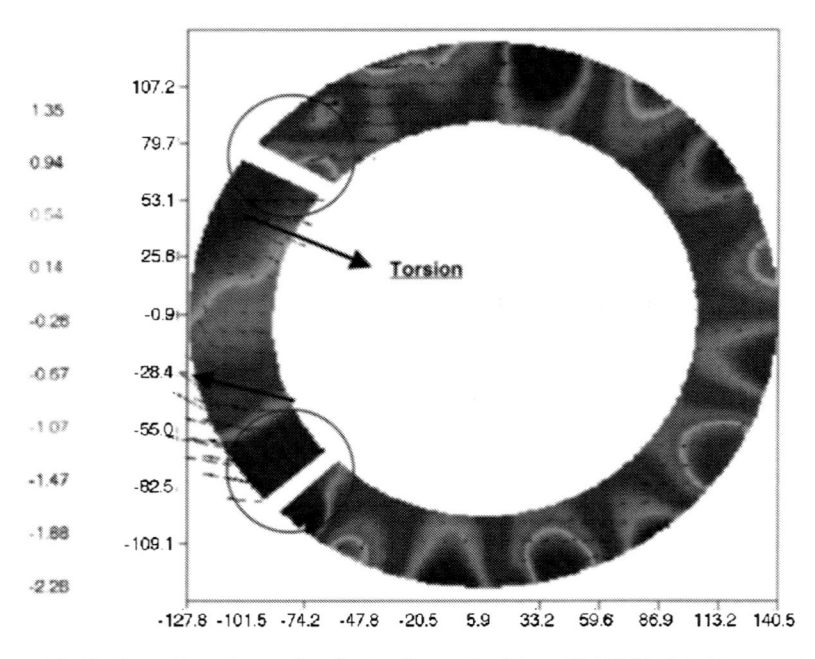

Figure 19: Deformation of squealing brake disc and calliper (f=12050 Hz). (Automatic acquisition and measurement)

4. CONCLUSION

It was demonstrated that ESPI is a powerful tool for analysis in highly dynamic processes, such as brake squeal events. The instantaneous full field three-dimensional measurement can especially help in the understanding of transient events. The ESPI technique allows easy access to the complete three-dimensional behaviour of the squealing brake disc. The system allows independent or fully automatic operation, for complete testing versatility. For example, recent developments have lead to the determination of both the phase and amplitude of harmonically vibrating discs. Combined with modal or strain/stress analysis software, the ESPI technique can be widely used in industry or research institutes for analyzing highly dynamic processes.

REFERENCES

1. R. Jones, C. Wykes, Holographic and Speckle Interferometry, ch.3-5, Cambridge University Press, Cambridge, New York, New Rochelle, Melbourne, Sidney, 1989
2. R.Krupka, T.Walz, A.Ettemeyer, New Techniques and Applications for 3D-Brake Analysis, Dr. Ettemeyer Application Report No. 04-00, April 2000.
3. Z. Wang, A. Ettemeyer, Some Applications of Pulsed ESPI to Break Squeal Analysis, Dr. Ettemeyer Application Report No. 03-98, March 1998.
4. Measurements with thanks from pulse holography laboratory of AlliedSignal, Reinbek/Germany with the generous support of Mr. Abendroth, Dr. Giese and Mr. Wernitz.

Mechatronic Brake Systems

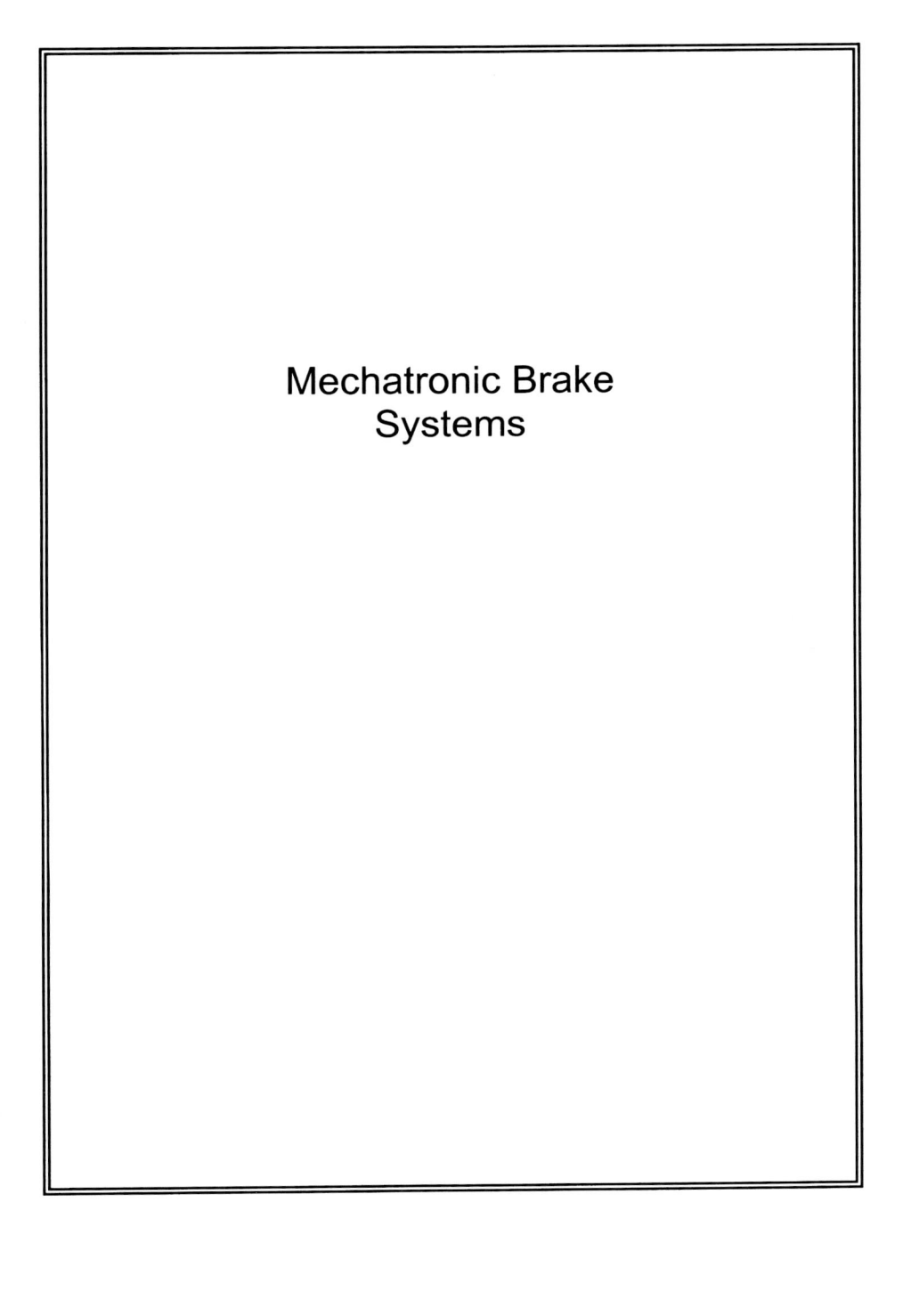

A regulatory approach to complex electronic control systems when sharing life with the braking system

A J MENDELSON
Department for Transport, Local Government, and the Regions, London, UK

SYNOPSIS

Over the last 5 years the number of on-board electronic systems had significantly increased and national government experts have become more concerned about their effect on vehicle safety. The interaction of these new electronically controlled functions with existing primary safety systems has raised questions over their compatibility and operation in the event of system failure.

Government and industry experts have worked together to produce a common understanding on the process that must be undertaken during design and the information that must be presented at Type Approval.

This presentation is aimed at emphasising to system designers the need to work with colleagues in other disciplines in order to develop technologically advanced motor vehicles.

1 INTERNATIONAL LEGISLATION

The design and performance criteria for the braking system used on motor vehicles and their trailers are laid down in European Directive 71/320/EEC, as amended by Directive 98/12/EC, and in United Nation ECE Regulation 13.

These regulations are continually being up-dated to ensure that vehicles are designed to meet a minimum guaranteed safety level and to cater for advances in new technology. It is also important that the regulations are not design restrictive and designers are free to be creative in their work.

All the work to develop the regulation is carried out under the control of the United Nations. Biannual meetings are held between national experts and representatives bodies of different areas of industry. At these meetings proposals can be put forward to accept new system features where these have a substantial influence on currently regulated functions such as braking.

Discussion at an early stage can lead to the setting down of essential design and performance rules, which may be of benefit to the subsequent common understanding. The meetings also permit a level of flexibility, which accepts, after negotiation, some adjustment of the rules so as to make them more workable in practical terms for full production.

Once agreement has been reached at this level the proposed amendment is passed to the World Forum for Harmonisation of Regulations (WP.29) for final modification and agreement.

The European Community has been happy not to duplicate the work carried out by the UNECE, but unfortunately this has resulted in the Directive being several levels and many years behind the Regulation. Generally it is more difficult to amend a Directive due to the procedures involved but it is expected that in due course the Directive will be up-dated to reflect the contents of Regulation 13. The European Community is a signatory to UNECE Regulation 13 in its own right.

2   COMPLEX ELECTRONIC CONTROL SYSTEMS

The capability of electronic controls on motor vehicle and trailers is advancing at an accelerating rate as sophisticated new sensors and processing systems are being developed at a cost which can be accepted by the industry. The original anti-lock braking system (ABS) has spawned a fully electronically controlled braking system (EBS). EBS can provide additional functions that may shorten the brake response time, improve compatibility and brake force distribution between tractors and trailers, balance brake lining wear on individual vehicle axles and simplify the installation process for the vehicle manufacturers.

Electronic control systems are available to provide increased traction when the vehicle is accelerating, greater stability under cornering and adaptive speed control when driving. All these can become complex systems when integrated with the braking system to provide the appropriate control of the vehicle. Electronically assisted steering is now common on smaller passenger cars and in time may give way to fully steered by wire systems. On its own a steered by wire system would not be a complex electronic system but if it was designed to use data generated by other functions e.g. ABS speed sensors, to perform corrections to the steering without the invention of the driver then it would become a complex electronic system.

Within the vehicle systems there will be a hierarchy of control in which a controlled function may be over-ridden by a higher level electronic control function. When the higher-level control function detects changes in the vehicle behaviour it can instigate corrections by commanding variations in the normal function of a vehicle control system. This is illustrated by a stability control system that detects excessive lateral acceleration at the rear of the vehicle and consequently applies one of the rear brakes to counter the force.

3 UNECE REGULATION 13 ANNEX 18

3.1 Creation of Annex 18

Over the last 5 years the number of on-board electronic systems had significantly increased and national government experts have become more concerned about their effect on vehicle safety. The interaction of these new electronic controlled functions with existing primary safety systems has raised questions over their compatibility and operation in the event of system failure.

Vehicle and brake system manufacturers have worked together with national experts to produce a common understanding on the steps that must be undertaken during the design process and the information that must be disclosed to the type approval authority at the time of type approval.

This work has resulted in the preparation of a new annex which, has been written in non-specific terms so that it can be incorporated directly into any regulations. In the case of Regulation 13 it has become Annex 18.

3.2 Contents

The New Annex introduces a definition for electronic control systems, which are subject to a hierarchy of control in which a controlled function may be over-ridden by a higher level electronic control system. This allows the higher-level control system to assess data available and then modify the vehicle behaviour by commanding variations in the normal control function. Examples of this are vehicle stability control systems, which apply the brakes automatically in order to alter the behaviour of the vehicle.

The Annex does not specify the performance criteria for the relevant system, this is covered in the main body of the regulation, but covers the methodology applied to the design process and the information which must be disclosed to the technical service for type approval purposes.

3.3 Vehicle Type Approval

At the time of Type Approval the vehicle manufacturer must declare the complex systems fitted and supply the technical service with a package of documentation which gives access to the basic design of the system and the means by which it is linked to other vehicle systems or by which it directly controls these other systems.

The documentation must provide evidence that the design and development has had the benefit of expertise from all fields, which are involved. This documentation will be taken as the basic reference for the verification of the performance function of the system under failure and non-fault conditions.

Any additional material and analysis data provided by the manufacturer to support their application for type approval would not be retained by the approval authority.

3.4 Safety Concept
In accordance with the new annex the vehicle manufacturer will declare the systems that may be considered to be complex systems and explain how the systems interact with each other. Particular attention is required to identify the steps taken to ensure the safe operation of linked systems in the event of complete or partial failure of one of the systems.

The manufacturer will be required to provide an explanation, sufficient to satisfy the approval authority, and possibly a demonstration of how safe operating conditions are maintained under fault conditions. This may have resulted in the inclusions of a separate back-up system.

Since modern systems use digital computers, signal processors and data transmission in schemes of distributed and inter-linked control, a wide ranging safety strategy has to be provided. To achieve this effectively, the vehicle manufacturer and systems suppliers nowadays have to collaborate more closely than ever.

The documentation presented at type approval must be supported by an analysis of how the vehicle will behave in the event of one of the functions within the complex electronic systems failing. The supporting information may be in the form of a Failure Mode and Effect Analysis (FMEA) or a Fault Tree Analysis (FTA) or any similar process.

In the event of a failure the driver must be warned that the vehicle may not benefit from that particular safety device. This can be made by a warning signal or message display and shall remain as long as the fault exists.

3.5 Verification and Test
At the time of type approval the technical service has the right to test the vehicle in a fault condition. Faults may be induced into the vehicle by changing sensor outputs to electrical units or mechanical components. The results shall correspond to the failure analysis supplied.

3.6 Approval
The Approval Authority will study the documentation supplied by the manufacturer as well as the results of tests performed by the Technical Service when assessing the vehicle. Ultimately they must be confident that vehicle safety remains at the highest level.

3.7 Additional Definitions Within Regulation 13
In association with the introduction of Annex 18, Regulation 13 has been amended to include two new definitions: Automatically Commanded Braking (ACB) and Selective Braking (SB).

ACB is a function within a complex electronic control system where the operation of the brakes is intended to reduce the speed of the vehicle but their control may or may not be under the control of the driver.

Again SB is a function within a complex electronic control system where the primary purpose of the application of individual brakes is not to retard the vehicle but to alter the vehicle behaviour.

4 THE FUTURE ROLE OF THE DESIGNER

Modern designs are pushing the boundaries of technology further and further ahead as more and more electronic systems are introduced into motor vehicles. Today it is becoming increasingly difficult for a system designer to be solely devoted to "braking". Brakes are being asked to provide a method for other systems to control the operation of the vehicle, this may vary from simple traction control to complex stability control systems.

System designers have to work with colleagues in other disciplines to ensure that their systems are compatible. Steps have to be taken to ensure the correct operation of the vehicle whilst providing additional safety features. Suspension engineers who develop an advanced stability control system may rely upon the brakes to amend the vehicle behaviour but this could not be done without input from the brake engineer. After all it is the brake engineer who knows the spare capacity of the brake system and the situation that may arise in the event of the brakes being over worked.

Vehicle manufacturers have had to take on the role of system integrators, planning the whole installation and often, when dealing with multiple data networks, have to arbitrate in the settlement of signalling and functional control priorities.

5 LOOKING AHEAD

It is possible to envisage future systems in which all the traditionally individual vehicle mechanical control functions are integrated electronically and combined with external signals so as to produce features of higher level control. These revised features may ease the task of driving whilst providing increased operational safety through tireless monitoring of vehicle and traffic behaviour.

Development of linear solenoid valve for brake-by-wire system

T MIYAZAKI, A SAKAI, A OTOMO, E NAKAMURA, A SHIMURA, and Y TANAKA
Toyota Motor Corporation, Aichi, Japan
T MATSUNAGA
Fluid Dynamics and Heat Transfer Laboratory, Toyota Central R & D Laboratories Inc., Aichi, Japan
S NIWA
Chassis System Development Division, Toyota Motor Corporation, Shizuoka, Japan

ABSTRACT

This paper reports on the structural feature and the electrically controlling technology of a linear solenoid valve, which is suitable for the Brake-by-wire system. And the brake system controls the friction brake force and the regenerative brake force cooperatively to reduce exhaust gas further and improve fuel consumption of a hybrid vehicle.

1. INTRODUCTION

The recent growing need for global environment and energy preservation requires the automotive industry to further reduce vehicle exhaust gas emissions and improve fuel efficiency. The hybrid vehicle, which is one immediate solution, uses motor(s) not only to drive the wheels but also for regeneration. For its brake system, the demand has been increasing for blending control of the friction and regenerative brake forces. The brake-by-wire system, one of the most effective means of meeting this demand, separates the driver-operated input unit from the wheel cylinder hydraulic pressure, and electrically controls wheel cylinder hydraulic pressure. This brake system therefore enables free control of wheel cylinder hydraulic pressure without adverse influence on brake pedal operation. For the brake-by-wire system to work effectively, smooth control of wheel cylinder hydraulic pressure is essential. For a key technology in achieving this, we developed a linear solenoid valve that enables smooth hydraulic pressure control for the brake-by-wire system. This paper reports on the structural features and electrical control method of the developed valve. (The linear solenoid valve self-excited vibration analysis described in this paper was carried out jointly with Toyota Central R & D Laboratories, Inc.)

2. HYBRID SYSTEM AND BLENDING OPERATION OF REGENERATIVE BRAKE AND FRICTION BRAKE FORCES

2-1. 4WD Hybrid System
We have developed a 4WD hybrid passenger car that is equipped with driving motors for front and rear wheels respectively. The developed hybrid system combines parallel hybrid and series hybrid characteristics. In the parallel hybrid characteristic, the front wheels are driven by the motor, as well as by the engine. In the series hybrid characteristic, the front motor is driven by the engine as a generator, and the rear wheels are driven by the other motor (Fig. 1).

During braking operation, the hybrid system uses the front and rear motors as generators to maximize energy recovery. Energy thus recovered and stored in the battery is reused at start and during ordinary driving, to reduce fuel consumption and exhaust gas emissions. Regenerative brake force is used to replace a part or all of the friction brake force and to replace the engine brake force generated when the accelerator pedal is released.

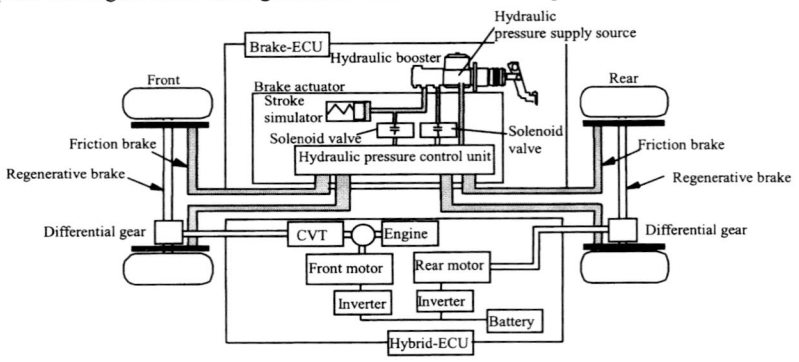

Fig.1 Hybrid System and Brake System

2-2. Blending Operation of Regenerative Brake and Friction Brake Forces
As shown in Fig. 2, the brake force of the developed hybrid system is divided into two portions: engine brake force generated when the accelerator pedal is released, and service brake force generated when the brake pedal is depressed. The engine brake force is determined by vehicle speed and accelerator pedal operation. The service brake force is applied when the driver depresses the brake pedal, as in conventional vehicles. The service brake force is the sum of friction (hydraulic) brake force and regenerative brake force. The

blending control of these forces is performed by the Brake-ECU and the Hybrid-ECU. The Brake-ECU controls the service brake force, while the Hybrid-ECU controls the regenerative brake force in response to commands from the Brake-ECU. To enable as much energy regeneration as possible during service brake, the Brake-ECU optimally controls the percentages of friction and regenerative brake forces.

Fig. 3 shows a specific control example. In the early stage of braking, the vehicle speed is high, and the regenerative brake force is low due to motor characteristics. Therefore, the Brake-ECU increases the percentage of the friction brake force to secure the total required brake force. As the vehicle speed decreases, the regenerative brake force gradually increases, the Brake-ECU accordingly decreasing the percentage of the friction brake force.

Since the regenerative brake force is zero when the vehicle is stationary, the Brake-ECU performs hydraulic control so that when vehicle speed is decreased to a certain level the regenerative brake force gradually begins to decrease while the friction brake force is gradually increased to secure the brake force intended by the driver. To ensure efficient high energy-regeneration, quick shifting between the friction brake force and the regenerative brake force is necessary. To this end, highly responsive friction-brake-force adjustment, i.e., highly responsive hydraulic control, is essential.

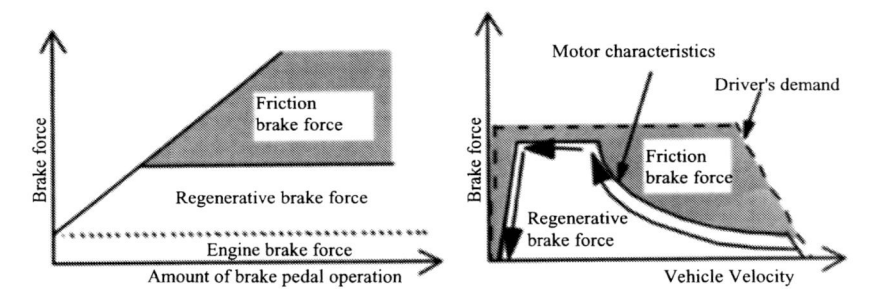

Fig.2 Allotment of Brake Force   **Fig.3 Change in Brake Force**

Fuel efficiency improvement resulting from the regenerative blending operation with the developed hybrid system is about 20% in comparison to the conventional vehicle in the Japanese 10-15 mode driving testing.

3. CONFIGURATION AND FUNCTIONS OF BRAKE-BY-WIRE SYSTEM

Fig. 4 shows the brake-by-wire system configuration incorporating the developed technology. We call this brake system the "Electronically Controlled Brake (ECB) system." Major system components are as follows:
(a) Input unit senses the driver intention and creates optimal pedal 'feel.'
(b) Hydraulic pressure control unit controls the hydraulic pressure of each wheel cylinder independently.
(c) Hydraulic pressure source supplies hydraulic energy for brake force.

(d) ECU centrally controls service brake (regenerative braking), ABS, Vehicle Stability Control System (hereinafter referred to as "VSC") and Traction Control System (hereinafter referred to as "TRC").
(e) Sensors for detecting vehicle running condition, which are wheel speed sensors, an acceleration sensor, a yaw rate sensor, and a steering sensor.

The brake-by-wire system comprising these components provides the following functional features:
(A) Since the driver-operated input unit is separated from the hydraulic pressure control unit, wheel cylinder hydraulic pressure can be controlled freely without influencing brake pedal operation.
(B) The hydraulic pressure source can be used to control the hydraulic pressure of each wheel cylinder without being influenced by negative vacuum pressure fluctuation at engine stop or brake pedal operation.
(C) The hydraulic pressure control unit is provided with the function to adjust the pressure smoothly and individually for each one of four wheels. So the hydraulic pressure adjustment of wheel cylinders, which is necessary for kinds of braking devices such as the regenerative brake blending operation, ABS, VSC and TRC, can be integrated.

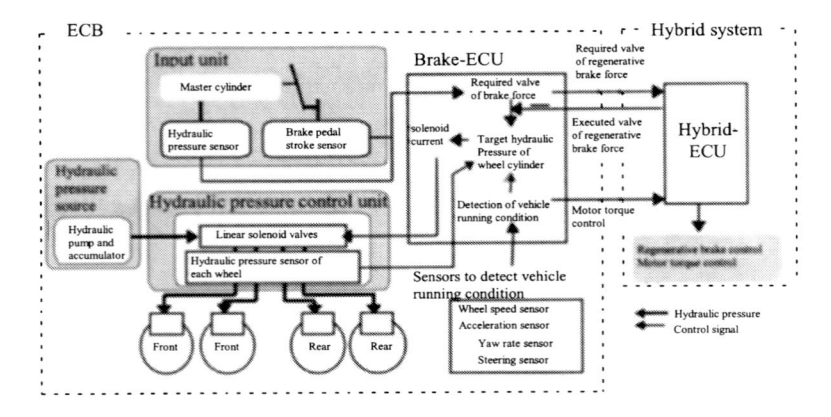

Fig. 4 ECB Composition

3-1. Hydraulic Brake System Configuration
Fig. 5 shows the hydraulic circuit diagram of the ECB system. This hydraulic circuit comprises the following components in addition to the hydraulic circuit of the conventional brake system:
(a) Stroke sensors(PSS1 and PSS2) for detecting pedal stroke, and hydraulic pressure sensors for detecting master cylinder hydraulic pressure(PMC1 and PMC2)
(b) Master cylinder cutoff valves(SMC1 and SMC2), which separate the brake pedal to be operated by the driver, and the wheel cylinder
(c) Stroke simulator to produce brake pedal reaction force and stroke
(d) Linear solenoid valves for augmentation and reduction, which adjust hydraulic pressure for each wheel independently

(e) Hydraulic pressure sensors(PFR, PFL, PRR, PRL, PACC1 and PACC2) for detecting hydraulic pressure of respective wheel cylinders and of hydraulic pressure source

3-1-1. Wheel Cylinder Hydraulic Pressure Adjustment for Brake Control

The Brake-ECU constantly monitors signal outputs from pedal stroke sensors (PSS1 and PSS2) and master cylinder hydraulic pressure sensors (PMC1 and PMC2). Upon detecting the depressed brake pedal, the Brake-ECU closes the master cylinder cutoff valves (SMC1 and SMC2) to prevent master cylinder hydraulic pressure transmission to each wheel cylinder, so that the wheel cylinder hydraulic pressure can be controlled without affecting pedal stroke. Simultaneously, master cylinder hydraulic pressure is transmitted to the stroke simulator, which then produces pedal reaction force and pedal stroke. The Brake-ECU also closes the valves SC1 and SC2 to separate the left and right wheels, thereby establishing an independent hydraulic line for each wheel.

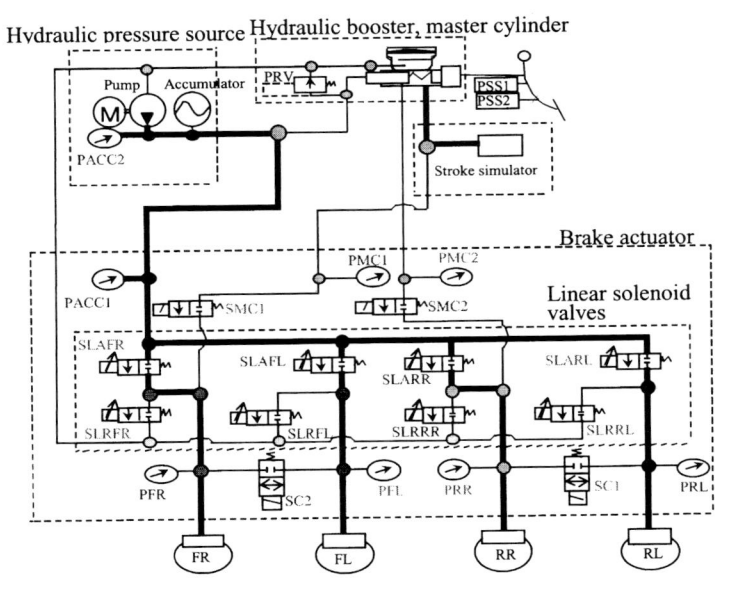

Fig. 5 Hydraulic Circuit Diagram (System Control Mode)

Fig. 6 shows the hydraulic control system configuration for wheel cylinders. The brake system comprises the following components:

(a) Accumulator (hereinafter referred to as "ACC") as a hydraulic pressure source in which hydraulic pressurized by a hydraulic pump is accumulated; and an ACC hydraulic pressure sensor
(b) Pressure-augmenting linear solenoid valve (hereinafter referred to as "SLA")
(c) Pressure-reducing linear solenoid valve (hereinafter referred to as "SLR")
(d) Wheel cylinder hydraulic pressure sensor
Components (b) through (d) are provided for each wheel.

Using these components, the hydraulic brake system controls wheel cylinder hydraulic pressure as described in the following.

3-1-1-1. Augmenting Wheel Cylinder Hydraulic Pressure
Wheel cylinder hydraulic pressure is augmented by the following method:
(i) SLR is closed.
(ii) SLA is gradually opened, allowing brake fluid to flow from ACC to wheel cylinder. Signal output from wheel cylinder hydraulic pressure sensor is sent to ECU to calculate SLA opening.

3-1-1-2. Reducing Wheel Cylinder Hydraulic Pressure
Wheel cylinder hydraulic pressure is reduced by the following method:
(i) SLA is closed.
(ii) SLR is gradually opened, allowing brake fluid to flow from wheel cylinder to reservoir. Signal output from wheel cylinder hydraulic pressure sensor is sent to ECU to calculate SLR opening.

3-1-1-3. Maintaining Wheel Cylinder Hydraulic Pressure
SLA and SLR are closed to maintain wheel cylinder hydraulic pressure.

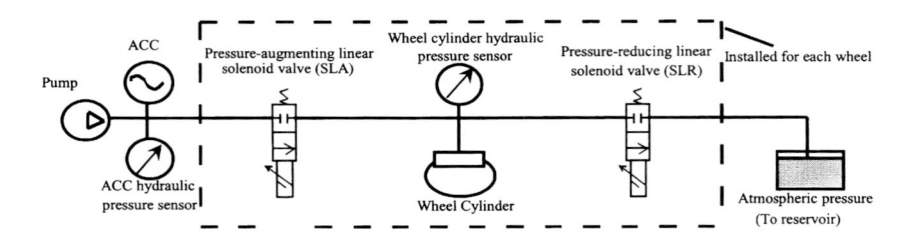

Fig. 6 Hydraulic Control System

4. LINEAR SOLENOID VALVE

4-1. Outline of Linear Solenoid Valve
The ECB system uses linear solenoid valves to perform above-mentioned hydraulic pressure control for service braking (regenerative braking), ABS, VSC, TRC etc. For this purpose, the linear solenoid valve must be:
(a) Smooth in adjusting wheel cylinder hydraulic pressure;
(b) Highly responsive in adjusting wheel cylinder hydraulic pressure;
(c) Quiet, since it is operated frequently (each time brake pedal is pressed).

The linear solenoid valve we developed in view of these requirements has the following characteristics:

(a) Smooth control of valve opening. → See Section 4-3.
(b) High responsiveness based on ABS technology.

(c) Functions to prevent self-excited vibration that causes noise and hydraulic pressure oscillation. → See Section 4-5.

4-2. Linear Solenoid Valve Type

The linear solenoid valve used in the brake system is of the poppet type, as shown in Fig. 7. As a drawback, the poppet valve tends to experience self-excited vibration, causing oscillation and noise. Use of this valve therefore requires self-excited vibration preventive technology, which will be detailed in Section 4-5.

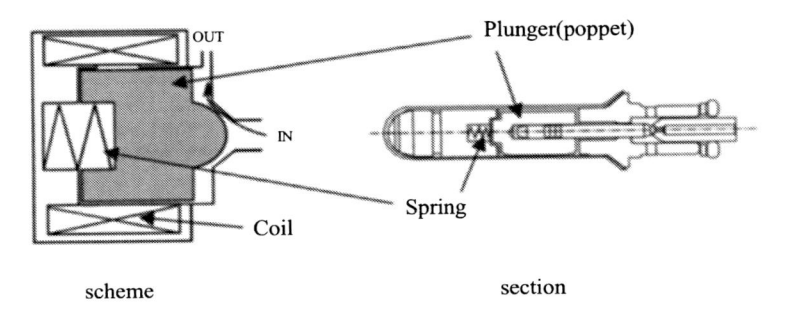

Fig. 7 Linear Solenoid Valve

4-3. Principle of Linear Solenoid Valve Operation

The opening angle of a linear solenoid valve is equivalent to the displacement of the plunger (shaded area in Fig. 8). Brake fluid flows from the inlet (IN) side through this opening to the outlet (OUT) side. Plunger displacement (valve opening) is determined by solenoid electromagnetic force, spring force, friction force, hydraulic pressure, and fluid force, as described in the following.

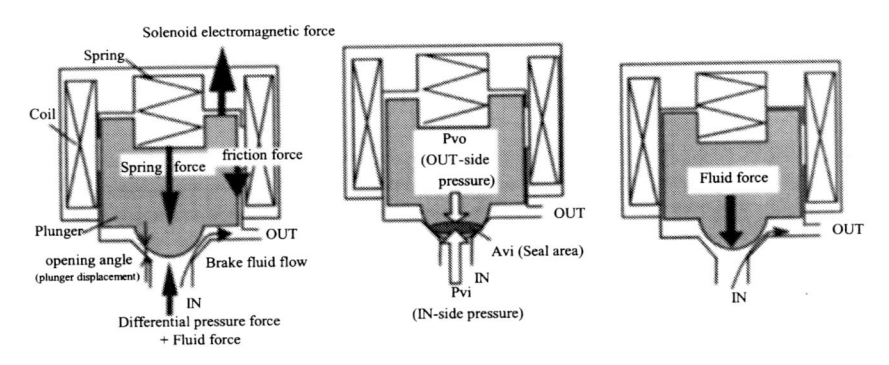

Fig. 8 Forces Acting on Plunger

Fig. 9 Differential Pressure Force

Fig. 10 Fluid Force

Fig. 8 shows the forces acting on the plunger of linear solenoid valve.

(a) Differential pressure force is caused by the difference in hydraulic pressure on both sides of seal area Avi, as shown in Fig. 9. Assuming that the IN- and OUT-side pressures are Pvi and Pvo, respectively, differential pressure force (Pvi - Pvo) · Avi acts on the seal in the valve-opening direction.
(b) Fluid force is the force of fluid flow in the valve that acts on the plunger in the valve-closing direction as shown in Fig. 10.
The resultant forces (a) and (b) (Differential pressure force + Fluid force) were surveyed through experimental and simulation analyses.
(c) Spring force increases proportionally with plunger displacement.
(d) Change in solenoid electromagnetic force which is determined by electric current flowing through the coil is small regardless of plunger displacement.
(e) Friction force which acts on the plunger in the opposite direction of plunger movement.

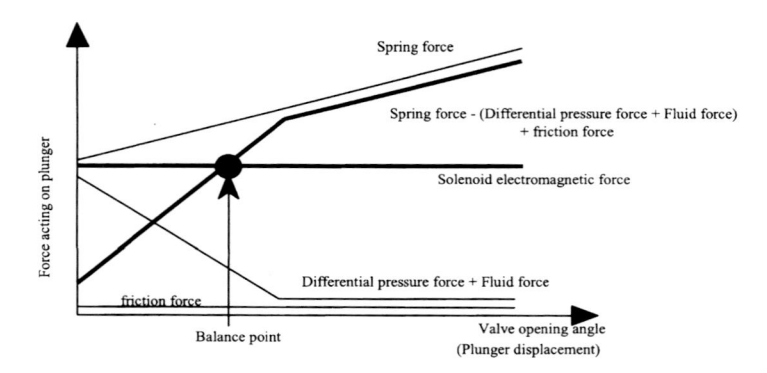

Fig. 11 Balance of Forces Acting on Plunger

Fig. 11 schematically shows the balance among the resultant of (differential pressure force + fluid force), the spring force, the solenoid electromagnetic force, and the friction force when the plunger moves in the valve-opening direction. Plunger displacement is set so as to satisfy the following equation:

Spring force - (Differential pressure force + Fluid force) + friction force
=Solenoid electromagnetic force   --- (1)

Therefore, valve opening can be smoothly controlled by varying the solenoid current to control the solenoid electromagnetic force.

4-4. Wheel Cylinder Hydraulic Pressure Control by Linear Solenoid Valve
The controller of wheel cylinder hydraulic pressure using the linear solenoid valve has a feedforward/feedback control loop with two degrees of freedom, as shown in Fig. 12, to simultaneously satisfy both responsiveness and stability requirements. It is necessary to enhance feedforward accuracy and decrease feedback gain so as to ensure stable hydraulic pressure control without sacrificing responsiveness.

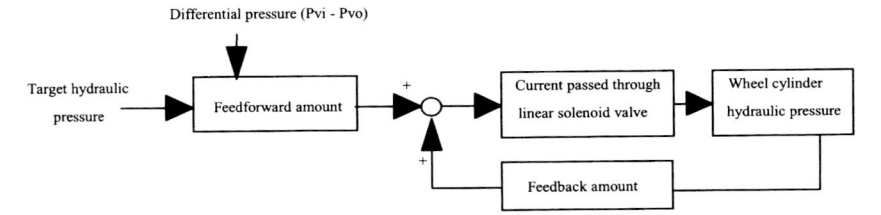

Fig. 12 Hydraulic Pressure Control System of Linear Solenoid Valve

Feedforward amount is calculated from the differential pressure (Pvi - Pvo) and the target hydraulic pressure. In principle, the feedforward amount is equal to the current value (hereinafter referred to as "valve-opening current") at which the linear solenoid valve begins to open. When the valve-opening current is allowed to flow in the linear solenoid valve, the fluid force = 0. So on the basis of the equation (1),
Solenoid electromagnetic force = Spring force - Differential pressure force + friction force
The Brake-ECU learns and memorizes this valve-opening current against some hydraulic pressure differences, to ensure that an accurate feedforward amount is sent to the linear solenoid valve.
Next, feedback amount is calculated using PID control; the sum of feedforward and feedback amounts is obtained as control output and supplied in the form of current to the linear solenoid valve.

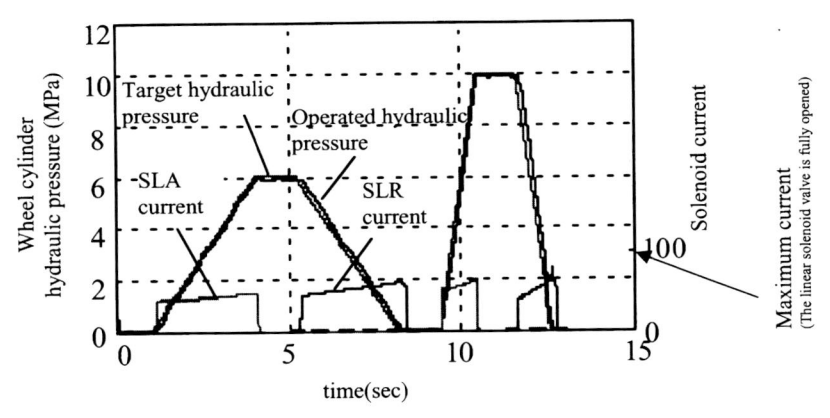

Fig. 13 Wheel Cylinder Hydraulic pressure Waveforms

Fig. 13 shows the waveforms of wheel cylinder hydraulic pressure controlled by the linear solenoid valve. The hydraulic pressure changes smoothly, complying well with the target pressure.

4-5. Self-excited Vibration Analysis for Linear Solenoid Valve

As mentioned earlier, a poppet valve tends to experience self-excited vibration, which can cause noise and poor hydraulic pressure control performance. Regarding this, some study reports are available[1] on a hydraulic system that controls hydraulic pressure upstream of the valve. However, there are no study reports on hydraulic systems that control hydraulic pressure downstream of the poppet valve. We therefore carried out self-excited vibration analysis on a hydraulic system for controlling downstream hydraulic pressure.

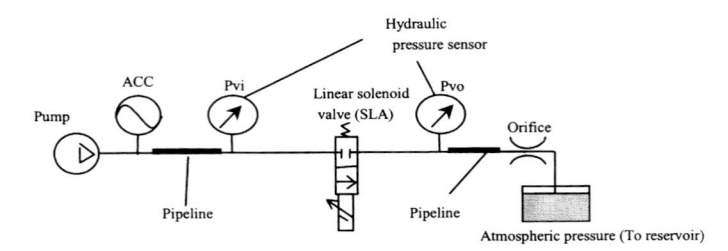

4-5-1. Analytical Hydraulic System
Analysis was carried out through experiment and simulation. Fig. 14 shows the hydraulic system used in the analysis. To simplify complex phenomena, this analytical hydraulic system differs from the system in Fig. 6 in the following points:
(a) Orifice is used in place of the wheel cylinder.
(b) SLA alone is connected, with SLR omitted.

Fig. 14 Analytical Hydraulic System

4-5-2. Self-Excited Vibration State
Fig. 15 shows example self-excited vibration waveforms observed in the hydraulic system of Fig. 14. The hydraulic pressure oscillation frequency variation ranges from approximately 400 Hz to approximately 900 Hz, depending on pipeline length on the upstream and downstream sides of the linear solenoid valve, and differs from the natural frequency (approx. 200 Hz) of the plunger spring mass system in the valve. This implies that self-excited vibration is a combination of the vibrations of the linear solenoid valve and the hydraulic system (including pipelines). In modeling each component, therefore, the influence of hydraulic system's frequency response must be taken into account.

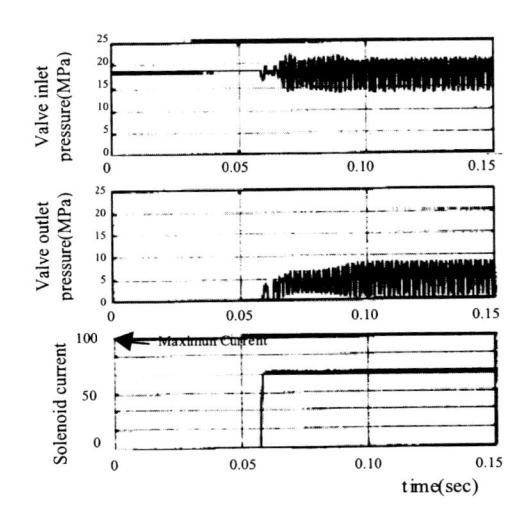

Fig. 15 Hydraulic pressure Oscillation Waveforms (Example)

4-5-3. Simulation Model
The following paragraphs describe the modeling of each component of the system shown in Fig. 14.

4-5-3-1. Modeling of Linear Solenoid Valve
Various forces acting on the plunger are modeled in more detail. Fig. 16 shows the forces. Hydraulic pressure upstream of the linear solenoid valve (hereinafter referred to as "valve upstream pressure"), Pvi, acts on area Avi, and hydraulic pressure downstream of the linear solenoid valve (hereinafter referred to as "valve downstream pressure"), Pvo, acts on area Avo. Hydraulic pressure in the fluid chamber of the linear solenoid valve (hereinafter referred to as "in-chamber hydraulic pressure"), Pvb, acts on area (Avi + Avo).

Equation (2) is the dynamic equation for the plunger, in which Ff is the fluid force, and Fsol is the solenoid electromagnetic force. Equation (3) shows the speed of the hydraulic pressure change in the fluid chamber. In the equation, K is the fluid bulk modulus, V is the fluid chamber capacity, and Q is the fluid flow rate into the fluid chamber.

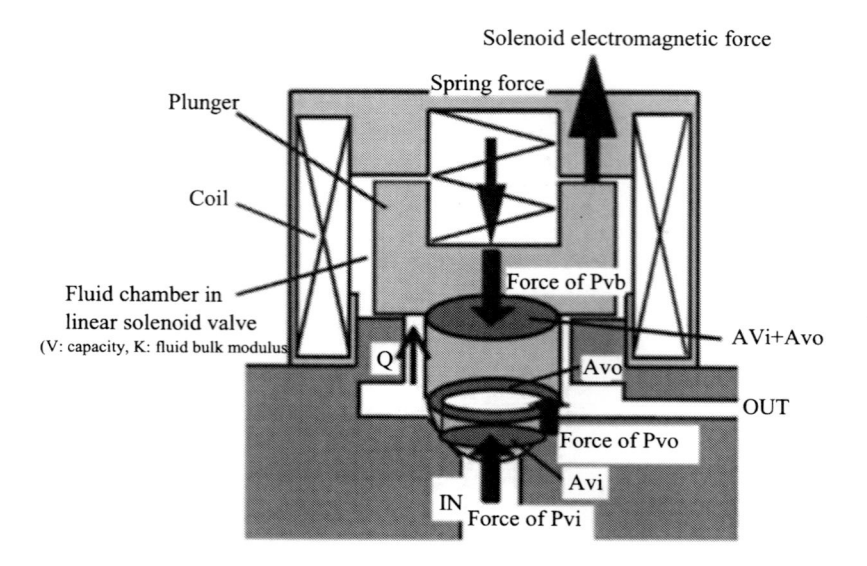

Fig. 16 Forces Acting on Plunger

$$m\ddot{X} + c\dot{X} + kX = \{PviAvi + PvoAvo - Pvo(Avi + Avo) + Ff + Fsol\} \quad \text{---} \quad (2)$$

$$\frac{dp}{dt} = \frac{K}{V}Q \quad \text{---} \quad (3)$$

4-5-3-2. Modeling of Pipeline

To reproduce pipeline resonance, an impedance model was adopted as the pipeline model. The relation between flow rate(Q) and hydraulic pressure(P) at each end of the pipeline is expressed by equation (4), in which Zc is the characteristic impedance and Γ is the transmission operator. Modal approximation was carried out to express this relation in the form of the sum of polynomials[2]. Equation (5) is an example result of modal approximation. A_i and B_i can be obtained by calculating the residues at the poles of 1/cosh Γ. ξ_i and ω_{ni} can be expressed as the functions of pressure waveform transmission speed, pipeline length, and pipeline inside diameter. Since experiments revealed that the pipeline is less susceptible to higher-order vibration, the fourth order was considered maximum order n for the approximation. The pressure waveform transmission speed in the pipeline was measured using an actual system.

$$\begin{bmatrix} P_1 \\ Q_1 \end{bmatrix} = \begin{bmatrix} \cosh\Gamma & Zc\sinh\Gamma \\ Zc^{-1}\sinh\Gamma & \cosh\Gamma \end{bmatrix} \begin{bmatrix} P_2 \\ Q_2 \end{bmatrix} \quad \text{---} \quad (4)$$

$$\frac{1}{\cosh\Gamma} = \sum_{i=1}^{n} \frac{A_i + B_i}{s^2 + 2\xi_i\omega_{ni}s + \omega^2} \quad \text{---} \quad (5)$$

4-5-3-3. Modeling of Accumulator (ACC)

Based on linear approximation of the filler gas status change, the relation between ACC pressure and flow rate of fluid out of the ACC can be obtained by applying equation (6) to bulk modulus K of equation (3). In equation (6), n is the polytropic index and P_a is the ACC pressure.

$$K = n P_a \qquad \text{--- (6)}$$

4-5-3-4. Modeling of Orifice

Equation (7) was applied to the orifice and linear solenoid valve (relation between valve opening and fluid flow rate) of Fig. 14. In equation (7), C is the discharge coefficient, A is the opening area and ρ is the fluid density. Values measured with an actual system containing a linear solenoid valve were used for discharge coefficient C.

$$Q = CA\sqrt{\frac{2\Delta P}{\rho}} \qquad \text{--- (7)}$$

4-5-4. Model Validation

Simulation results were compared with experimental results to confirm that each model properly reflects experimental values.

4-5-4-1. Comparison by Pipeline Length and Self-Excited Vibration Frequency

Figs. 17 and 18 compare the experimental and simulation results regarding the oscillation frequency of hydraulic pressure Pvo at the time of self-excited vibration, with upstream and downstream pipeline lengths, respectively, as parameters. The simulation result coincides well with the experimental result.

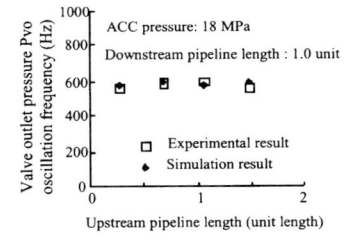

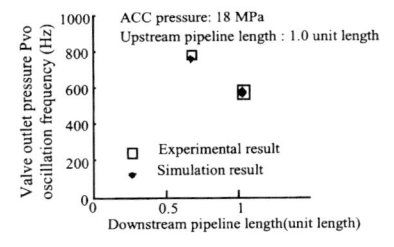

Fig. 17 Upstream Pipeline Length and Oscillation Frequency

Fig. 18 Downstream Pipeline Length and Oscillation Frequency

4-5-4-2. Comparison by Self-excited Vibration Occurrence Area

Fig. 19 compares the experimental and simulation results concerning the self-excited vibration occurrence area of ACC pressure and current input to the linear solenoid valve. Clearly, the simulation reproduces the experimental result at high accuracy. In this diagram, change in solenoid current value (horizontal axis) represents change in linear solenoid valve opening angle under varying ACC pressure (vertical axis).

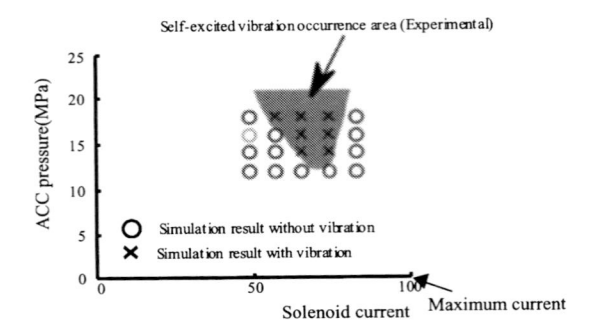

Fig. 19 Comparison by Self-excited Vibration Occurrence Area

These comparisons confirmed that each model constructed for the analysis accurately reproduces the self-excited vibration phenomenon.

4-5-5. Self-Excited Vibration Analysis by Simulation
Analysis using the above-mentioned simulation model focused on plunger vibration, which causes hydraulic pressure oscillation. The result revealed the contributions of the following parameters to self-excited vibration.

4-5-5-1. Influence of Pipeline Length
Fig. 20 shows the relation between downstream pipeline length and linear solenoid valve plunger vibration amplitude. The self-excited vibration in question is a combination of vibration of the linear solenoid valve itself and that of pipelines connected upstream and downstream of the valve, and is therefore affected largely by the frequency response of the pipelines.

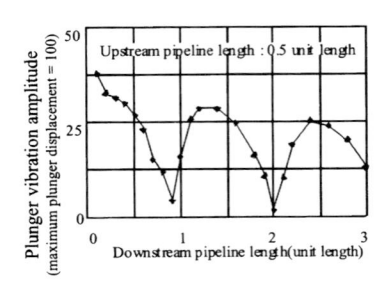

Fig. 20 Downstream Pipeline Length and Plunger Vibration Amplitude

Fig. 21 Plunger Displacement and Vibration Frequency

In addition, plunger vibration frequency at the time of self-excited vibration reflects the resonance frequencies of the upstream and downstream pipelines, as indicated in Fig. 21. In the range where plunger displacement is small, the action force of IN-side pressure (Pvi · Avi) is greater than that of OUT-side pressure (Pvo · Avo), and plunger vibration frequency is equal to the resonance frequency of the upstream pipeline. On the other hand, in the range where plunger displacement is large, the action force of IN-side pressure is smaller than that of OUT-side pressure, and the plunger vibrates at the resonance frequency of the downstream pipeline. As plunger displacement increases, there is a range where upstream pressure oscillation differs from downstream pressure oscillation. This phenomenon was also observed in the experiment.

4-5-5-2. Influence of Fluid Bulk Modulus in Linear Solenoid Valve Fluid Chamber

Fig. 22 shows the relation between fluid bulk modulus in the linear solenoid valve fluid chamber (hereinafter referred to as "fluid bulk modulus") and self-excited vibration. Fluid bulk modulus influences both plunger vibration frequency and amplitude. Delay in transmission of in-chamber hydraulic pressure Pvb, compared to transmission of valve outlet pressure Pvo, as shown in Fig. 23, is the cause of plunger vibration. This transmission delay varies depending on the fluid bulk modulus, and significantly influences the vibration amplitude, i.e., the probability of self-excited vibration of the plunger.

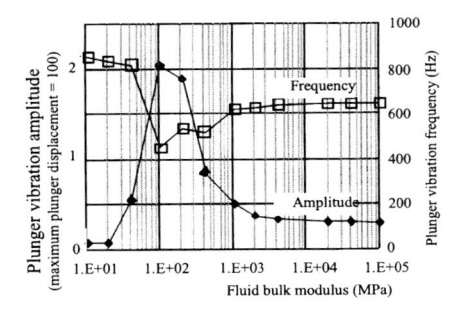

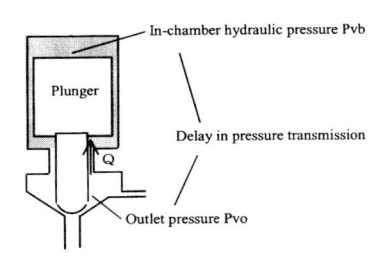

Fig. 22 Fluid Bulk Modulus and Vibration

Fig. 23 Delay in Pressure Transmission

4-5-5-3. Influence of Spring Constant

Fig. 24 shows the relation between linear solenoid valve spring and state of plunger vibration. Plunger vibration frequency is relatively constant irrespective of the spring constant, indicating that self-excited vibration is a combination of plunger spring mass point resonance vibration and the hydraulic system around the plunger. Plunger vibration amplitude is affected significantly by the spring constant. As the spring constant increases, the plunger vibration amplitude decreases. Self-excited vibration is therefore less likely to occur at larger spring constant. In other words, self-excited vibration can be suppressed by increasing the spring constant.

Based on the above-mentioned analytical results, we have developed a linear solenoid valve with high fluid bulk modulus by eliminating impurities in the chamber and high spring constant that enables stable braking operation (see Fig. 25).

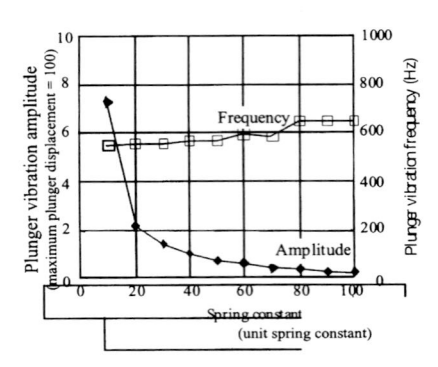

Fig. 24 Spring Constant and Plunger Vibration

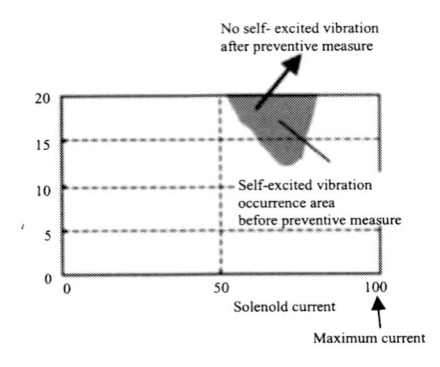

Fig. 25 Result of Effect Verification Study

5. CONCLUSION

We have developed a linear solenoid valve for use in the brake-by-wire system, and a wheel cylinder hydraulic pressure control technology using the developed valve. The developed technology enables the brake-by-wire system to smoothly and quietly control wheel cylinder hydraulic pressure, realizing a natural brake 'feel' that properly reflects the driver's intention and vehicle dynamic control of the next generation.

REFERENCE

1) Hayashi, Ohi; "Oscillation Phenomenon in the Poppet Valve Line," Paper for JSME, 56-532, C1990, 25
2) C. Y. Y. Hsue, D. A. Hullender; "Modal Approximation for the Fluid Dynamics of Hydraulic and Pneumatic Transmission Line," Fluid Transmission Line Dynamics, Special Publication for the ASME Winter Annual Meeting, Boston, Massachusetts, 1983, pp. 51

Design and control of an electric parking brake

R LEITER
TRW Automotive, Koblenz, Germany

Synopsis

The combination of the growing need for comfort and the increase in safety requirements has resulted in the development of the power window, the electronic seat and mirror adjustment and ultimately, in the development of the electric parking brake.

While the basic concept remains the same for all vehicles, functionality and operational concepts will continue to serve as the criteria for differentiating between the vehicle classes and vehicle manufacturers. The product development described in the following pages offers a multitude of possibilities in this regard. As in the case of all new electronic systems, a cost benefit cannot be directly anticipated at the beginning of the series introduction; in the long run, however, the product is shown to demonstrate adequate potential.

1 Introduction

Anyone who has ever driven a car is already familiar with that uncertain feeling one gets, when he or she is forced to park the vehicle at an uphill grade during an Alpine holiday. A gear is engaged, the front wheels are turned towards the curbstone and stones are even placed under the wheels of the car. Seasoned alpinists always carry one or two wheel chocks with them. The reason for this? No confidence in the parking brake; a fact, which is also confirmed by the statistics of the TÜV (German Technical Inspection Agency).

Many of us are familiar with the struggle to shift the hand brake lever with both hands after it has been wrenched on by the hulk of the family the previous evening (unless the car is a Mercedes). And which Swedish driver parks his car during winter with the hand brakes engaged? Nobody, because the cables could freeze!

Vehicle manufacturers and suppliers have developed a variety of effective solutions with regard to this problem. In 1997/98 a team from TRW (then Lucas) has collaborated towards the development of an entirely new concept, which will go into serial production for the first time in 2002. At TRW, the product is called EPB, an abbreviation for electric parking brake.

2 System components

The actuating element and the brakes constitute a single modular unit in this system, i.e. the operating mechanism is integrated in the rear wheel brakes.
Both actuators are controlled through an separate electronic control unit. This control unit is connected redundantly to a power supply. The operating element (here, a rocker switch is shown) is directly connected to the control unit. Otherwise, the electronics receives all the necessary signals from the CAN - Network.

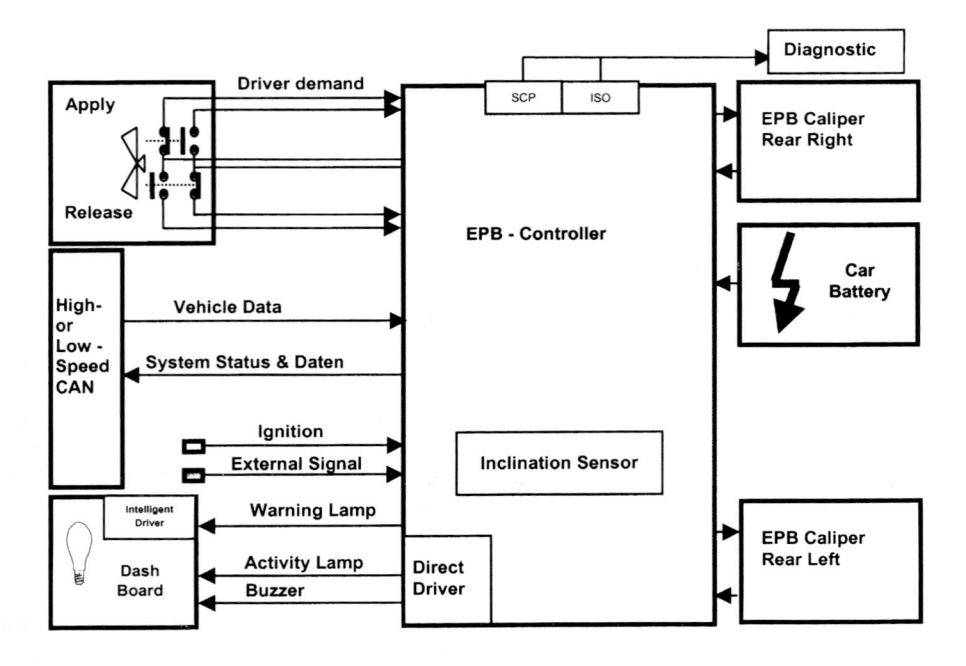

Figure 1: System design and components

3 System design

The longer service life of the vehicle and possibly, higher number of operations (since the operation is simpler and requires no effort) have prompted our customers to require up to 100,000 load cycles.

At 12V, 20°C ambient temperature and 200°C disc temperature, the system will reach a clamp force which guaranties standstill at 30% incline in 0.8s. The release time is defined, as the time from the starting of the motor up to the time for zero clamp force; the clearance has no impact. A typical release time of the TRW-System is 0.5s.

In the course of the development process, the prevailing international specifications have continuously been developed. Probably the most crucial specification is the requirement concerning the parking of a fully loaded vehicle at a hill with an 8% gradient, in the event of a failure of the parking brake system (excluding the power supply). Engines with little or no compression (i.e. electric valves, which are open in case of an electrical failure) are not capable of delivering the needed braking torque at standstill. This results in a limited use of parking brakes with a single-motor concept and a considerable disadvantage in the long term, as compared to the double-motor concept followed by TRW. Although the faulty battery is meanwhile a condition of use (especially in the FMEA), the legislation does not address conditions regarding the energy supply.

The TRW-EPB-System complies with the legal requirements pertaining to mechanical locking (after power off), and the maximum force difference of 20% between left and right, as well as with the requirement on child lock during switched-off ignition (release is not possible). In accordance with ECE 13 the parking brake must be independent of the service brake; in the TRW-System, this is achieved through the movable spindle nut in the brake piston. This means that the parking brake still overrides the service brake.

A number of drivers would like to freewheel down the garage driveway with the ignition switched off and then brake in front of the gate using the parking brake. In garages, the vehicles are pushed by hand and then decelerated again by using the hand brake. Consequently, the implementation of dynamic braking was part of the specification.

4 Definition of the mechanical and electrical interfaces

For the development, the following interfaces have been defined:

Maximum clamping force:	17 kN
Brake to actuator:	plug in, two fastening screws
Power consumption actuator:	Max. 20 A
Input voltage ECU:	Min. 9 V
Actuator-feedback:	Hall-effect sensor

This was necessary in order to be able to start working with the in-house-design, subcontractors and electronic-suppliers simultaneously.

Figure 2: Brake and actuator assembly

5 Design of the vehicle brake

The structure of the brake corresponds to that of a conventional rear wheel brake, which utilizes the well-known „ball in ramp principle" from TRW.
Instead of the automatic adjustment device, a spindle was used, on which a spindle nut is positioned, and this is either pushed against the piston head or is again retracted from it. The thread is self-locking and therefore conforms to the legal requirement pertaining to mechanical locking.

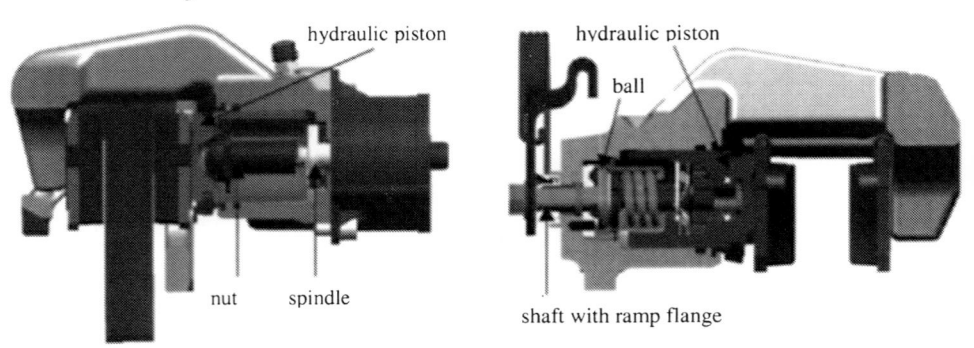

Figure 3 : EPB applied, compared with actual Ball & Ramp Brake

Probably the most important specification is the magnitude of the maximum clamping force.

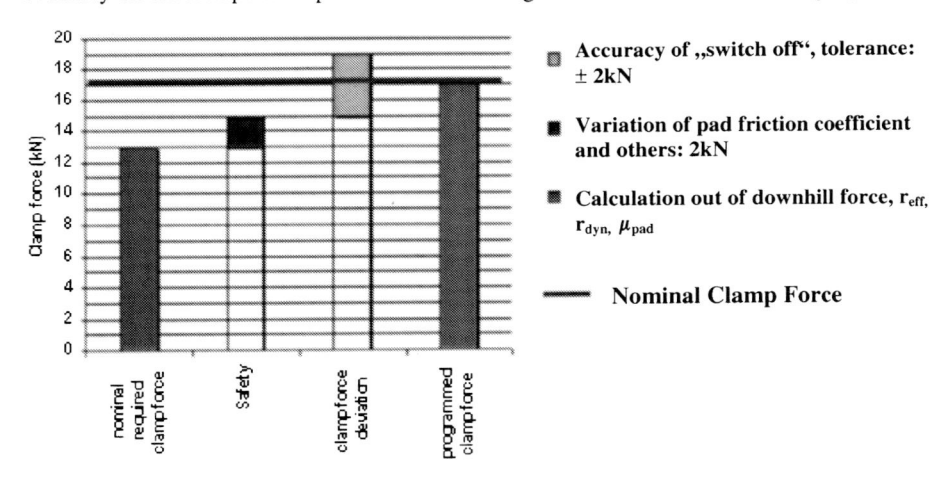

- Accuracy of "switch off", tolerance: ± 2kN
- Variation of pad friction coefficient and others: 2kN
- Calculation out of downhill force, r_{eff}, r_{dyn}, μ_{pad}
- Nominal Clamp Force

[Chart: Clamp force (kN) vs categories: nominal required clampforce, Safety, clampforce deviation, programmed clampforce]

Figure 4: Schematic design EPB clamping force

Modern parking brakes are designed for a temperature range of up to 300 °C, thereby defining the nominal friction coefficient. Moreover, the gradient of the slope (30%), the effective friction radius and the mass of the fully loaded vehicle are likewise taken into consideration in the calculation of the clamping force.

It must be presumed that in such a slope, the driver also uses the service brake. For all calculations of service life, therefore 40 bar service brake pressure have been added.

Since there are no accepted load cycles for electric parking brakes, and the software engineers were not able to develop any convincing strategies for the reduction of the clamping force, it was necessary to design a substantial bridge size, especially for aluminum rear wheel brakes. The required increase in bridge thickness can reach up to 5 mm. It is preferred to manufacture the caliper from gray cast iron, to facilitate the utilization of narrow installation spaces (distance between brake disc and rim).

In the end, we designed the parking brake for a maximum clamping force of 17 kN. Measurements in the brakes have shown that, already at this clamping force a "slip-stick" effect occurs in the thread of the spindle. Control in this area is achieved by the selection of the reduction gear, which generates a torque oscillating around a mean value; but this will be discussed in more detail later.

Another important discussion is about the emergency release options. At the lowest level, we have created the possibility of releasing the brake by means of a Torx-wrench. It is a garage solution, but still a possibility, if nothing else works.

The spindle is supported at the collar by means of an optimized thrust bearing at the base of the brake housing. The spindle itself is a cold-formed part with a rolled thread. The functional areas are fine machined prior to the annealing process. The design of the self-locking, sawtooth thread was a demanding iteration between production technology, output data (clamping force) and service life.

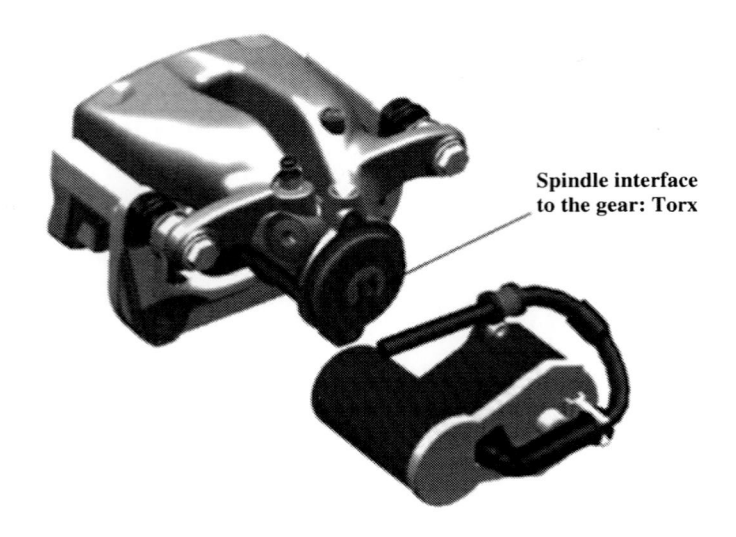

Figure 5: Manual emergency release through Torx

In contrast to the standard screws, this is a screw drive mechanism with a continuous alternation of the supporting threads. The pitch of the thread is equivalent to 1mm. This means that the garages may not use the standard piston resetting devices. There has been developed a new piston retraction tool, available via the Aftermarket Organisation. A limit stop is fitted at the collar and serves to prevent the twisting of the spindle nut against the collar. This is necessary for the manual, as well as the electrical pad change. The spindle nut is positioned on the spindle and transmits the clamping force via the conical cup into the brake piston.

The piston interior is of a form that prevents the spindle nut from turning relative to the piston. The counter torque of the anti-twist stop is generated by means of the piston seal, as well as the friction force between piston and pad back plate during application. In order to avoid any seizing tendency, the counterpart of the screw drive mechanism has been manufactured out of hardened steel. In this case it is the spindle nut, since the full thread length of the spindle is utilized due to the pad wear, while the spindle nut has only approximately 10mm of supporting thread.

6 Electric Actuator

The costs play a significant role in the selection of the electric motor. A 12V-direct current motor was selected, one that is already manufactured in large quantities.

Future 42V electrical systems will have no effect on the size of the motor; however, the peak current (at present, up to 20A) will be clearly reduced (target is below 5A). This will facilitate cost savings with respect to the power stages in the control unit.

The speed of the motor is dependent on the voltage and from this, it can be correctly concluded that the application times of the parking brake are likewise dependent on the voltage, i.e. the brake operation will be slower with the ignition switched off (approximately 12V) than with the engine running (approximately 14V).

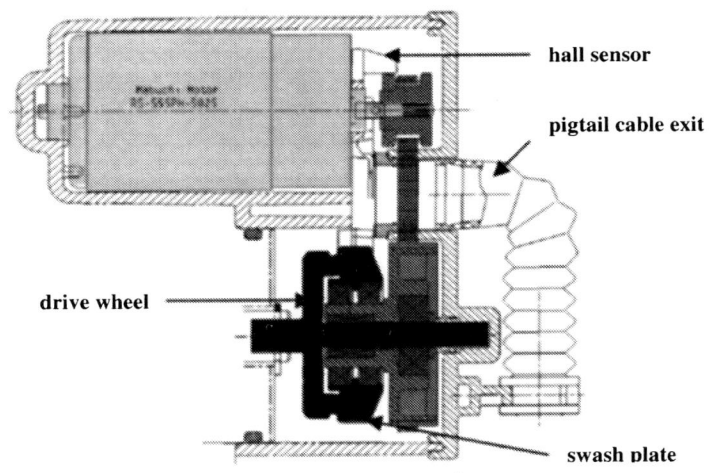

Figure 6: Actuator with motor and gear

The standard torque used is about 150 mNm. Due to the change in the internal resistance of the motor winding which is dependent on the temperature, one must take care that there is sufficient reserve for the upper range of temperature during the selection of the winding. While temperatures of up to 180°C occur at the caliper, the plastic housing allows the motor to operate in the considerably more favorable temperature range. The highest temperature that we have measured at the motor was 85°C, following high-performance multiple stops.

Efficiency and clamping times impact the selection of the subsequent ratios of reduction. The motor drives a toothed belt, which on the other side drives the drive wheel of a swash plate gear. Toothed belts have a very favorable efficiency within the nominal load range and decouple the noise of the fast-running direct current motor. By means of the integrated drive torx, the swash plate gear is fitted directly onto the spindle of the brake. Its design features allow relatively large position tolerances for the thrust bearings. The direct contact with the brake led to the necessity of manufacturing the gear wheels from temperature-resistant plastic; PEEK was selected.

The slightly angled drive shaft forces the swash plate to roll on the driving wheel. In our application, the swash plate has 50 teeth and the driving gear has 49 teeth. After one rotation of the drive wheel, the driving wheel has moved by one tooth in the opposite direction, when the swash plate is secured against twisting by the two brackets (as shown in Figure 6).. The brackets run in lubricated guides in the housing, which are reinforced with steel plates. The rest of the gear is virtually free of any lateral force. Through the oscillation of the brackets in the guides, varying supporting lengths of the brackets will result with the rotational speed, thus producing a harmonic wave on the torque curve. Consequently, this facilitates a relatively uniform and controllable rotation of the spindle, up to the point of seizure.

The toothed belt drive and swash plate gear have an overall ratio of approximately 1:150, i.e. for one rotation of the spindle (equivalent to 1mm clamp distance) the motor has to rotate 150 times.

The actuator housing is made of glass-fiber reinforced plastic with 3mm wall thickness. Its main functions include sealing (the system is hermetically sealed), protection against gravel impact, as well as heat insulation between the hot brake and the electrical components in the actuator.

An optional hall-effect sensor is fitted at the motor pinion to be able to determine the absolute position of the spindle nut for the display of the pad wear. The motion of the motor is monitored via voltage/current measurement. The clearance is adjusted by voltage dependent time control after reduction of clamp force.

Depending on the installation conditions in the vehicle, an emergency release device can be integrated in the lid, which is situated in the extension of the motor shaft (an operation which can be carried out with an Allen key, using very low force). However, even then, it is still necessary to either dismount the wheel, or to open a slot at the wheel arch from inside.

The relatively substantial mass of the entire wheel suspension keeps the level of dynamic stress at the actuator within limits; the actuator itself, can even withstand peak loads of up to 50g without any problems. More than 210,000 cycles have been attained with the actuators in reliability tests.

As in the case of ABS-sensors, the critical points to be considered are the alternating lateral cable bending, as well as the sealing of the cables. By using a suitable corrugated pipe material (PA12) and cable material 70 million alternating bending cycles during the endurance tests conducted, in which low temperature conditions were also included as a part of the tests. The classical type of cable sealing (single-wire sealing through sealing cushion, soldering in the lead frame, silicone sealing of the lead frame against the lid) selected here will be 100% tested in manufacturing with low pressure.

If possible, the cable should be installed parallel to the brake hose and routed into the passenger compartment and from there it is connected to the harness. An even better alternative would be, the installation of the cable on the lower wishbone, in the direction of the center line of the vehicle, because the cable is subjected to alternating bending only through slight angles and the portions of the flexing work will be reduced.

7 Electrical supply

The EPB system is supplied by two voltage supplies, each separately protected with a 30A fuse. The control unit is supplied through diodes from both voltage supplies, so that it will still operate, even if one of the voltage supplies fails. Each voltage supply is assigned to an actuator, i.e. if a short circuit triggers the fuse on one side, the other actuator can still be operated without any restrictions.

During the preparation of the harness diagram, the possible voltage drop between the energy source and the load has to be taken into consideration. After all, the required clamping force will only be attained at a current of up to 20A. The number of connections and the conductor cross sections are both taken into account in the calculation. TRW therefore requires at least a 4mm^2 conductor cross section for the live cables (supply and ground), up to the connector of the actuator. Under certain circumstances, i.e. extremely long cables, a conductor cross section of up to 10mm^2 may be required.

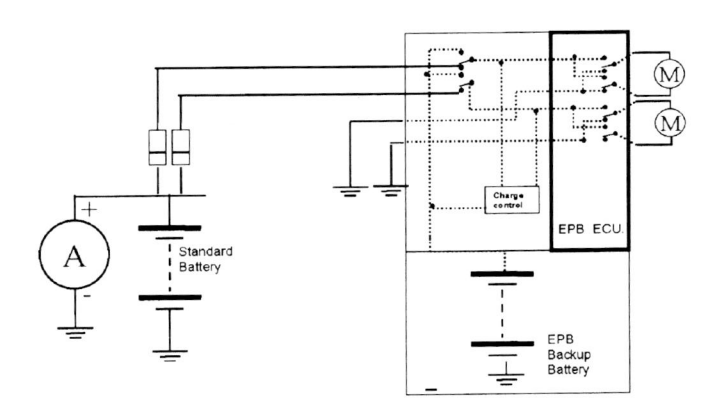

Figure 7: Secondary battery connection diagram

The most elegant form of emergency release is the installation of a secondary battery. It helps to increase the system availability.

Sudden power failure while driving is a criterion. The service brake and its emergency function, as well as engine braking are available. However, subsequent to this, safe parking is no longer possible, when engine is stalled and the engine compression can not provide the 8% requirement (fully laden). Crucial are vehicles with electrical gear shift, which also does not work if the energy supply has failed.

The more probable case is a power failure of an already parked vehicle. Since it would not be possible to start the vehicle anyway, the owner may wish to move the vehicle using muscle power, if it obstructs traffic, for example, and to park it again afterwards.
Both situations therefore, do not deal with normal operating conditions, but with the parking of a vehicle which has already broken down, i.e. a defective vehicle, a task which is possible with the mechanical parking brake systems.

The EPB requires approximately 500 watts for the application process, which takes about 1s, and around 300 watts for the release process, which also takes 1s. In order to provide this energy even at –40°C, a high-current resistant battery with approximately 2 Ah, is required. TRW has developed a solution with a charge regulator integrated in the control unit, but this has been dismissed due to cost reasons.

The better solution is the installation of a terminal, in which energy for the EPB can be supplied, as needed, by means of an external supply (e.g. through the use of a battery jumper cable).

It is certainly unfair, if the electric parking brake alone would have to bear the consequences of an archaic energy system within the vehicle. No doubt that there are other parties who will be interested in the option of a secondary battery, and they can be recognized by the fact that they are connected to the positive battery terminal (e.g. Keyless Go systems, alarm circuits) and most of them have an internal backup battery. Modern systems such as Steer by wire, Brake by wire or even the Joystick control will not operate without a redundant electrical system.

8 Control unit

Connectors with eight power-terminals and a few signal inputs were not available on the market. This prompted a development for a connector as well for a new ECU-housing in cooperation with AMP/Tyco.

The processor selected was the HC08, an 8-bit-processor from Motorola. The smallest version of this processor with 16K ROM was considered oversized at the beginning of the project. Software development has since revealed that three times as much memory space is needed. Each of the motors is operated via a corresponding FET.

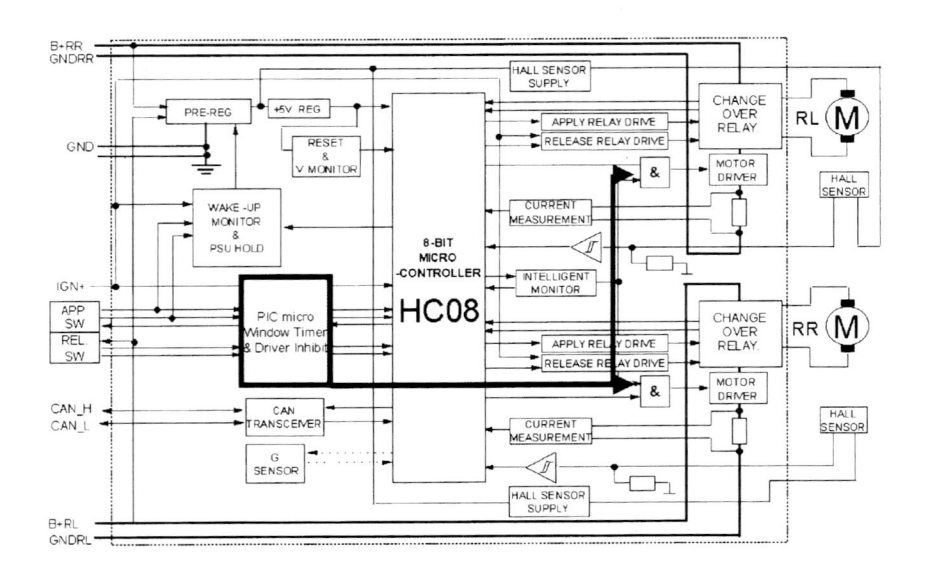

Figure 8: ECU Block diagram

The change of rotation is achieved by means of two relays. With an individual shunt, the real current flow, as well as the voltage level and polarity are measured.

The control unit can contain up to four power stages for the direct control of indicator lamps; preferably, these should be operated via the CAN.

During ignition on, the control unit is in the active-mode and continuously performs self-tests. After ignition off and a post-running time, the control unit switches to an energy-saving sleep-mode (approximately 200 µA) and will be activated again, either by actuation of the switch or through ignition on.

Originally, the EPB system was regarded as a comfort system and designed for the Low-Speed CAN. Hence, it's spartan but nevertheless effective safety features with one "intelligent monitor". The output signal of the double integrator is constantly read back and from this, the processor calculates the length of the required drive pulse, in order to hold the output signal between two voltage thresholds monitored by hardware. If the monitor exits from the given window, a RESET will be generated. The code for the activation of the double integrator is spread throughout the whole program, so that CPU arithmetic problems, endless loops and non-operative program parts are detected.

The connection to the High-Speed CAN and the interactions with ESP and engine control required a second approach for the safety architecture. This led to a two-processor solution, which will enable the interception of possible bugs. Both processors must decide independently that both EPB-motors must be activated.

Diagnosis and End of Line configuration is provided by means of the K-Line. The protocol used is Keyword 2000. Alternatively it is also available via the CAN.

For vehicles with integrated drive-away assistant, an additional inclination sensor in the control unit is provided as an optional feature.

9 Control element

The significant advantage of the EPB is the flexible layout of the control element, not only with regard to the design, but also with respect to the installation location and functionality. Electrical operation allows the use of minimal forces, an advantage which is especially appreciated by physically weak and handicapped individuals.

Meanwhile nearly all conceivable versions of the control element have already been designed, built, and tested with varying results. The simplest and most cost-effective design is a push-button switch, which applies the brakes if actuated once and again releases it, if actuated again.

The variant preferred by TRW is a dual position push-button switch with a neutral position, which is mounted in the longitudinal axis of the vehicle. The front position serves to apply the brake, the rear position to release it. If not actuated, the switch remains in the neutral position. The control philosophy also plays a significant role in the selection of the control element. Should the driver be relieved of the responsibility for the adequate clamping of the parking brake, the simple „on/off" elements will suffice. But if the responsibility for the clamping force is left to the driver, then tangible control elements are better. However, for this purpose, the control strategy and the FMEA must be very well supported.

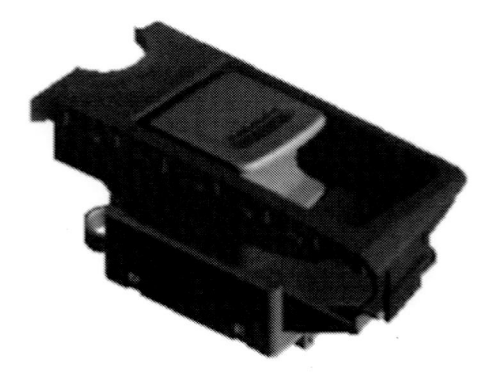

Figure 9: Dual position push-button switch

The single and dual position push-button switches will feature a redundant design, each with an normally open (NO) and an normally closed (NC) contact for each switch position. A capacitor will be connected to the NO-contacts, so that the NO-cable connections can also be monitored constantly with a pulsed test voltage. Therefore, in the event of a single fault, the action which the driver requests from the parking brake can still be recognized. This is also required in the more recent legislation on electronic brakes under the term, "control transmission".
Alternatively can the "Ignition off" signal can be used to apply the EPB, when main switch failure is recognised and the Drive Away Function used for the release.

10 Basic software functions

The software operates in a 20ms main loop, in which all modules are continuously addressed. It functions purely as a state machine, in which the exit from various modes can only be executed through certain events (Mode Control). Sub-routines, such as the combination of switch commands with other system signals, (Demand Calculator) or the motor control (Clamp Force Controller) are also designed as state machines.

The main function is the static application of the brake with ignition on. For this purpose, the switch must be actuated. The software checks, if the ignition is switched on (direct input from the ignition switch, but only as a signal), whether the CAN-bus is operating, if the CAN-bus ignition signal is OK and if the wheel speeds are zero. After 100ms (5 cycles), the command is accepted and passed on to the motor control.

This activates all operative motors in the clamping direction and also monitors the clamp current. At the same time, the voltage and the speed measurements are taken. During the take up of the clearance (current typically around 2A), the motor speed is determined and the motor temperature is estimated from this figure, by taking the voltage into account. Both voltage and temperature require a correcting value for the cutoff current, which will then be added to the idle current. A cycle counter provides another correcting value for the mechanical wear of the gear.

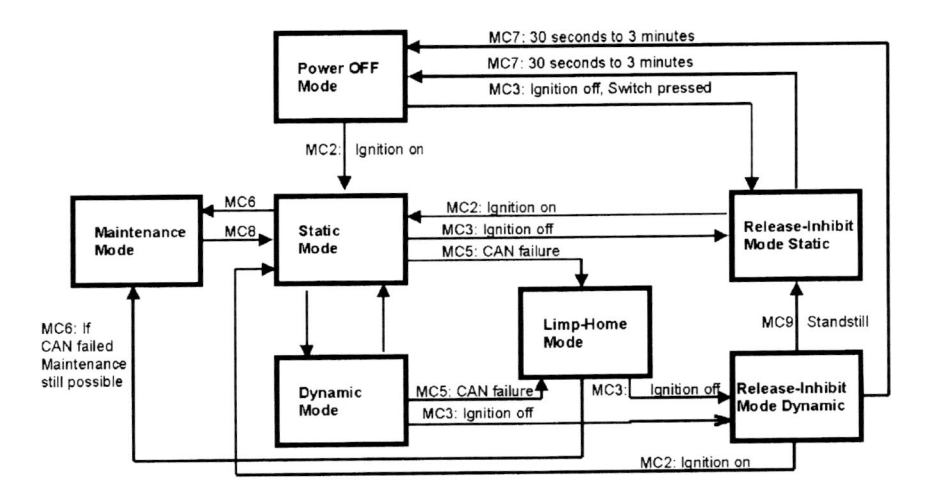

Figure 10: EPB-Functional states

With respect to the determination of the cutoff current, the tolerances in the system (spindle/spindle nut unit, manufacturing tolerances in the swash plate gear, rated power deviations of the electric motor and the tolerances during the current measurement) were taken into consideration. A fully loaded vehicle on a 30% uphill gradient is assumed.

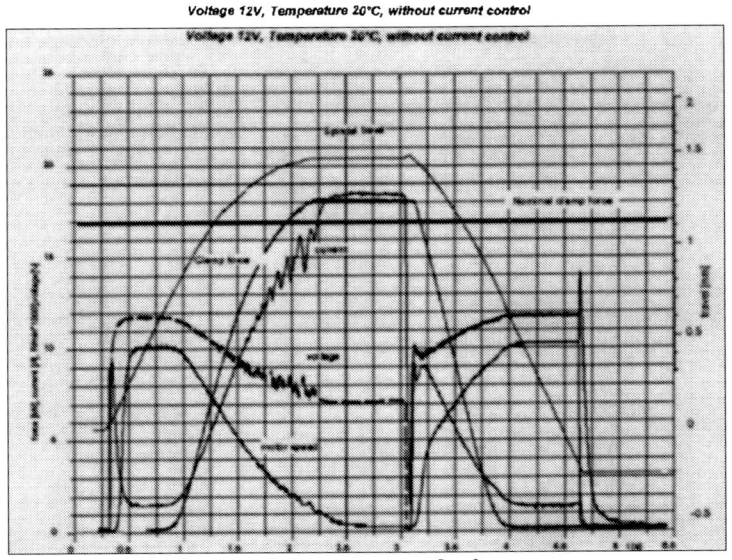

Figure 11: EPB clamping and release process

While driving, a temperature model similar to that of the Traction Control System is running simultaneously, in order to detect an excessively hot disc. Through a clamping repetition after appr. 3min, the system offers the possibility of achieving a single (cycle-related) rise in force which exceeds the nominal clamping force; this will cover the driver who wants to park his car on a 30% uphill gradient, following a series of high speed stops.

During the release of the brake, a lower current will be required than during the application; one-third less is typical. As soon as the current curve has fallen to the idle current level, this point will then be recorded as the new relative zero point and the motor is switched off, after a time-controlled period for clearance adjustment.
If the ignition is switched off and the switch is actuated, only the clamping of the parking brake is possible; a release however cannot be activated (child lock).

Crucial are Drivers who switch off the ignition when freewheeling downhill. There is no wheel speed information available via the CAN-bus. Stochastically distributed road irregularities continuously excite the body of the vehicle to produce pitching motions. With the help of a software band-pass filter, the freewheeling of the vehicle can be derived from the inclination sensor. Speeds of over 20 km/h will be reliably detected.

If the service mode is activated via the diagnosis cable while the ignition is switched on, both spindle nuts will retract up to their limit stop. After that, the brake pads can be replaced. The removal of the service signal and an application operation will result in the exit of the service mode and a re-calibration of the system (it will be assumed that the brake pad is new).

Of course, the parking brake is also functional while driving. In order to avoid a locking of the rear axle, a software which is similar to the ABS automatically controls the clamping force in accordance with the modified individual wheel control principle at the maximum adhesion limit. However, the electric motors have a far higher level of reaction time than ABS-valves, so that the closed loop control which runs at approximately 0.5 to 1 Hz, can not reach the high level of the ABS. Nevertheless, the driving stability is maintained to a great extent, even under split-μ conditions. If the ABS or ESP were already activated prior to the application of the parking brake, the dynamic braking will not be carried out as long as the activity flags of the closed loop control systems are set.
As an alternative, the dynamic deceleration can also be transferred to the service brake system by means of a CAN-message, and the braking will be executed by the service brake system, either through the electronic booster or the ESP-system.

11 Additional software functions

The electronic activation capability of the parking brake enables our customers to take advantage of a multitude of new functions.
For instance, an automatic application command can be issued, if the ignition is off when the vehicle is stationary. If this function is not desired in individual cases, then it must be canceled through the use of appropriate devices (e.g. on-board computer).

Another possible function would be an automatic application command, when the driver leaves his seat while the vehicle is stationary (e.g. recognized through the seat-occupant detection or door contact).

Additionally, a mode which applies the EPB at any standstill and then releases for drive-away could be activated by a "Stop and Go" switch for use in slow moving traffic. This function will be switched off automatically when vehicle speed reaches i.e. first time 30 kph.

Future systems, which decelerate until standstill, have one problem: The electric valves in the boosters or in the ESP systems cannot be continuously supplied with current in order to maintain the hydraulic pressure and there are no reliably sealing valves available for a reasonable price available (pressure loss up to 2bar/s). Here the electric parking brake offers advantages through its mechanical locking feature and activation by electronic means.

If the alarm system transmits an alarm signal on the CAN-bus during standstill, the parking brake could be applied (and release inhibited), in order to make it difficult for the potential thief to drive the vehicle away.

The extend of pad wear can be determined through the continuous counting of the pulses produced by the hall-effect sensor. Whether the two additional sensors and the wiring justify the expenditure, is according to customers preference.

Probably the most pleasant function of the electric parking brake is the automatic release during drive-off. The required engine torque is determined through the integrated inclination sensor and the brake is released upon reaching this threshold. The "fine art" of the Drive-Away function is the drive-off on a flat road. The inaccuracies, which result in the calculation of small engine-torque values is known, especially to the TC-developers; in certain cases, the deviations can be up to 60%. The change in the engine speed gradient allows the recognition of the clutch engagement and the brakes will be released as soon as a sufficiently high level of torque is reached. This algorithm does not function satisfactorily for the whole range of drivers, and this prompted us to introduce a parallel algorithm: during drive-off, the body of the vehicle lowers slightly. The inclination sensor recognizes and the brakes will be released. In the end, both algorithms wait for a vehicle reaction, so that a slight but acceptable jerk will always be noticeable. If the information about the imminent clutch engagement process is available to the system, i.e. through monitoring of the clutch pedal speed, a completely jerk-free drive-off can also be attained with a manual transmission.

Vehicles with automatic and semi-automatic transmissions will hardly have a problem here; the torque characteristic is dependent on the actuation of the accelerator pedal. Even vehicles with automatic transmissions already roll back when driving uphill grades (12%) due to constantly lower creeping torques. If the parking brake is not used, vehicles with semi-automatic transmissions will generally roll back during shifting. These vehicles will also derive considerable benefit from the drive-away assistant function.

12 Safety concept

Aside from the standard techniques such as the continuous ROM-Test, RAM-Test, intelligent monitor as extended watchdog and monitoring of the voltage supply, the redundant control element, the fault lamp and the actuators are also monitored for proper operation. Through a constant current source (approximately 2A), the power output stages are regularly checked for switch on and switch off. Owing to their service life, only the polarity reversing relays can be driven once, and only during ignition on, to facilitate the monitoring of the electric circuits to the motors even without actuation.

If possible, all signals will undergo a plausibility check prior to their processing. If the ignition is switched on, the CAN must also be available; if the motor is switched on, the current must reduce timely closed after the initial peak current. When the brake is applied, this must be followed by a rise in current. If there is a substantial decrease in vehicle speed, this is either due to an uphill grade or the actuation of a brake switch. Changes in vehicle speed must be synchronous with the signals of the inclination sensor, a piezoelectric acceleration sensor with capacitive pick-off from the company, VTI Hämlin.

The possible faults are classified as:
a) minor faults (e.g. loss of hall-effect sensor), the EPB System continues to operate and displays the fault.
b) serious faults (e.g. communication with the control unit of the motor is interrupted), the function (here Drive Away) will be switched off, the EPB-System continues to operate and displays the fault.
c) fault of a channel (e.g. cable to the motor is defective), the EBP-System will try to release the brake at the next release command and will switch off the affected channel thereafter; the other channel continues to operate, the fault is displayed.
d) system fault (e.g. processor fault), the EPB-System switches itself off, the fault is displayed.

Recovered faults will be reset at ignition "on", i.e. Switching-off the fault lamp is also possible without a diagnostic device.

13 System availability

A special topic is the handling of undervoltage. In the case of a flat battery, the EPB-System itself can trigger off the undervoltage, when switching on the motors. In this case, we have decided to keep the motors running, even if the desired level of clamping force can no longer be attained.
At least the motors will continue to run as long as the control unit is active. The driver is informed by means of the flashing activity lamp and the fault lamp.

14 Driver information

Basically, there are two indicator lamps, which are available to the driver - the yellow fault lamp and the red activity lamp.

The fault lamp will be switched on for any irregularities recognized by the system (irrespective of the error category). With regard to CAN-faults, which have corrected themselves, the fault lamp will be switched off again without any intervention from the driver.

Other faults will be stored in the fault memory during ignition "off" and temporarily deleted at the next ignition "on", until the first cycle when the safety software has detected the fault once more; unless the fault has been recovered in the meantime. During ignition "off", the fault lamp will light for another 3 minutes, if a fault was present. Subsequently, the control unit switches to sleep-mode. If the control element is actuated at the same time, the control unit will be activated and the fault will be displayed again for another 3 minutes.

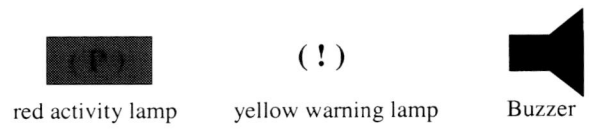

	red activity lamp	yellow warning lamp	Buzzer
Mode:	Brake released: off Brake applied: **ON** Brake active: **ON** Fault during Activation **FLASH**	no fault: off faults in History and not present: off any fault (incl. 3 min after "ignition off" **ON**	no fault: off any fault during activation: **ON** In dynamic mode: **ON**

CAN – EPB message contains information about system status for other systems

Figure 12: Driver information

In the event of an existing fault, each actuation of the control element leads to the activation of the buzzer, which is located in the interior of the vehicle. As a result, an increased level of attention will be received from the driver and any further driving with an existing fault will be made more uncomfortable.

The red lamp will be switched on, if the specified maximum clamping force is attained. The lamp will be turned off if the clearance is adjusted during the release. In contrast to the manual hand brake, wherein the activity lamp is already switched on at minimal lever travel, in the EPB-system, the lamp is only switched on if the vehicle is parked safely.

The activity lamp will flash, in the event that the brake does not attain the required clamping force (e.g. the voltage is too low). If the monitored fault lamp is defective, the red lamp will flash in order to catch the attention of the driver.

15 Cost considerations

At present, the system is considered to be more expensive than other comparable mechanical systems and their components. But it also offers advantages such as functional reliability, self-diagnosis and additional functions and flexibility of internal design.
At the start of the series production, materials have been selected, which fulfill the desired standard of functionality with a large safety margin. Housing material and wall thickness could be optimized. The very expensive PEEK of the gear wheels could be substituted, using suitable compound-materials.

Based on the criteria for the dimensioning of the clamping force detailed in the chapter, System Design, there is a need for optimization. It must be possible to calculate a reduced level of required clamping force based on the criteria, vehicle weight, temperature of the rear-wheel brake disc and slope. Through this, a load cycle collective can be established, and this will lead to a reduction of the currently required bridge size. However, this load cycle also requires the acceptance of our customers.

The increasing number of control units produced, as well as the use of ASIC´s will deliver the expected cost reduction, as long as our customers are satisfied with a generic control unit. The EPB-System can be introduced into large-scale production at a low additional price, in comparison to the present parking brake systems.

At the moment, TRW is currently investigating the feasibility of directly integrating the control unit in the actuator; at least the power output stages and its logic. The activation and communication processes would be carried out via the CAN or LIN bus. This would offer the possibility of achieving an equalization of costs in comparison to today's mechanical systems.

The new Range Rover – maximizing customer benefits from a slip control system

P THOMPSON
Slip Control Systems, Land Rover, Solihull, UK

SYNOPSIS
This paper describes the functions of the slip control system fitted to the new Range Rover. It illustrates how maximum benefit has been gained from the system by innovative use of sensor signals from all relevant vehicle systems. The unique challenges of developing a slip control system for a vehicle with such a broad range of abilities are discussed.

1. THE HISTORY OF THE RANGE ROVER
The Range Rover created the segment that is now known as the luxury sport utility vehicle (SUV) when it was launched in 1970. As the market for 4x4 vehicles has expanded, the Range Rover has moved gradually upmarket, but has retained its class leading off-road reputation. Accompanying the increasing sophistication of the Range Rover have been a number of firsts in the field of slip control systems. In 1990 the Classic Range Rover was the first vehicle to be fitted with all-terrain ABS. This was followed by the introduction of off-road traction control in 1993, and four wheel traction control in 1999.

The new Range Rover, announced in November 2001 continues this trend, by offering a state of the art slip control system, incorporating a number of innovations.

2. NEW RANGE ROVER TARGETS
At the outset of Project L322, targets to define the performance of the new Range Rover were carefully considered. The figure shows Land Rover's interpretation of trends in the SUV market - a gradual shift away from off-road performance, and an increasing emphasis on on-road dynamics. The Land Rover brand values demand class-leading off-road performance, therefore the challenge for the new Range Rover was to produce a vehicle with the broadest range of abilities of any vehicle on sale. On-road benchmarks were the latest generation of road-biased 4x4s, while the off-road target was to match or beat the performance of the 1999 model year Range Rover across the whole spectrum of off-road surfaces.

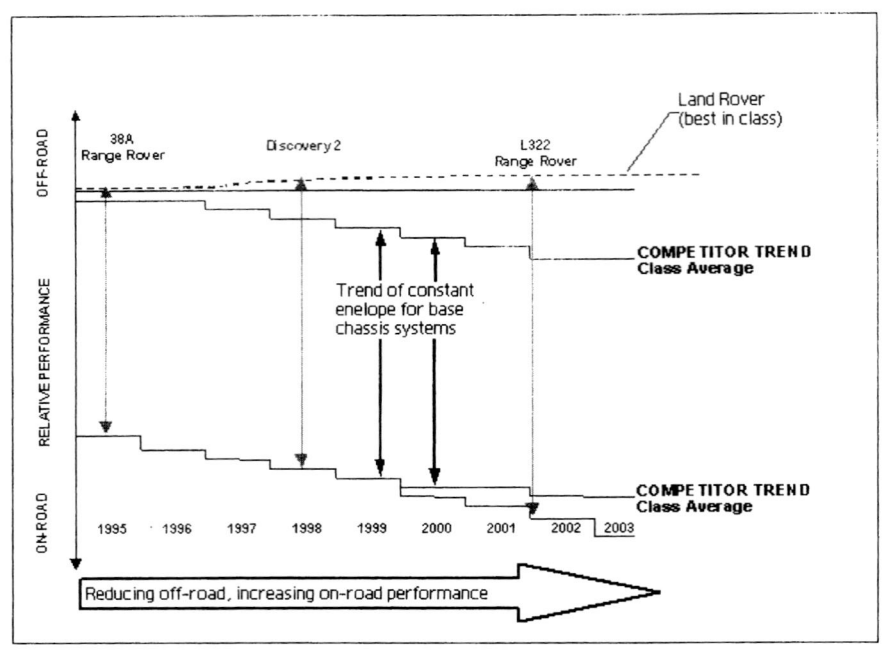

Figure 1. Performance trends – full size 4x4

The key to the appeal of the Range Rover off-road is not just the possibility to drive across the most demanding terrain, but the feeling of relaxed control for the driver whilst doing so. This remained a focus throughout the development of the slip control system.

3. NEW RANGE ROVER ARCHITECTURE
Reaching the performance targets of the new Range Rover demanded a radical change in the architecture of the vehicle. Key features include:
- Extremely stiff steel monocoque construction.
- Three massive subframes supporting front and rear suspension, and transfer gearbox.
- Four wheel independent suspension (long travel font struts, double wishbone rear).
- Four wheel electronically controlled air suspension, with switchable side-to side linking to improve axle articulation.
- 4.4l V8 petrol or 3l turbo diesel engine.
- 5 speed automatic gearbox, with Steptronic function.
- Controller Area Network (CAN) bus linking powertrain, chassis and electrical systems.

4. NEW RANGE ROVER SLIP CONTROL SYSTEM
One specification of slip control system is used in all models of the new Range Rover. It is a Bosch DSC 5.7 system, developed specifically for the application to include the following functions: Dynamic Stability Control (DSC), Electronic Traction Control (ETC), Anti-lock Bracking System (ABS), Emergency Brake Assist (EBA) and Hill Descent Control (HDC).

Throughout the development of the system, the requirement for the vehicle to perform in an extremely wide variety of environments on and off-road led to a number of innovations in the functionality of the slip control system. Much attention has been paid to the possible side effects on the sensor inputs of driving in varied terrain, to ensure a robust performance from a potentially sensitive system.

The integration of so many functions in one slip control system clearly requires careful handling of combinations of or conflicts between different controllers. This is handled by each controller outputting a target level of wheel slip (in the case of DSC, ETC and ABS) or brake torque (HDC, EBA) and using pre-determined priorities to calculate an overall target slip and hence target brake torque for each wheel.

Integration of so many controllers also demands that the various sub-systems share common values for vehicle reference speed, instantaneous brake pressure, brake disc temperature etc.

The slip control system consists of the usual components:
- modulator with attached ECU
- 4 x active wheel speed sensors (Hall effect)
- pressure sensor (built in to modulator)
- steering angle sensor
- lateral acceleration sensor
- yaw rate sensor (integrated with lateral acceleration sensor)
- pre-charge pump

Additionally the slip control system communicates with the following vehicle systems using the CAN bus:
- engine
- automatic gearbox
- transfer gearbox
- air suspension
- instrument pack.

Key

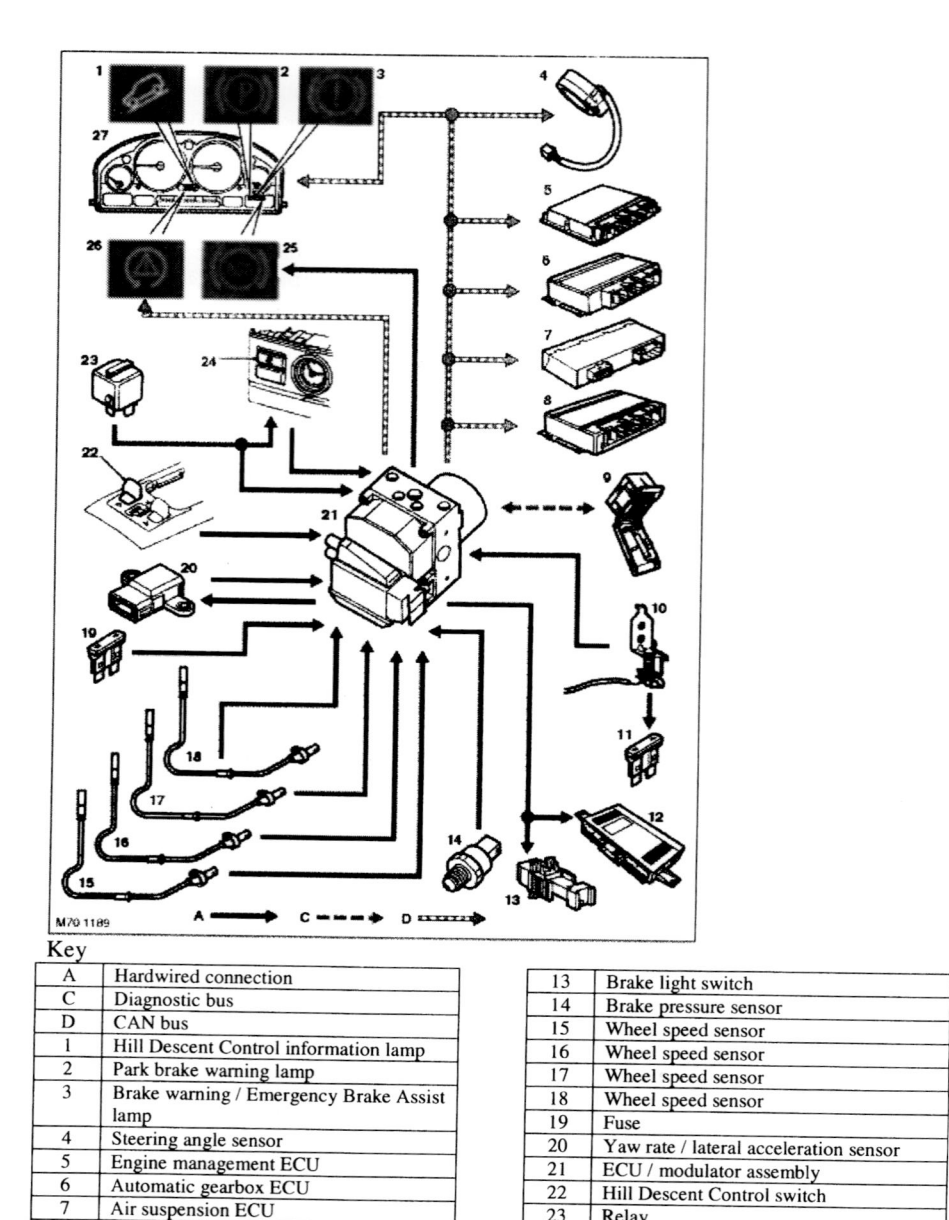

A	Hardwired connection
C	Diagnostic bus
D	CAN bus
1	Hill Descent Control information lamp
2	Park brake warning lamp
3	Brake warning / Emergency Brake Assist lamp
4	Steering angle sensor
5	Engine management ECU
6	Automatic gearbox ECU
7	Air suspension ECU
8	Transfer gearbox ECU
9	Diagnostic socket
10	Park brake switch
11	Fuse
12	Light control module

13	Brake light switch
14	Brake pressure sensor
15	Wheel speed sensor
16	Wheel speed sensor
17	Wheel speed sensor
18	Wheel speed sensor
19	Fuse
20	Yaw rate / lateral acceleration sensor
21	ECU / modulator assembly
22	Hill Descent Control switch
23	Relay
24	Dynamic Stability Control switch
25	ABS warning lamp
26	Dynamic Stability Control warning lamp
27	Instrument pack

Figure 2. Slip control system interactions.

5. SUB-SYSTEMS
5.1. Dynamic Stability Control (DSC)

The operations of ETC and DSC are closely related and the slip controller often moves seamlessly from one to the other. For the purpose of this description, any intervention related to vehicle stability is classified as DSC. ETC functionality is restricted to the use of the vehicle brakes to improve vehicle traction by optimising the distribution of drive torque.

The DSC performs the usual function of such systems – to recognise and correct unwanted vehicle behaviour in the form of under- or oversteer. Also included is the correction of instability induced by wheel spin during harsh acceleration on slippery surfaces. The response of DSC to these events is to cut engine torque and if necessary, to brake individual wheels to correct the vehicle's yaw attitude.

A DSC system tuned purely for on-road use will inevitably prove intrusive in terms of unwanted engine intervention when a vehicle is driven on rough tracks or off-road. To overcome this issue, the wheel speed signals are analysed. Road surface irregularities cause frequent changes in individual wheel speeds; when a counter of irregularities exceeds a threshold limit, and the vehicle is travelling below a predetermined speed threshold, rough or off-road driving is detected. The DSC system responds by raising the thresholds which trigger an intervention. The outcome is a vehicle which can be driven on loose surfaces with the feeling of security offered by DSC, but without the irritation of regular unwanted interventions by the system.

Detection of rough roads offers very good off-road performance without the need for a high level of skill and experience from the driver. However, in certain extreme driving conditions such as deep soft sand, mud or snow, forward motion is best achieved by maximising vehicle momentum at the expense of stability. The fitment of a DSC switch enables a skilled driver to temporarily de-select most of the stability enhancing functions of DSC, leaving the traction optimising function of ETC active. Use of this function as a temporary measure only is encouraged by illumination of the DSC warning lamp after de-selection, and automatic reactivation of DSC at the next ignition cycle. The purpose of DSC de-selection using the switch is to offer maximum traction, therefore, while the driver is pressing the brake pedal the full stabilising functions become active.

5.2. Electronic Traction Control (ETC)

ETC performs a key role in meeting the targets of effortless off-road performance with minimum driver intervention.

Although it is generally a characteristic of traction control systems that they provide automatic support without the need for driver intervention, the speed and refinement of the response greatly affect the driver's perception of the system.

A further challenge is the requirement to provide timely control of wheel slip in critical conditions, without developing an over-sensitive system which could cause irritation in everyday driving.

Brake intervention traction control systems work on the principle that a normal open differential sends an equal torque to each output shaft. Therefore a spinning wheel not only reduces the torque transmitted by that wheel, but also the torque transmitted by the second wheel sharing its differential. In a 4x4 vehicle with a differential in each axle, and one in between the axles, this effect is multiplied. By applying a brake torque to a spinning wheel, ETC allows the equivalent amount of engine torque to be applied to the wheel on the other side of the differential. Contrary to popular belief, ETC does not transfer torque to the other wheel. Similarly if both wheels on an axle have lost grip, applying a brake torque to both of them allows the equivalent drive torque to be applied to the other axle. The new Range Rover uses two measures to maximise the benefits of these principles.

Firstly, the maximum drive torque available at the non-slipping wheels is directly related to the brake torque applied at the slipping wheels. The combination of large diameter disc brakes and the ability of the ETC to generate an exceptionally high level of pressure (up to 150 Bar) ensures that even on steep gradients with more than one wheel without grip, ETC is able to function effectively. The potential for the high brake torque to overwhelm the drive torque is overcome by the combination of high torque engines and low gear ratios.

Secondly, the combination of ETC and a Torsen differential enhances the performance of both components. The torque biasing effect of a Torsen differential is limited by the torque it is outputting. In the case of a centre differential, if grip is lost at both wheels of an axle, the Torsen output falls to zero, and its ability to bias torque is lost. The application of brake torque to the slipping wheels by ETC restores the ability of the Torsen to function. In turn, the action of the Torsen speeds up the response time of the vehicle, by reducing the level of pressure needed to allow torque to be transmitted to the non-slipping axle.

In order to resolve the conflicting demands of over sensitivity in normal driving and the quick response to loss of traction in critical off-road conditions, it is clearly necessary to recognise certain conditions. ETC uses the rough road detection described above to eliminate unwanted interventions on loose surfaces, where small amounts of wheel slip go unnoticed by the driver. When driving in slippery conditions such as snow and ice on-road, the driver is much more sensitive to slip and ETC intervenes as soon as wheel slip is initiated.

Very low speed off-road driving, with the vehicle crawling over severe obstacles puts particular demands on a traction control system. In such conditions, where the vehicle's wheels can lose contact with ground, or the surface conditions can momentarily offer very little grip, optimum traction and controllability are achieved by reducing wheel spin to an absolute minimum. The response of a standard traction control system can lead to shunting of the vehicle in such conditions. Inevitably, traction control cannot begin until wheel slip is recognised; normally traction control activity (that is braking of a slipping wheel) finishes as soon as the wheel is turning at the same speed as the other wheels. Normally this is because the wheel has regained grip. In severe circumstances the wheel that is being braked only slows down because of the traction control activity, but still has no grip. Reduction of the traction control braking effect in this case leads to further wheel slip, and a pause in forward progress of the vehicle until another cycle of traction control activity begins.

Two measures have been incorporated into new Range Rover's ETC to recognise such conditions, and a unique system response has been developed. Recognition of driving conditions in which extreme articulation causes one or two (diagonal) wheels to lift is achieved by analysing the outputs from the EAS height sensors which are available from the CAN bus. Alternatively low speed crawling conditions are recognised by a combination of a manually selected low gear (using the Steptronic function of the gearbox) and low speed driving, again using signals available from the CAN bus. The response of the ETC system when either of these conditions are met is to greatly reduce the rate at which brake pressure is reduced when a spinning wheel comes under control. This holding of brake pressure continues until the suspension height sensors indicate that the braked wheel has regained contact with the ground, or until the vehicle has travelled a set distance. The effect for the driver is that smooth control of the vehicle is possible, even in extreme circumstances, without the need for manual engagement of differential locks, or other traction aids.

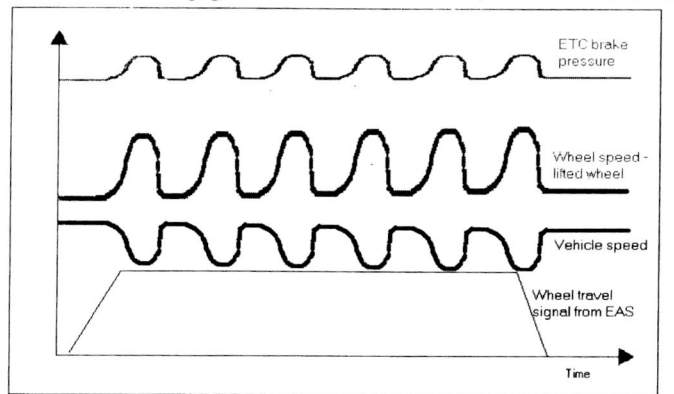

Figure 3. Standard ETC response – wheel lifted due to high articulation

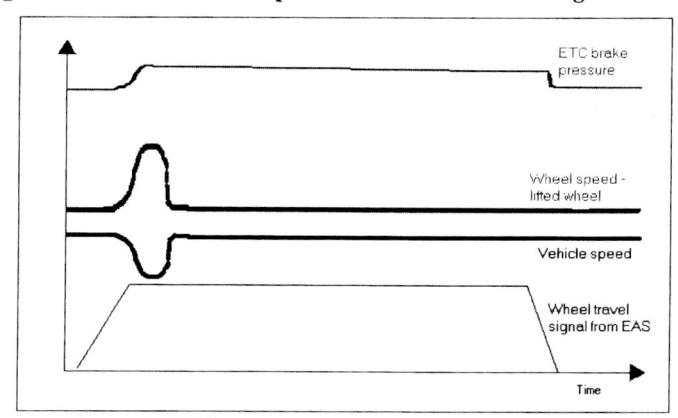

Figure 4. ETC response using wheel travel information

5.3. Anti-lock Braking System (ABS)

As for other aspects of the slip control system, the performance of the ABS system in the full range of off-road conditions has been given much emphasis. Again, the availability of information from the DSC sensors has enabled the development of strategies to tailor the response of the ABS system to the driving conditions.

The ABS system uses the following signals to assess the driving conditions and driving style:
- Rough road detection (as used by DSC and ETC)
- Steering wheel angle
- Vehicle speed
- Vehicle acceleration
- Brake pressure
- Engine speed / torque
- Transmission ratio

For example, a comparison of the actual vehicle acceleration with the expected vehicle acceleration for a given engine output and transmission ratio can give an indication of driving on a gradient, or on a surface with high rolling resistance.

Combinations of the various sensor outputs trigger different measures in the ABS software. These influence the ABS controller output by adjusting the target slip levels and by altering the rate at which pressure is released from the brake of a locked wheel. The aims of the measures are to optimise stopping distances on a range of surfaces and to prevent over-sensitivity of the ABS system on rough roads.

The key measure for improving stopping distances on loose surfaces is to allow much deeper wheel slip than normal, in order to allow the building of a wedge of material in front of the tyres. The risk of this strategy decreasing the stability of the vehicle is eliminated by the reduction of target slip to normal levels when the driver turns the steering wheel, and by the constant supervision of the DSC system.

5.3.1. Electronic Brake-force Distribution (EBD)

The ABS system includes the almost universally fitted (EBD). This system offers particular advantages over pressure reduction valves in a vehicle with a wide variation in front to rear weight distribution resulting from a relatively high centre of gravity, and high payload potential.

5.3.2. Corner Brake Control (CBC)

CBC is intended to enhance vehicle stability when the driver brakes during cornering or lane-change manoeuvres. Unlike ABS and EBD, which react to wheel slip, CBC becomes active before excessive wheel slip has developed, by responding to the lateral acceleration signal. If the driver brakes while the vehicle is cornering above a lateral acceleration threshold, or if the vehicle's lateral acceleration exceeds the threshold during a braking manoeuvre, CBC becomes active. Vehicle stability is enhanced in two ways by limitation or progressive reduction of braking pressure at the inside rear wheel. Firstly, reduction in braking force at this lightly laden wheel improves the ability of the tyre to transmit cornering forces. Secondly, the effect of a higher brake torque on the outside of the vehicle than the inside counteracts the tendency of a vehicle to oversteer when decelerating in a corner.

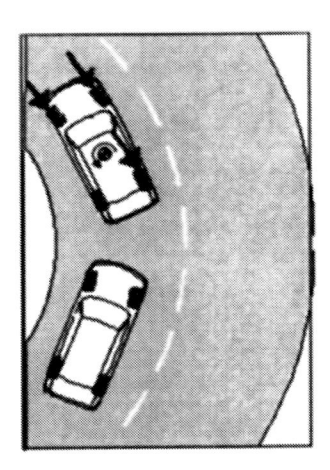

Figure 5. CBC – brake forces counteract yaw moment

5.3.3. *Emergency Brake Assist (EBA)*
The new Range Rover features two functions to ensure that maximum deceleration is attained when the driver's actions indicate it is required. In common with most EBA systems, a combination of brake pressure increase gradient, speed of transfer from throttle to brake, vehicle speed and brake pressure are used to determine a panic braking situation. A second trigger for EBA is included, with the intention of overcoming the tendency of drivers to stop increasing pedal effort as soon as ABS activity is detected through the pedal. Irrespective of the brake pressure increase gradient, when driver braking is sufficient only to cause the front wheels to enter ABS control.

Once triggered, EBA automatically increases the level of deceleration, by increasing brake pressure using the ABS recirculation pump, until all wheels are under ABS control. This level of deceleration is maintained until the pressure sensor indicates that the driver has started to release the brake pedal. From this point, the EBA target deceleration is reduced in line with the ratio of current driver demand to maximum driver demand. Eventually, EBA assistance fades to zero. This target deceleration strategy ensures that the driver's normal instinctive use of the brake pedal effort to control deceleration remains applicable.

5.4. Hill Descent Control (HDC)
Land Rover's patented Hill Descent Control system is included for the first time in a Range Rover. The system has been further developed since its introduction in Freelander and Discovery. The principle of HDC is to assist the driver in making severe descents when driving off road, or in difficult conditions on-road, by automatically applying the brakes to control vehicle acceleration and maintain a low, constant descent speed.

Traditionally, low speed descents were made using only the engine braking effect in a very low gear ratio. HDC, in combination with a suitable gear selection offers the following advantages over engine braking alone:

- Continued braking at remaining wheels when one wheel loses grip (engine braking effect lost when a single wheel loses grip)
- Braking bias towards downhill axle
- Even lower descent speed
- Consistent descent speed regardless of gradient and vehicle loading
- ABS over-ride to retain steerability in slippery conditions.
- Brake lights lit to warn following drivers of decelerating / slow moving vehicle

HDC functions, when selected by the driver, by automatically applying all four brakes if engine braking alone is not sufficient to control the vehicle to a previously defined descent speed. The descent speed is dependent on the gear selected by the driver: the lower the gear, the lower the descent speed, down to a minimum of approximately 3.5 km/h. If the driver selects HDC with the gear-selector in the "D" position, rather than the manually selectable Steptronic position, the gearbox switches to a unique HDC shift-map, which ensures that the most appropriate gear is selected.

The driver is able to influence the target descent speed in three ways:
- By selecting a different gear.
- Using the brake pedal: if the driver presses the brake pedal while HDC is active, the brakes operate as normal, over-riding HDC control. The speed at which the vehicle is travelling when the brake pedal is released (down to a minimum of approximately 3.5 km/h) is selected as the new target descent speed.
- Using the throttle pedal: if the driver presses the throttle pedal while HDC is active, the system responds in two stages. For approximately the first 20% of throttle pedal travel, the target descent speed increases in proportion to the throttle pedal position up to a maximum of 35 km/h. If the throttle pedal is pressed further, HDC activity is temporarily suspended. When the throttle pedal is released, the target descent speed is set to the vehicle speed, or the default speed for the selected gear, whichever is lower. The response to the throttle inputs has two functions: to allow the driver more control over the target descent speed, and to prevent HDC from fighting against the engine when the driver wishes to accelerate the vehicle without first de-selecting HDC.

The HDC system incorporates a number of features to discourage its use at higher speeds on road. This is for two reasons. Firstly, if the vehicle is driven with system inadvertently selected (for example by a previous driver), unexpected braking of the vehicle could confuse the driver and/or other road users. Secondly, if a driver learns to use HDC as a convenient way of decelerating the vehicle in traffic, it could lead to an unusually high level of usage of the system, possibly leading to durability issues for the modulator. The following measures aim to emphasise that HDC is a low speed system, and prevent its inadvertent use on-road:
- Selection of HDC is only possible below 35 km/h.
- When driving in high range, HDC is automatically de-selected at 60 km/h.
- If the vehicle ignition is switched off for more than 6 hours, HDC is automatically de-selected when the vehicle is started. Automatic de-selection is not carried out sooner in case the engine is re-started after stalling during a hill climb, or after a short pause during off-road driving.

6. CONCLUSIONS
Innovative algorithms using all available sensor signals have enabled the slip control system of the new Range Rover to contribute to a vehicle performance which exceeds the original targets. In contrast to a standard road vehicle slip control system, the system provides robustness in off-road driving conditions and optimised performance in a large range of conditions by interpreting sensor signals to recognise driving surface conditions. The success of this approach justifies the current industry trend towards closer integration of vehicle control systems.

© Land Rover

Regenerative braking for all-electric vehicles

N SCHOFIELD, C M BINGHAM, and **D HOWE**
Department of Electronic and Electrical Engineering, University of Sheffield, UK

ABSTRACT

Electric vehicles are set to play a prominent role in addressing the energy and environmental impact of an increasing road transport population by offering a more energy efficient and less polluting drive-train alternative to conventional internal combustion engine (ICE) vehicles. The recovery of vehicle kinetic energy during braking, which is more readily implemented with electric vehicle formats than pure ICE configurations, can yield an improvement in vehicle energy conversion and will influence the design of friction braking systems of the future. However, the high battery current demands experienced with energy source (battery) electric vehicle systems during acceleration and braking has limited the full exploitation of vehicle kinetic energy recovered during braking in the past.

The incorporation of a peak power buffer into an electric vehicle drive-train can significantly reduce the peak power load on the primary energy source, while at the same time, the response of the vehicle is more consistent when accelerating and braking, since it is independent of the traction battery state-of-charge. However, the implementation of dual power and energy source systems requires detailed analysis of the vehicle drive-train in order to optimise component specifications and energy management strategies.

This paper discusses the detailed simulation and test evaluation of a laboratory based electric vehicle traction system incorporating a supercapacitor peak power buffer, valve-regulated-sealed lead-acid traction battery and high efficiency permanent magnet traction machine and power conversion electronics. The results illustrate the functionality of the peak power buffer for acceleration and regenerative power transients, thus improving battery energy utilisation and recovery of vehicle kinetic energy during braking.

1. ELECTRIC VEHICLE ENERGY CONSIDERATIONS

Research and development in the field of high power, high efficiency traction systems for electric vehicles has been reported world-wide over the last three decades and such systems are reaching a technical maturity with cost optimisation in volume manufacture now the primary target area[1]. However, a major area for further research is in improving the specification and integration of system components and management of energy flow to realise optimum energy utilisation from the drive-train.

A major operational problem with electric vehicle traction batteries is their poor performance at high current charge/discharge rates, typical of vehicle acceleration and braking demands. For example, a high power demand from a lead-acid traction battery, even for a short duration (2-3s), chemically limits the remaining energy capacity of the battery due to lowering of the available active mass and effective surface area of the battery plates[2]. In addition, high cyclic operation reduces battery life adding a significant premium to the operating costs of an electric vehicle. The incorporation of a peak power buffer into an electric vehicle drive-train can reduce the peak power load on the primary energy source, resulting in a significant improvement in battery energy utilisation due to levelling of the battery energy demand. Additionally, the response of the vehicle is more consistent when accelerating and braking, and the recovery of vehicle kinetic energy during braking enhanced, since the drive-train energy management is independent of the traction battery state-of-charge or terminal voltage.

For example, the power requirements of a small urban, 1.5 tonne, family vehicle during typical urban mission profiles are highly dynamic, being characterised by a relatively low mean continuous power of ~2-4kW, but high peak transient powers of ~45kW, the latter equating to acceleration of the vehicle from 0-50km/h in 8 seconds. Similarly for sub-urban driving, where a 4:1, peak-to-average power is typical, as illustrated in Fig. 1, again showing power requirements of a small 1.5 tonne vehicle, over an enhanced ECE15 (urban) and sub-urban velocity mission profiles, the specifics of which are detailed in Table 1.

For the mission profiles illustrated in Fig. 1, regenerative braking can yield an improvement in vehicle wheel drive energy requirement of 20% and 12% for the urban and sub-urban parts respectively, which is not unrepresentative. Similar potential gains can also be shown for public transport vehicles (buses) that have power demands predominantly characterised by high acceleration and regenerative braking, over low mean energy consumption. Goods delivery vehicles operating in urban areas also exhibit similar trends.

However, recovery of vehicle kinetic energy does not fully translate to battery energy reduction, with only marginal benefits <5% being reported[3]. The shortfall can be attributed to:
- the poor response of most readily available chemical battery technologies (in-particular lead-acid) and their associated management systems, to pulsed power demands, and
- inefficiencies in the battery-to-wheel drive-train, although some of the energy dissipated in the drive-train components during regenerative braking, i.e. traction machine and inverter, can be utilised for in-cab heating via their respective coolant circuits.

Regardless of how the recovered kinetic energy is expanded, i.e. in terms of losses in the electrical transmission or regenerated to the vehicle battery or peak power buffer, the energy equates to a reduction in friction brake requirements, which could lead to lower service specifications for such components.

Consequently, the incorporation of a peak power buffer, chosen for high specific power as opposed to energy density, into an electric vehicle drive-train can significantly reduce the peak power loads placed on the primary energy source. However, the use of dual power and energy sources requires a detailed analysis of drive-train operation in order to optimise component specifications and develop energy management strategies.

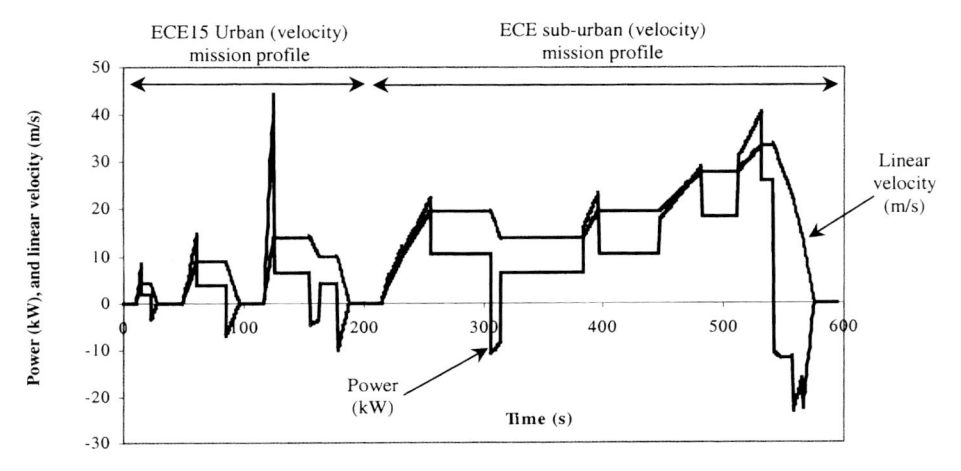

Fig. 1. Electric vehicle power requirements over urban and sub-urban mission profiles.

Table 1. Typical average and peak wheel-drive power requirements for a 1.5 tonne urban vehicle.

	Time duration (s)	Average power		Peak power (kW)
		With regen. braking (kW)	Without regen. braking (kW)	
Enhanced ECE15	0 – 195	2.72	3.39	44.5
Sub-urban cycle	195 - 595	10.17	11.56	40.6

2. VEHICLE SIMULATION

Whilst, in general, top-level models of many vehicle drive-train components, such as gear-stages and traction machines, can be based on efficiency maps derived from tests or manufacturers data, these provide little insight into how their design parameters/characteristics influence overall vehicle performance. Therefore, since the choice of power electronic converter topology, semiconductor device technology, pulse-width-modulation switching strategy, current control strategy, nominal battery voltage, and traction machine design parameters all affect the system efficiency, to a greater or lesser degree, detailed modelling of key components was undertaken.

(a) Top level model of electric vehicle traction system

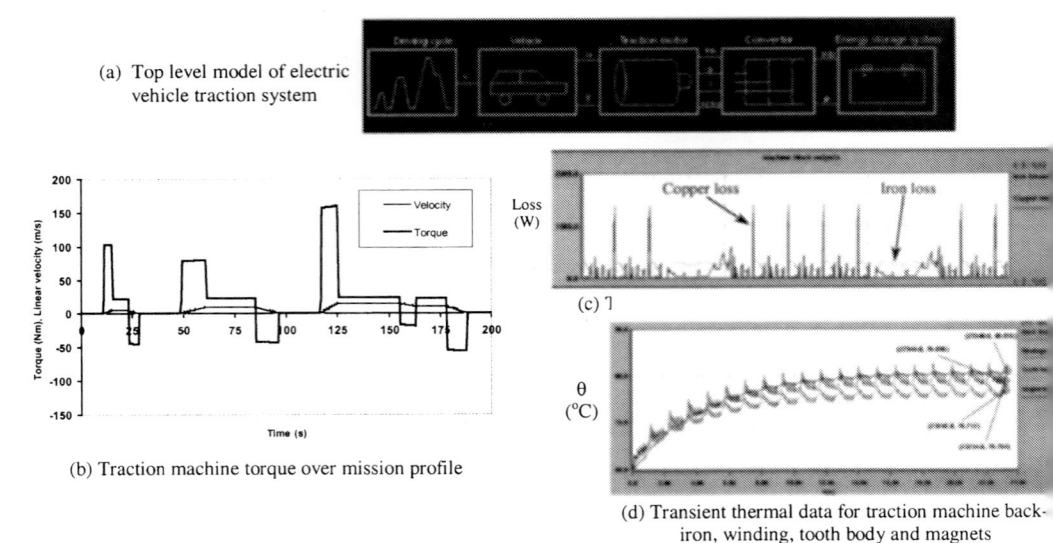

(b) Traction machine torque over mission profile

(c)

(d) Transient thermal data for traction machine back-iron, winding, tooth body and magnets

Fig. 2. SABER™-based simulation tool.

Time domain simulations, using the SABER™ simulation platform and incorporating experimentally validated macro-models of the various vehicle system components, provided a detailed insight into the influence of the drive-train architecture, the ratings of the energy and peak power buffer sources, and the energy management strategy on the overall vehicle performance under prescribed driving cycles. A comprehensive suite of macro-models of various traction system components has been established, and subsequently employed in top-level SABER™-based simulations of a benchmark urban electric vehicle, an example of the software graphics and results being shown in Fig. 2. The top-level model, Fig. 2(a), comprises of a number of component models with pre-specified inputs/outputs. Vehicle simulation is facilitated by the interconnection of the component models thus allowing a high degree of design flexibility whilst optimising the drive-train system topology. From inputs of specified driving cycles (e.g. standard ECE, FTB, CARB, etc. or other user defined), the torque-speed requirement at the driven wheel(s) was calculated from a model of the vehicle kinematics, Fig. 2(b). This was translated to a power demand from the battery, or a power input when regeneratively braking, via models of the transmission and reduction gearing, the traction machine and the power electronic drive. Hence, from the estimated battery loss and its state-of-charge, the range of the vehicle could be determined, as well as other performance data, such as component losses, Fig. 2(c), and temperature rise, Fig. 2(d).

However, whilst time-domain simulation tools such as SABER™ are capable of solving large mixed signal systems, containing detailed models of semiconductor devices, analogue/digital controllers and electromechanical devices, etc., the computational effort to simulate a vehicle over representative driving cycles whilst accurately accounting for operation of each switching device would be prohibitive. Hence, a hierarchical approach to system modelling was employed, in which a top-level vehicle model such as that in Fig. 2(a) was supplied with data calculated off-line from detailed models of the relevant traction system components[4].

3. DUAL SOURCE SYSTEMS

The developed simulation models have been used to evaluate alternative dual power and energy source traction system architectures that combine a supercapacitor peak power buffer with the battery energy source. For example, Fig. 3. illustrates two possible structures for a dual power source system, showing (a) the traction battery directly supplying the traction inverter DC link voltage and the power buffer connected to supplement the traction peak power via a dc:dc converter, and (b) the peak power buffer directly connected to the traction drive DC link and the batteries connected via a dc:dc converter.

Fig. 3(a) represents a conventional electric vehicle traction system to which a power buffer and DC:DC converter interface has been added to provide system transient loading. The DC:DC converter typically interfaces between a low voltage, high current supercapacitor system (chosen due to concerns over cell charge and voltage management) and a traction battery voltage of nominally 240V. As such, the DC:DC power transients are in the order of ~45kW, which is demanding in terms of silicon and passive component (i.e. inductor) volumes.

(a) Battery on DC link   (b) Power buffer on DC

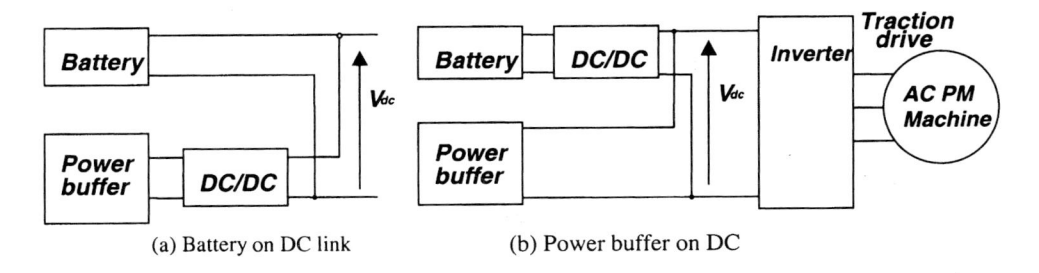

Fig. 3. Two possible dual source structures.

Since the terminal voltage of a traction battery will vary considerably between fully charged and discharged states, the traction drive has to be rated to cater for a wide range of operating voltages without loss in performance. Therefore the change from the more conventional approach to one where the DC link is allowed to continually vary, as in Fig. 3(b), has minimal impact on the traction drive design. Further, a direct connection between the peak power source and traction system is more energy efficient since the DC:DC converter power handling requirements are much reduced, ideally, only transferring the average vehicle power from the traction battery. Consequently, the DC:DC converter can have a reduced silicon rating, an important commercial consideration.

Fig. 4 shows simulation results for a small, 1.5 tonne, urban passenger vehicle with a traction system configured with the variable DC link dual source structure of Fig. 3(b), and a 180Wh supercapacitor as the peak power buffer. Note, other peak power buffer technologies, e.g. flywheels, could provide this function. The vehicle is simulated over 4x enhanced ECE15 mission profiles and an additional suburban cycle, as shown in Fig. 4(a). For illustration, a simple energy management scheme is implemented within the model such that (i) the peak power buffer normally supplies the peak power, (ii) the battery supplies the average power, (iii) when the buffer unit is fully charged any surplus regenerative energy is diverted to the battery, and (iv) when the DC link falls below a minimum set level, all further power requirement is drawn directly from the battery, as shown in Fig. 4(b). The resulting supercapacitor current is highly dynamic as shown in Fig. 4(c), while the battery current (d) is, for this scenario, unipolar and with a significantly reduced cyclic profile. Note, for the case of no peak power buffer, the battery current would be the summation of the currents of Fig. 4(c) and (d).

The simulation results for battery current demonstrate that the use of a 180Wh supercapacitor power buffer, suitably integrated in a vehicle drive-train augmenting energy management, can significantly reduce the peak battery current magnitude and cyclic loading during urban driving, with the consequential improvement in vehicle range and lifetime from batteries in electric vehicles.

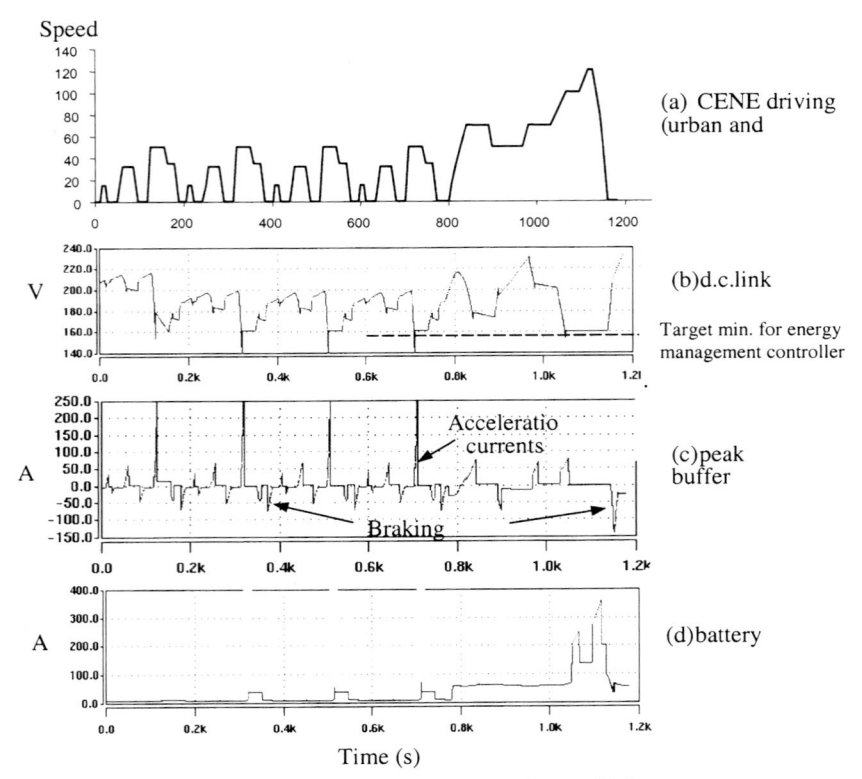

Fig. 4. Simulation results for 1.5 tonnes urban vehicle over four urban and one sub-urban mission profile.

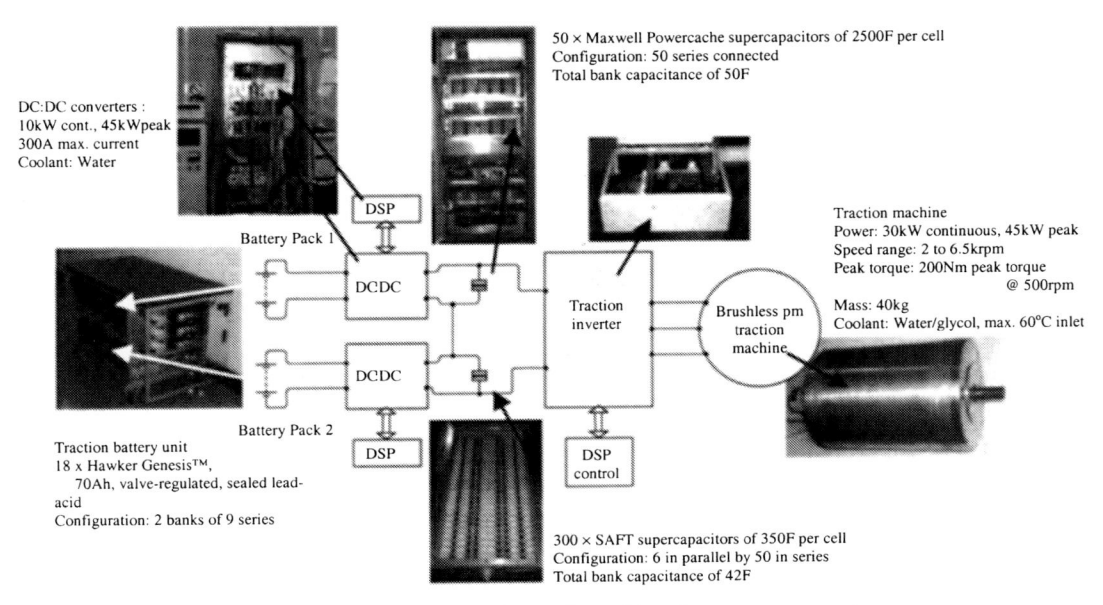

Fig. 5. **Brass-board traction system schematic and hardware components.**

4. BRASS-BOARD DEMONSTRATOR.

To experimentally validate the vehicle simulation tools, address the technicalities of drive-train component integration and develop the energy management control, electric vehicle drive-train components were assembled into a laboratory based 'brass-board' demonstrator comprising of 2 supercapacitor banks, a 135V bank comprising of 50x Maxwell cells of 2500F, and a 135V bank comprising of 300x SAFT cells of 350F. The supercapacitor banks were series connected to provide the dc link for the traction system with individual DC:DC converters (to account for the variation in supercapacitor specifications) interconnected to 2 Hawker sealed lead-acid battery packs, as illustrated in Fig. 5. The dc link supplies a brushless permanent magnet traction machine and inverter, the mechanical output of which is loaded via a dynamometer test facility. The Maxwell supercapacitors, protection fuse and switchgear, plugs and sockets for interconnection, are housed within a 19" rack unit, likewise the DC:DC converters, DSP control platforms for the DC:DC converters and traction inverter, and energy management unit (EMU), as illustrated in Fig. 5. The EMU interfaces to the power converter DSP's via a CAN2b link. Similarly, information from cell voltage and temperature monitors on the Maxwell and SAFT supercapacitor banks were fed to a local display PC via CAN.

One of the features of supercapacitor technologies perceived to be technically limiting is their low cell voltage, typically max. 2.7V, which fosters parallel connection (and hence high current) for cell balancing and load share. However, such low voltage systems are not conducive to vehicle traction systems handling peak powers of ~45kW, essentially the silicon and component interconnections would not be cost effective. Hence, a high voltage (270V) supercapacitor configuration was implemented for the project brass-board demonstrator. This necessitated the design of supercapacitor cell balancing and monitoring circuitry to ensure safe and effective operation. An important exploitation outcome of the research was the applications experience gained during 'brass-board' tests. Results showed that once the series connected supercapacitor cells had been fully charged and the cells balanced, i.e. a 'forming' procedure, the series configuration was then relatively insensitive to dynamic load variations, as illustrated in Fig. 6(a), showing individual cell voltage variation as the Maxwell supercapacitor bank is cycled over the ECE15 mission profile. A simple and low energy 'forming' process being implanted via the vehicle EMU during periods of system rest.

Fig. 6(b). illustrates measured data from the brass-board test facility showing system dc link voltage, traction machine current, supercapacitor and battery currents over one ECE15 mission cycle. The results illustrate the predicted reduction in battery current magnitude, how the energy management utilises battery current in addition to regenerated current to recharge the supercapacitors, and the essentially unipolar nature of the battery current.

Repetitive cycling of the traction battery over ECE15 cycles demonstrates a clear improvement in battery loading, better energy utilisation and hence range for the dual source scenario, as illustrated in Fig. 6(c), showing measured battery terminal voltage for a system operating with and without a supercapacitor power buffer and energy management unit. The improvement in battery energy utilisation equates to an increase in vehicle range of 56% for this case study, the increase in energy usage per cycle for the battery and supercapacitor case being due to the conversion efficiency of the DC:DC converter, Table 2.

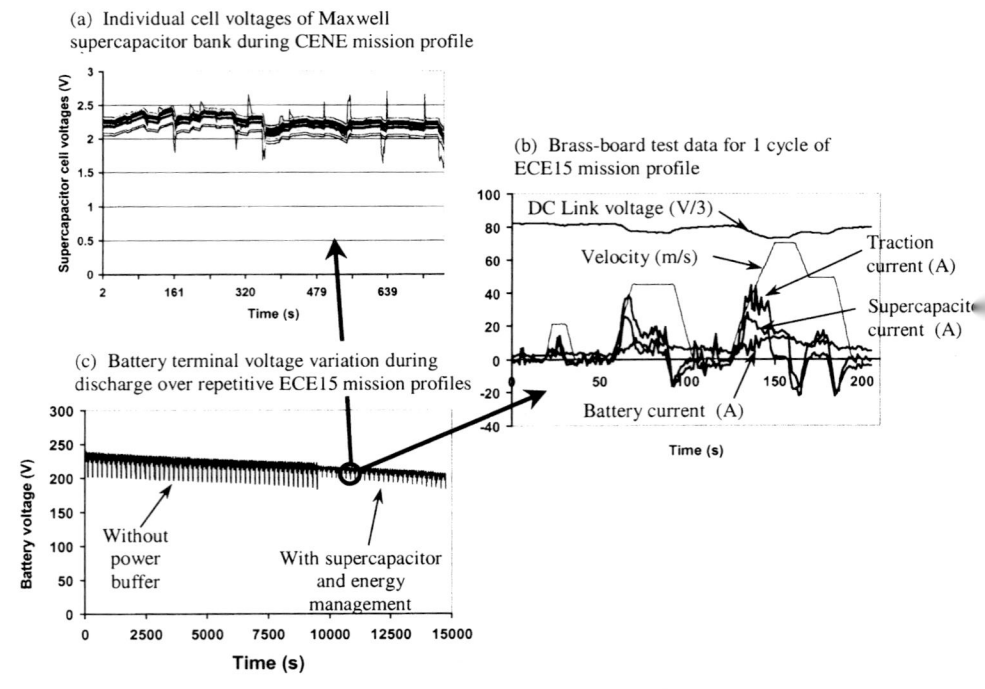

Fig. 6. Brass-board traction system test data.

Table 2. Lead-acid traction battery test data.

	Battery alone	Battery supercap. and EMU
Nominal dc link voltage (V)	216	216
Test duration (s)	9,483	14,748
ECE15 cycles completed	48.6	75.6
Range (km)	54.9	85.4
Battery performance (Ah)	33.5	55.6
Battery performance (kWh/cycle)	0.145	0.156
Total energy (kWh)	7.05	11.77

5. SUMMARY

The integration of vehicle system components and the development of energy management philosophies are significantly enhanced by the facility of a detailed and re-configurable simulation based design tool. The use of a peak power buffer can significantly reduce peak currents drawn from the traction battery of an electric vehicle resulting in a marked improvement in vehicle range and battery lifetime. A peak power buffer is an essential drive-train element if vehicle kinetic energy is to be fully exploited during regenerative braking. Additionally, with appropriate energy management of the vehicle dual energy and peak power sources, the response of the vehicle is more consistent when accelerating and braking, since it is independent of the traction battery state-of-charge.

The variable DC link architecture, where the power buffer is directly connected to the traction drive, yields promising results for both supercapacitor and flywheel systems. However, careful energy management is required to ensure effective use of the limited energy storage capacity of the peak power buffer whilst maintaining system voltage stability.

6. ACKNOWLEDGEMENTS

The authors wish to acknowledge the support of the UK EPSRC and the European Commission under the Framework IV RTD Programme[3].

7. REFERENCES

[1] Brusaglino, G., Ravello, V., Schofield, N. and Howe, D.: 'Advanced drives for electrically propelled vehicles', Proc. 33rd Int. Symp. on Automotive Technology and Automation (ISATA), Dedicated Conf. on Electric, Hybrid, Fuel Cell and Alternative Fuel Vehicles, Paper 00ELE045, pp.293-300, Dublin, 25-27 Sept. 2000.

[2] Brodd, R. J., Kordesch, K. V.: 'Lead-acid batteries', book, Wiley-Interscience Publication, ISBN 0-471-08455-7.

[3] 'Energy optimised traction system for electric vehicle (OPTELEC)', Project funded by the EC under Framework IV, Contract No. BRPR-CT97-0499, Project No. BE97-4502.

[4] Mellor, P.H., Schofield, N., Brown, A.J. and Howe D.: 'Assessment of supercapacitor/flywheel and battery EV traction systems', Proc. 33rd Int. Symp. on Automotive Technology and Automation (ISATA), Dedicated Conf. on Electric, Hybrid, Fuel Cell and Alternative Fuel Vehicles, Paper 00ELE044, pp.235-242, Dublin, 25-27 Sept. 2000.

Development of 'ECB' system for hybrid vehicles

M SOGA, M SHIMADA, and **Y OBUCHI**
Vehicle Dynamics Control Engineering Department, Toyota Motor Corporation, Aichi, Japan

ABSTRACT

In anticipation of the increased needs to further reduce exhaust gas emissions and improve fuel consumption, a new brake-by-wire system called an "Electronically Controlled Brake" system (hereafter referred to as "ECB") has been developed. With this brake system, which is able to smoothly control the hydraulic pressure that is applied to each of the four wheel cylinders on an individual basis, functional enhancements can be added by appropriately modifying its software. This paper discusses the necessity of the ECB, the system configuration, and the results of its application on hybrid vehicles.

1 INTRODUCTION

The development of hybrid vehicles is being promoted out of concerns for environmental protection. In this regard, the Estima Hybrid, which was announced in June 2001, has adopted the world's first brake-by-wire system called ECB[*1], in order to achieve the three requirements at high levels: environmental technology, safety, and driving enjoyment. Also high brake performance is satisfied. Through the use of the Vehicle Dynamic Management, which was developed under a new concept to comprehensively control the braking and driving functions of the hybrid system, a high level of dynamic performance has been realized. This paper gives an outline of the ECB system, which manages comprehensive control, as well as the improvements realized in the vehicle dynamics and environmental performance through the adoption of the comprehensive Vehicle Dynamics Management system.

[*1]ECB: Electronically Controlled Brake System

2 OUTLINE OF THE ECB SYSTEM

2.1 Functional Requirements of a Brake System

This hybrid vehicle uses a regenerative brake system that recovers the braking energy in order to improve fuel consumption. As Figure 1 shows, this vehicle responds to the driver's brake force requirement by utilizing the maximum possible amount of regenerative brake force and resorting to the friction brakes for the amount of brake force that is lacking.

To make this possible, the system must:

- Be able to linearly control the hydraulic brake force in the normal operating range.
- Be able to generate the required brake force, without allowing the driver to feel the coordination functions of the regenerative and friction brakes.

Furthermore, out of concerns for preventive safety, the system is required to independently control the brake force of each wheel in a highly responsive and precise manner.

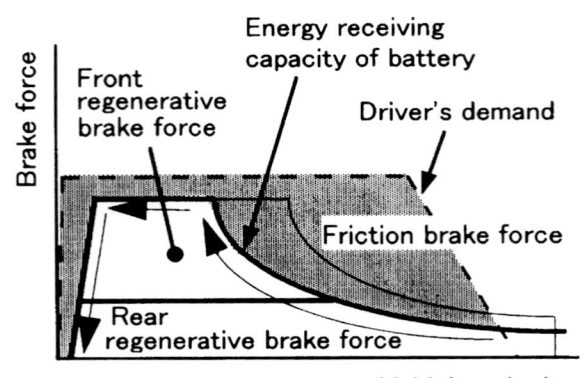

Figure 1: Example of coordination functions of the regenerative and friction brakes

2.2 ECB System Configuration

To satisfy the aforementioned requirements, the ECB is configured as shown in Figure 2. It consists of mainly two parts, that electrically detects the pedal operation of the driver, and that controls the hydraulic pressure to the wheel cylinders. They are hydraulically separated in normal operation, in order to achieve a so-called by-wire system that electrically controls the hydraulic pressure that is applied to the wheel cylinders. Furthermore, the ECB uses linear valves that are arranged in pairs, to control the hydraulic pressure of the brakes in all ranges, for normal braking or vehicle dynamic control, such as the Vehicle Stability Control (VSC).

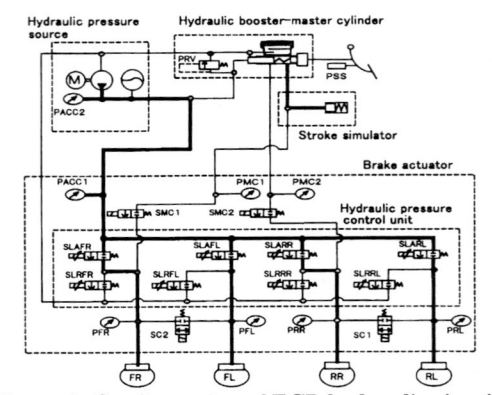

Figure 2: Configuration of ECB hydraulic circuits

2.3 Linear Hydraulic Pressure Control

Figure 3 shows the basic configuration of the linear hydraulic pressure control of the ECB.

The linear hydraulic pressure control consists of the difference between the actual hydraulic pressure and the target hydraulic pressure (which is added to the feed-back term) and the valve-opening current that varies with the difference in pressure upstream and downstream of the valve (which is added to the feed-forward term).

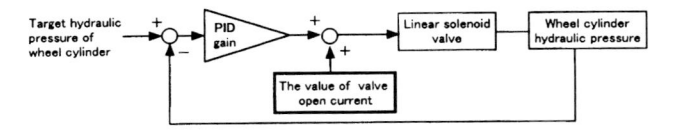

Figure 3: Configuration of hydraulic pressure control

Figure 4 shows a control example of the linear hydraulic pressure control. It controls the responsiveness and the controllability of the actual hydraulic pressure against the target hydraulic pressure within a range that satisfies the requirements of the various types of control applications, which will be discussed later.

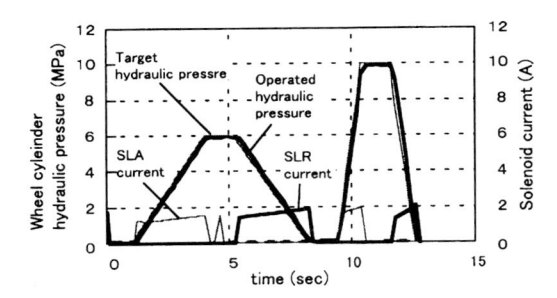

Figure 4: Example of linear hydraulic pressure control

3 OUTLINE OF THE VEHICLE DYNAMICS MANAGEMENT SYSTEM

The configuration of the entire system, which comprehensively controls the braking and driving functions through the combination of the ECB system and the hybrid system, and an outline of the control processes, are discussed in the following section.

3.1 Hardware Configuration of the System

Figure 5 shows the hardware configuration. In contrast to the aforementioned ECB system, the hybrid system is the drive system that contains front and rear drive motors to drive the respective axles. In the front unit, the engine and motor are laid out parallel to each other, and the drive force is transmitted via the CVT. The rear unit consists of only a motor and a reduction unit. Thus, it is an independent four-wheel drive system that does not require a transfer unit or a propeller shaft.

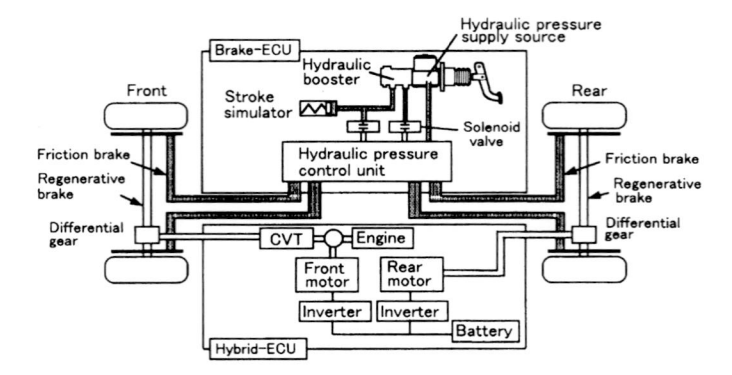

Figure 5: Hardware configuration of the system

3.2 Configuration of the Comprehensive Control Process

Figure 6 shows the control process of this system. The system detects the requirements of the driver through a sensor that detects the pedal input and a steering angle sensor, and computes the vehicle's dynamic targets. At the same time, the system detects the driving conditions of the vehicle through a wheel speed sensor, yaw rate sensor, and acceleration sensor. In accordance with the conditions of these sensors, the system selects and executes the respective control modules in the comprehensive braking and driving control logic. The commands for the brake force and the driving force, which are required by the wheels, are directed to and executed by the ECB hydraulic pressure control module and the hybrid computer, respectively.

The advantages of this configuration are described below. A new concept of comprehensive control that expands the control range from the critical limit of Vehicle Stability Control (VSC) to the normal operating range, this configuration improves the driver's comfort and the environmental performance of the vehicle. This system is hereafter referred to as "Vehicle Dynamics Management".

- On one hand, this configuration enables the system to compute the required brake force and driving force from the normal operating range represented by the regenerative coordination control, in order to maximize the regeneration of energy. On the other hand, this configuration facilitates the execution of continuous control by regulating the brake hydraulic pressure and the driving torque near the critical limit.

- The condition assessment modules, which were previously computed and executed on a module-by-module basis by the ABS and VSC brake control modules, have been integrated and organized. By implementing the control targets through the wheel cylinder hydraulic pressure, the actuator driving modules could be separated from the respective software applications in order to effect comprehensive control. As a result, the quality of the software has been improved and the application time has been shortened.

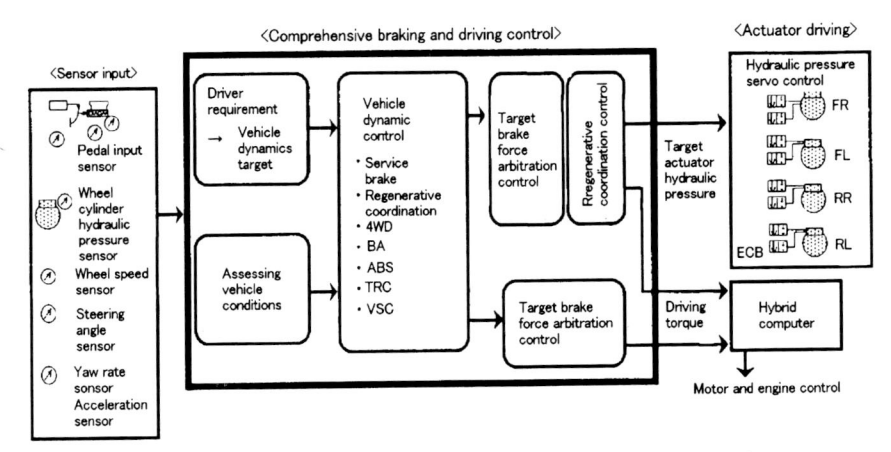

Figure 6: Vehicle Dynamics Management System Processing

4 VEHICLE DYNAMICS CONTROL THROUGH VEHICLE DYNAMICS MANAGEMENT

This section gives specific examples of the improvements in the vehicle dynamic performance and environmental performance that have been realized through the aforementioned system and control.

4.1 Service Brake Control and Regenerative Coordination Control

The service brake, which controls the longitudinal deceleration of the vehicle, detects and computes the driver's brake application. Then, it distributes the target deceleration to the hydraulic pressure brake and the regenerative brake. For the purpose of maintaining the distribution ratio of the front-rear brake force at a constant level, both the front and rear motors of this vehicle prioritize fuel efficiency while effecting the regeneration of energy. Figure 7 shows an example of the actual regenerative and hydraulic pressure coordination control of the front brakes, and Figure 8 shows the rear brakes. The coordination control alone improves the vehicle's fuel efficiency by approximately 20 percent, as verified by the 10-15 mode fuel consumption measurement. Brake performance is also improved, especially in high deceleration range (Figure 9).

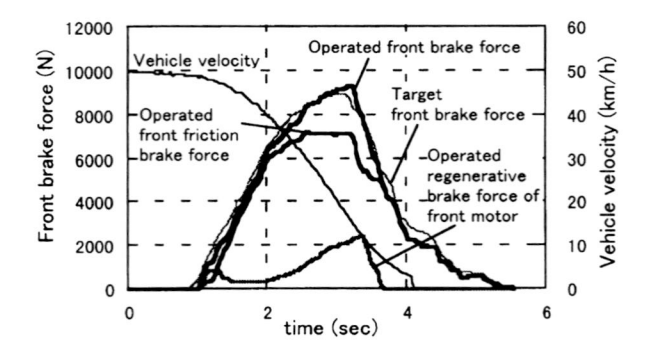

Figure 7: Example of hydraulic-regenerative brake control of the front wheels

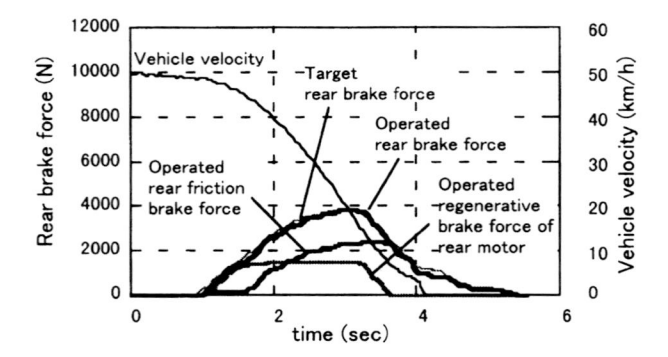

Figure 8: Example of hydraulic-regenerative brake control of the rear wheels

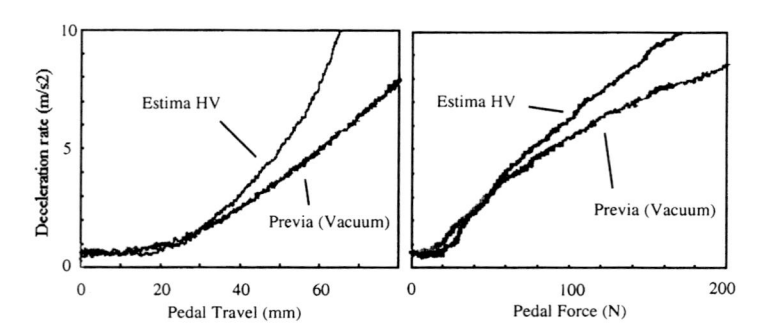

Figure 9 : Brake Performance of Estima HV

4.2 Fundamental Brake Load

Fundamental brake load is lowered as a effect of regenerative coordination, since regenerative brake absorbs brake energy partially. Figure 10-a shows comparison of brake pad temperature in case of regenerative brake active and inactive under R13 fade test pattern(Figure 10-b).

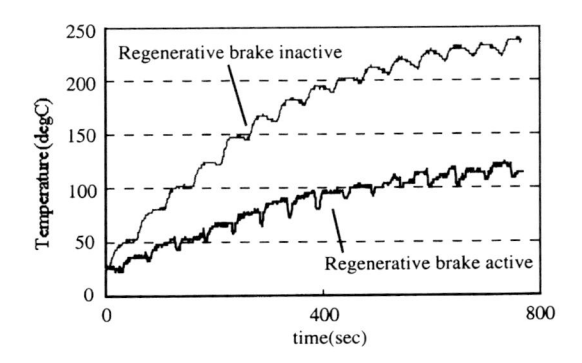

Figure 10-a: Comparison of Brake Pad Temperature, with Regenerative brake active and inactive

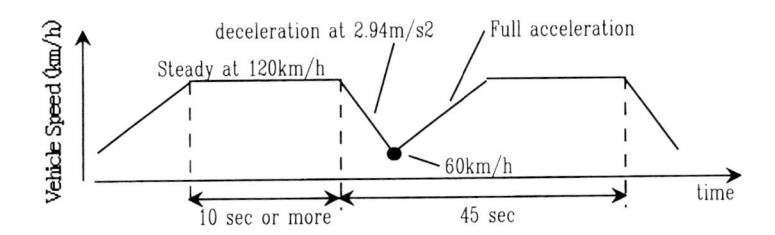

Figure 10-b: Test pattern

4.3 Cornering and Startoff Performance

As part of dynamic performance improvement technology that extends from the normal to the limit range, the driving performance during cornering and the startoff performance while driving on low-μ surfaces are described below.

4.3.1 Cornering Performance

Figure 11 shows a comparison between the Vehicle Dynamics Management (VDM) and the previous Vehicle Stability Control (VSC), with respect to the vehicle behavior and the driver's steering input, while the driving locus is constrained to the following conditions: $\mu = 0.35$, R = 150m, and initial speed = 19.4 m/s.

In the previous VSC control (Figure 11-a), the hydraulic pressure control, when control steps in near the limit and when control is operating, is effected by an ON/OFF valve through duty cycle control. In contrast, the control that has recently been realized by the VDM gradually applies hydraulic pressure to the wheel cylinders from before the limit. This improves the follow-up of the hydraulics to the target yaw rate and reduces the amount of steering correction made by the driver (Figure 11-b). Furthermore, because the amount of control itself is minimized and the control converges quickly (Figure 11-c, -d), smooth driving is realized, while the deceleration of the vehicle that occurs when the control steps in is kept small.

Figure 11-a: Steering angle and target yaw rate deviation with the previous VSC control

Figure 11-b: Steering angle and target yaw rate deviation with the Vehicle Dynamics Management

Figure 11-c: Comparison of control hydraulic pressure of the inner front wheel

Figure 11-d: Vehicle deceleration rate

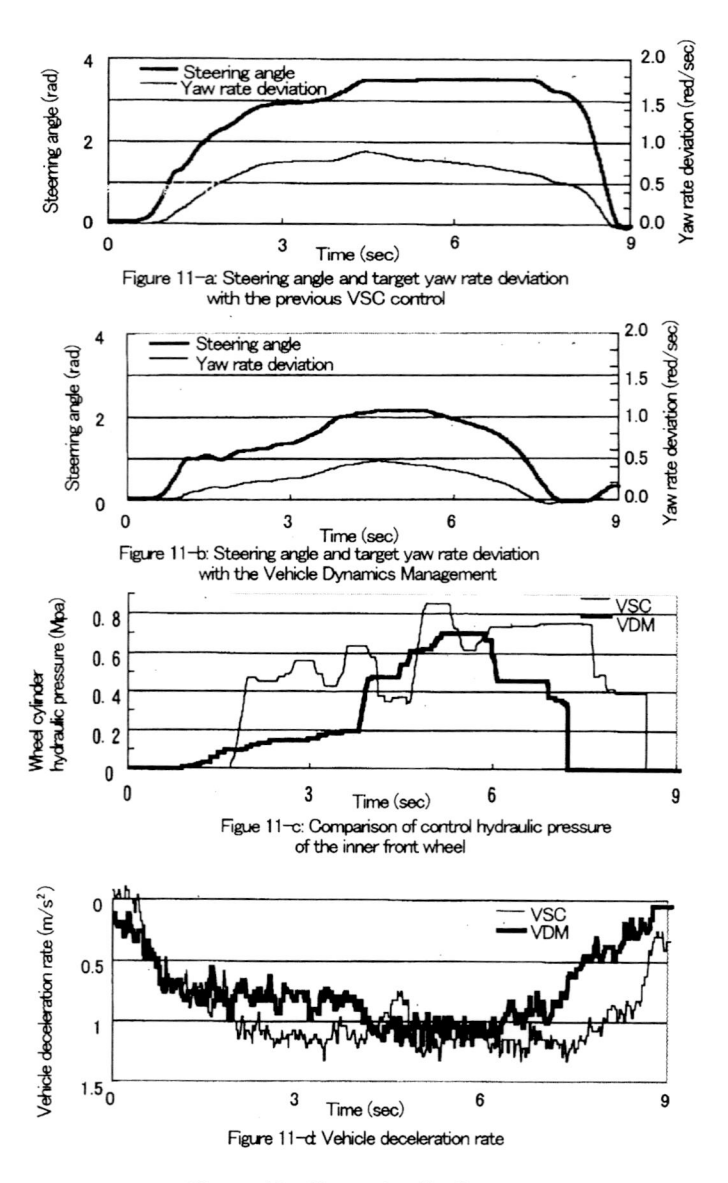

Figure 11: Cornering Performance

4.3.2 Cornering Startoff Performance on Extremely Low-μ Surfaces

When a vehicle starts off at an extremely low-μ surface intersection to turn right or left, its startoff performance is improved if it is a four-wheel-drive vehicle. However, if the front wheels slip, the drive torque applies to the rear wheels, causing the occurrence of the so-called "push-under" phenomenon, which the rear drive force pushes out the vehicle.

With respect to the vehicle behavior simulating a left turn at an intersection with a $\mu=0.35$ surface, Figure 12 shows the following:

[1] The actual vehicle data using the new system and control (Figure 12-a)

[2] The responsiveness equivalent to gasoline engine, with the control logic remaining as is (150 ms lag, Figure 12-b)

[3] Only the VSC control effected, without the drive force control (Figure 12-c)

The results are shown in Figure 12-d, according to the amount of deviation of the actual yaw rate in comparison to the target yaw rate per the aforementioned three pieces of data. Although the previous system can effect drive torque control in accordance with the deviation between the target yaw rate required by the driver and the actual yaw rate that is generated by the vehicle, the new system realizes a smoother cornering startoff performance by controlling the motor (whose torque can be controlled more responsively and linearly than in vehicles with ordinary engines) starting at the normal range.

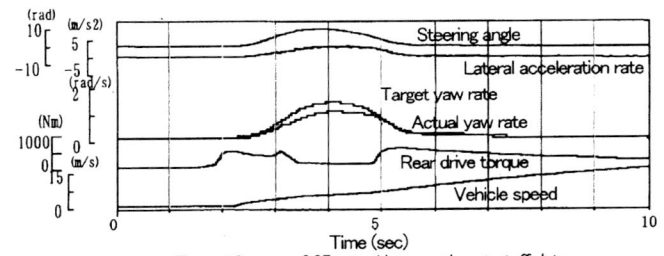

Figure 12-a: $\mu = 0.35$ actual intersection startoff data

Figure 12-b: Same as above (150 msec lag)

Figure 12-c: Same as above (no drive force control)

Figure 12-d: Comparison of yaw rate deviations with defferent controls

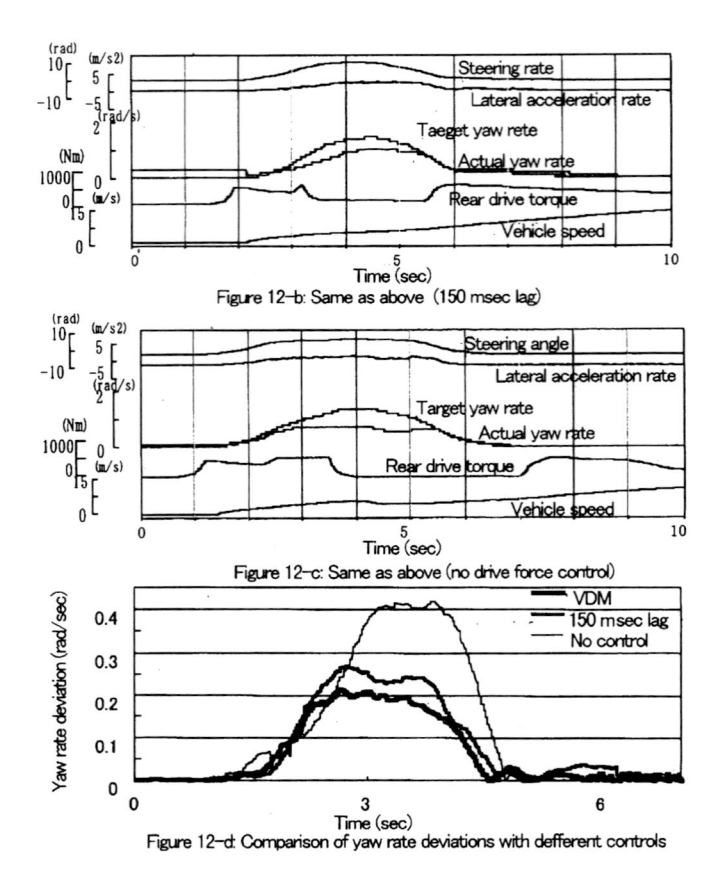

Figure 12: Cornering Startoff Performance on Extremely Low-μ Surfaces

5 CONCLUSIONS

For the purpose of improving the environmental performance (through the coordination with the regenerative brakes at the front and rear wheels), preventive safety, and vehicle dynamic performance, a brake-by-wire system called an Electronically Controlled Brake (ECB) system (which linearly controls the hydraulic pressure that is applied independently to the four wheels) has been developed, and the Vehicle Dynamics Management (which comprehensively controls the braking and driving forces) has been incorporated in this system. As a result:

- Excellent brake feel has been realized through regenerative coordination control, while improving the fuel consumption rate by approximately 20%.

- The driver comfort has been improved with respect to the vehicle behavior from the normal range to the limit range.

Because this technology can also be applied to ordinary vehicles in addition to hybrid vehicles, it is believed that this technology can contribute to enhancing the dynamics of the vehicles.

The authors wish to thank all the people who gave their valuable advice and cooperation to make the development of this technology possible.

REFERENCES

[1] W.D.Jonner et al,"Electrohydraulic Brake System-The First Approach to Brake-by-wire Technology",SAE960991

[2] A.Sakai,et al,"Toyota Braking System for Hybrid Vehicle with Regenerative System",EVS13

[3] A.T.Van Zanten,et al,"VDC Systems Development and Perspective",SAE980235

Enhancing road vehicle efficiency by regenerative braking

A M WALKER, M U LAMPERTY, and **S WILKINS**
Mechanical Engineering Department, Imperial College of Science, Technology, and Medicine, London, UK

ABSTRACT

Regenerative braking is an established feature of rail vehicles. The subject is under development in the road vehicle industry through recent developments in Hybrid Electric (HEV) and pure Electric Vehicles (EV).

Up to 30% of the overall energy demand can be satisfied by energy saved through regenerative braking, significantly improving a vehicle's overall efficiency. In addition, the brake force requirement on friction brake modules is reduced and, in-turn, their size can be reduced.

The paper discusses an overall approach to brake system, powertrain and energy storage system components. Computer codes have been developed to simulate component performance and examine control strategies, which demonstrate significant reduction in friction brake use. A notable finding has been the improved energy retention by the incorporation of ultracapacitors.

NOMENCLATURE

A	Vehicle front area	(m^2)
c_w	Coefficient of aerodynamic drag	
E	Energy	(W.h)
g	Gravity constant	(9.81 m/s^2)
h	Height	(m)
m	Mass	(kg)
t	Time	(s)
v	Speed	(m/s)
η	Efficiency	
μ	Coefficient of rolling resistance	
ρ	Density / Density of air	(kg/m^3)

1 INTRODUCTION

Hybrid Electric (HEV) and pure Electric Vehicles (EV) use more energy efficient technology than traditional vehicle technologies. They are also increasingly being seen as the answer to stringent emissions regulations in North America, Europe and Japan.

Most major automobile manufacturers are involved in HEV and EV development programmes, with a few early production platforms and countless prototype vehicles already on the road. Although the prime mover and powertrain configuration is unique to each model, a common feature of all these vehicles is regenerative braking.

The approach to vehicle energy efficiency, and incorporation of regenerative braking, requires that powertrain, brake, and energy storage components no longer be addressed separately.

2 REGENERATIVE BRAKING

2.1 Electrical Braking
Retaining the vehicle's inertial energy during deceleration, and saving for the next acceleration phase, is termed as "regenerative" braking. When returned, this energy reduces the demand on the power train. The vehicle's overall efficiency is improved such that up to 30% of the overall energy demand can be satisfied through energy recaptured during regenerative braking (1).

The level of energy recapture depends on the capacity and efficiency of the electric traction motor, which operates as a generator during deceleration, and the capacity and efficiency of the power control and energy storage system, connected to the motor. Only in exceptional circumstances is the motor unable to sustain the regenerative energy flow, due to insufficient cooling. The limiting component, in most cases, is the energy storage system, and its charge acceptance characteristic during heavy braking. Improvement, in braking performance, of a regenerative braking system can only be achieved by development, in terms of efficiency and response, of energy storage devices.

2.2 Friction Braking
Where a regenerative braking system is used, a friction brake system is still required, even if seldom employed. A brake control architecture is required for interaction of the components and communications. This brake control architecture might be similar to those in existence on rail vehicles (2), but is not addressed here.

As regenerative braking assumes the burden of most brake actuation, the brake force required from friction brake modules is reduced, and a reduction in physical size is possible.

2.3 Mechanical Integration
To integrate the friction brake and electric traction motor, particularly in axle-less vehicles (such as low floor buses), combination units are now under development (3), (4).

3 ENERGY STORAGE

3.1 Energy Conversion
Multiple energy conversions (e.g. $E_{kinetic} \rightarrow E_{electric} \rightarrow E_{kinetic}$) are present in vehicle operation. The vehicle's movement can be described as an energy storage mechanism – the energy of which is converted to electricity or other while the vehicle is at rest. Under this description, braking is the means by which the vehicle's kinetic energy is returned to the on-board electrical or mechanical storage devices. The powertrain is the means by which the on-board stored energy is converted into kinetic energy (i.e. vehicle movement), along with additional requirements to overcome losses.

3.2 Energy Density / Power Density
Energy stored chemically in a conventional vehicle benefits from a high energy density, as illustrated in Fig. 1, but suffers from a poor conversion process efficiency. Traditionally, only a small level of electrical energy storage is required in the vehicle for restarting the prime mover. EV technology depends fully on electrical energy storage, and these limitations have hindered vehicle range to date.

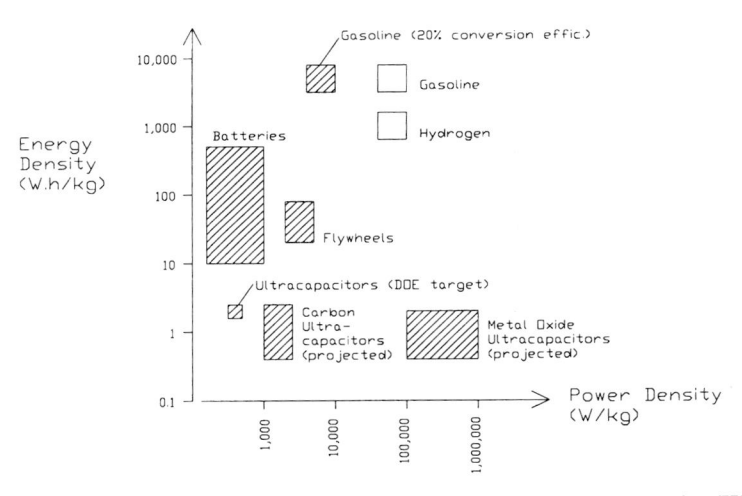

Fig. 1 Region plot of Energy Density (W.h/kg) versus Power Density (W/kg) in mechanical, chemical and electrical forms (5). Hydrogen includes containment vessel.

3.3 Electrical Energy Storage Device

3.3.1 Battery Development
Practical EV development requires batteries that provide performance comparable with conventional vehicles, and at a comparable cost. Although cost effective, standard lead acid battery technology provides a very limited range.

Table 1 Battery Development Objectives for Vehicle Applications (6).

	Energy Density (W.h/kg)	Power Density (W/kg)	Useful Life (year)	Cost ($/kW.h)
Vehicle Application Requirements	200	400	10	100

With relatively high charge acceptance, lead acid technology has acceptable power density, but poor energy density. Lithium-ion polymer technology is closer to providing acceptable energy and power density. However, cost is still a significant issue (Lead acid 120-150 $/kW.h, Lithium-ion 200 $/kW.h, and Nickel/metal hydride 200-350 $/kW.h), although work is proceeding on this (7), (8).

Table 2 Operational Properties of Existing Electrical Energy Storage Devices (9).

Energy Storage System	Energy Density (W.h/kg)	Power Density (W/kg)	Energy Efficiency (%)	Cycle Life (Num of Cycles)	Self Discharge (%/48 h)
Lead acid (10)	35-50	150-400	>80	500-1000	0.6
Nickel/cadmium (10)	40-60	80-150	75	800	1
Nickel/metal hydride (10)	70-95	200-300	70	750-1200+	6
Lithium-ion (10)	80-130	200-300	>95	1000+	0.7
Ultracapacitor (11)	2.5	>1270 (12)	92-98	10,000hr* (12)	2 (12)

* Ultracapacitor cycle life is for average operating temperature 35 °C, and voltage 2.3 V

3.3.2 Charge Acceptance and Ultracapacitors

Further development of battery technology for vehicle applications has focused on high charge and discharge rates. The rapid, and efficient, charging of the battery, the number of cycles per life, and full operation on low state-of-charge is important. Electrochemical battery life is dictated by a limited number of charge-discharge operations.

An alternative storage medium for absorbing and releasing large amounts of energy quickly is the double-layer ultracapacitor. Ultracapacitors have a very high rate of charge acceptance and discharge characteristic (12), which is appropriate in the vehicle application for quick acceleration and energy absorption from braking. Energy storage based on 2000-3000F ultracapacitors assists regenerative braking better than batteries by buffering the energy generated (through high current charging). However, a high discharge rate cannot be sustained for a long period, rendering a low energy density. The energy in the powertrain (i.e. when the storage unit is used for lower power output) still requires optimising through a battery, although the million-plus cycle life of the ultracapacitor greatly extends the battery life.

Ultracapacitors cost 270 $/kW.h (from a projected $30 per 2700F unit). One city bus, with a dieselelectric powertrain and an ultracapacitor bank, has demonstrated 24% of energy requirement provided by regenerated brake energy, resulting in an equivalent fuel efficiency improvement (13).

4 SIMULATION

The selection of powertrain components, the power requirements and the optimisation of control systems involve complex trade-offs, such as performance versus energy consumption and cost (14), (15), (16), range, emissions and acceleration. Comprehensive, computer-based simulation models, which accurately represent the power train and power requirements, are helpful in examining the powertrain configuration as these parameters can be evaluated in a systematic approach and sensibly matched to the vehicle application.

4.1 Simulation Program
The programming approach is modular, such that components can be respecified and retested without rebuilding the entire system. Evaluation may be performed under any user defined driving cycle, or embedded schedules, such as the FUDS (Federal Urban Driving Schedule) or the FHDS (Federal Highway Driving Schedule). There are three modes of simulation.

4.1.1 Non-feedback Mode
The non-feedback mode is the most commonly used where the simulation starts from a given drive cycle (time step, speed, road gradient) and vehicle configuration (powertrain specifications, component sizes etc.). For every time step the energy demand is calculated and then the performance in terms of fuel consumption, driving range and emissions are computed.

4.1.2 Cycle Feedback Mode
Cycle feedback is similar to the non-feedback mode, except that the maximum capacity of the powertrain components dictates the vehicle performance, as in reality, in attempting to satisfy the simulated driving cycle. In this case, the vehicle speed-time profile will stray from the pre-defined profile, which can be seen by overlaying the speed profiles, see Fig. 2.

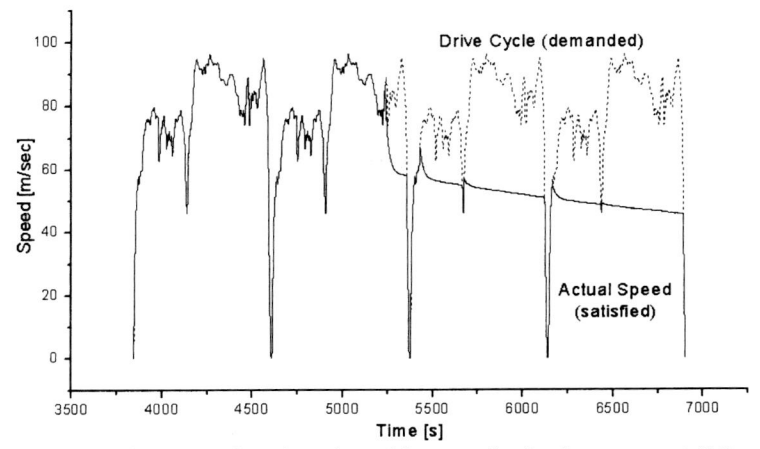

Fig. 2 Performance deterioration with an undersized component (14).

4.1.3 Powertrain Feedback Mode

The powertrain feedback mode is designed to determine the sizing of the powertrain components using feedback from the energy demand of the driving cycle. Hence, from a given vehicle arrangement and drive cycle, the program sizes the different powertrain components in order to match the requirements in terms of power and required range. This proves to be a particularly useful feature where parametric studies are required for a certain design.

4.1.4 Vehicle Dynamics Equations

For the simulation, the program uses vehicle dynamics equations for the calculation of the energy demand for every time step in the drive cycle. These equations include:

$$E_{tractive} = E_{Kinectic} + E_{Potential} + E_{rolling} + E_{aero} \quad [1]$$

where kinetic energy during a time step is

$$E_{Kinetic} = \frac{1}{2}m(v_E^2 - v_S^2) \quad [2]$$

change in potential energy is

$$E_{Potential} = mg(h_E - h_S) \quad [3]$$

losses due to the rolling resistance are

$$E_{rolling} = mg\mu(v_E - v_S)\Delta t \quad [4]$$

and losses due to aerodynamic drag, calculated by integrating over the time step, are

$$E_{aero} = \frac{c_w A \rho}{2} \int_{t-\Delta t}^{t} v(t)^3 \, dt \quad [5]$$

When $E_{tractive} > 0$, the powertrain provides energy to the motor to drive the wheels, given by

$$E_{Powertrain} = \frac{E_{tractive}}{\eta_{motor}} + E_{DriveTrain} \quad [6]$$

where $E_{DriveTrain}$ represents the energy losses due the inefficiency of the drivetrain and friction. In Equation [6] η_{motor} is the motor efficiency, which is a function of motor speed and torque, calculated from the motor map.

When $E_{tractive} < 0$, braking operates, and energy is recovered and fed back into the batteries, given by

$$E_{Powertrain} = E_{tractive}\eta_{regen} - E_{DriveTrain} \qquad [7]$$

where η_{regen} is the regeneration efficiency (also a function of motor speed and torque, determined from the motor map).

4.1.5 Battery Model
Batteries are modelled using a "hydrodynamic two-tank model" (17). This is analogous to the electrochemical reaction occurring within the battery, taking into account the battery time history to estimate the state of charge. Other energy storage devices are similarly modelled.

5 HYBRID H.G.V. EXAMPLE

Based on limited driving, the full traction and braking loads on a Hybrid H.G.V. truck have been investigated by simulation in a Non-Feedback Mode. This Hybrid vehicle has a Series Hybrid topology, with electrical rather than mechanical connection between the powertrain and the wheels, illustrated in Fig. 3(a). The Parallel Hybrid topology, illustrated in Fig. 3(b), with mechanical and electrical powertrain connection, has proportionately less electrical traction power. The Series topology benefits from electrical braking more than the Parallel.

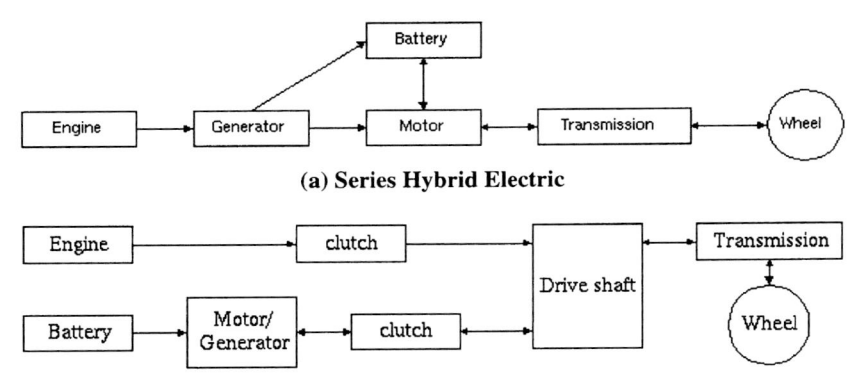

(a) Series Hybrid Electric

(b) Parallel Hybrid Electric
Fig. 3 Vehicle Systems Layouts

5.1 Drive Cycle
A drive cycle has been selected to be demanding on both traction power and braking power, as traction power is the determining factor in the vehicle configuration (18), (19). To date, no commonly used drive cycles exist, which demand heavy braking due to gradients, so a new drive cycle has been rendered. The selected route is part of Edinburgh's inner ring road – the streets Drum Brae North and Drum Brae South. Heavy braking was implemented due to the steep gradients involved, illustrated in Fig. 4. The cycle features an altitude climb of 50 m and descent over the 2 km street length, and has the same altitude at start and finish to ensure no overall energy change. The curvature of the Drum Brae streets is negligible, allowing even distribution on wheels.

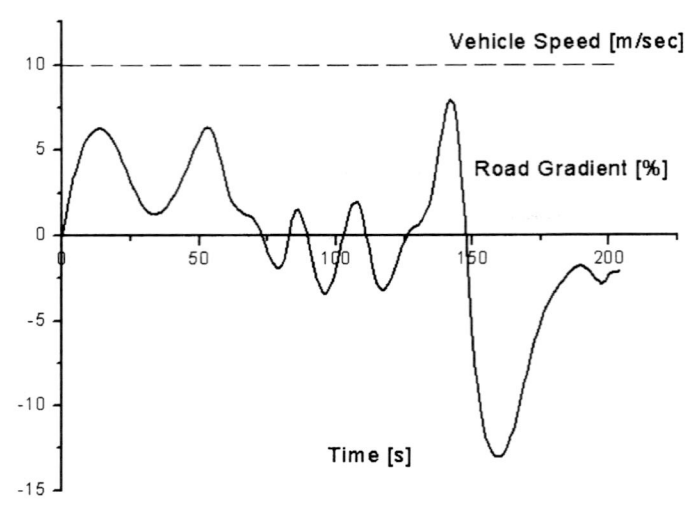

Fig. 4 Drive Cycle on Drum Brae North and South, Edinburgh (20).

5.2 Vehicle Speed
The vehicle speed was ramped to 10 m/s (36 km/h) prior to imposing the drive cycle. The speed was then maintained constant over the duration of the cycle, illustrated in Fig. 4. This ensured substantial, high-current discharge from the energy storage during ascent, and a rapid, high-current charge during descent.

5.3 Vehicle Configuration
Two configurations were implemented, with different electrical braking capacities. Availability of electrical braking depends on available charging capacity of the energy storage device. A "low charge acceptance" requires comparatively more use of the friction brake. A "high charge acceptance" allows more use of electrical braking and requires comparatively less use of the friction brake. This partially corresponds to the difference between Hybrid topologies. A high charge acceptance vehicle features more electric powertrain (the Series) than a vehicle with low charge acceptance (the Parallel).

5.3.1 Configuration A – "Low Charge Acceptance"
The general characteristic of "low charge acceptance" was achieved through a standard traction motor and a standard energy storage device in series. Without an alternative means for "dumping" excess regenerated energy (such as a brake resistor), electrical braking is effectively useless when the energy storage device is "full" (i.e. has 100 % state-of-charge).

5.3.2 Configuration B – "High Charge Acceptance"
"High charge acceptance" was achieved through a standard traction motor and a larger energy storage device, also in series. This energy storage device comprised a standard-sized battery

and, importantly, an ultracapacitor, in order to attain rapid, high current charging, which the battery was unable to sustain.

5.3.3 Control Strategy
In both cases, the control strategy used to implement powertrain optimisation was based on the state-of-charge of the energy storage device. The braking control strategy ensures that brake applications are achieved initially through electrical braking, with friction braking only employed where the electrical braking application is insufficient.

Table 3 Vehicle Configurations.

Configuration	Charge Acceptance	Motor Size (kW)	Energy Storage Device Size	
			Battery (Ah)	Ultracapacitor (F)
A	Low	90	180	–
B	High	90	180	50+

5.4 Results

Table 4 Total Brake Energy for Configuration A and B over Drive Cycle.

Configuration	Charge Acceptance	Total Brake Energy over Drive Cycle (W.h)	Electric Brake Energy (W.h)	Friction Brake Energy (W.h)
A	Low	600	285	315
B	High	600	415	185

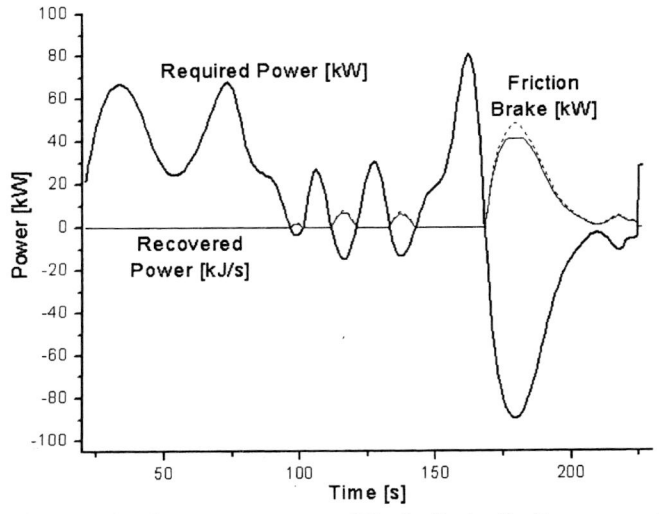

(a) Configuration A – the charge acceptance of the batteries limits recovered power (20).

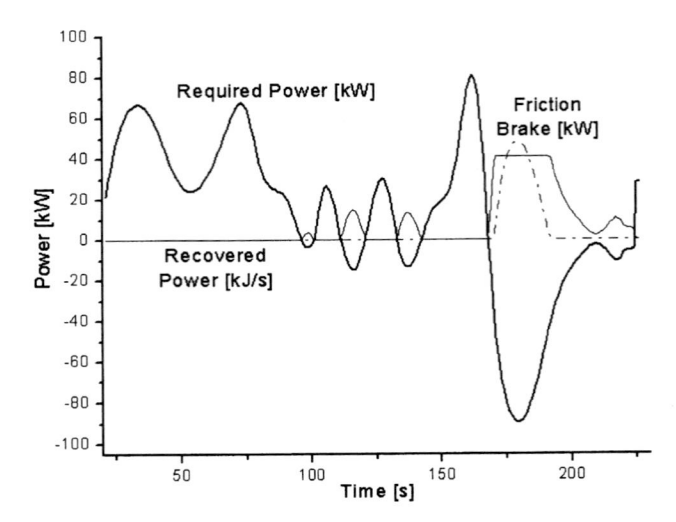

(a) Configuration B – the torque limit of the motor limits recovered power (20).
Fig. 5 Recovered power and friction brake use over drive cycle.

6 CONCLUSIONS

In the Series Hybrid Electric vehicle, braking through the electric motor is the norm, and the requirement for friction braking is reduced. The level of this reduction is dependent on the charge acceptance capacity of the energy storage device employed.

A specialised, heavy-braking drive cycle has been rendered to facilitate simulation of Hybrid Electric and pure Electric Vehicles. A Hybrid H.G.V. truck, with a Series Hybrid topology, has been simulated driving over this drive cycle at constant speed. With a low charge acceptance configuration, due to a standard energy storage device, the friction brakes were employed regularly. With a high charge acceptance level, due to the inclusion of an ultracapacitor with the standard battery, the friction brakes were rarely used, determined by the torque limit of the motor at the operating speed.

Therefore, friction brake modules on Hybrid Electric vehicles can be reduced in size to service a reduced requirement, and developed of a control architecture, optimising the combined use of electric and friction braking, is required.

7 REFRENCES

(1) Mitsubishi data sheet: "Motor Vehicle Brake Energy Conservation System – III", 2001
(2) Knorr-Bremse data sheet: "Distributed EP-Brake Controller EP2002", 2001
(3) Meritor data sheet: "Axles Designed for Ultra-Low Floor City Buses", 2001

(4) Irisbus data sheet: "Civis: New System of Electric Propulsion", 2001
(5) G. Zavitsanakis, P. Lambrou, N. Economidis, P. Lagias, "Energy Efficiency Simulation of a Hybrid Vehicle for Battery Use Optimisation", UG report Imperial College, 1996
(6) U.S. Advanced Battery Consortium, 1998, <http://www.uscar.org>
(7) E. Forouzan, B. Jay, "Improved, Low-Cost Lithium Polymer Battery System for Electric and Hybrid Vehicle Applications", Electrosource, Inc. promotion material, 2001
(8) D. Clerc, M. Fay, L. Wang, "High-Capacity Rechargeable Batteries Using Tin-Based Electrodes", T/J Technologies, Inc. promotion material, 2001
(9) A. Pal, D. Dareing, "Regenerative Braking Systems with Energy Capture for Heavy Duty Vehicles – A Research Objective", GATE Center, University of Tennessee, Knoxville, 2001
(10) D. A. J. Rand, R. Woods, R. M. Dell, "Batteries for Electric Vehicles", Taunton Research Studies Press, 1998
(11) Northeast Advanced Vehicle Consortium, 2000, <http://www.navc.org>
(12) Maxwell data sheet
(13) EPCOS data sheet: "Optimised Starting – Economic Driving", 2001
(14) C. Leontopoulos, M. R. Etemad, K. R. Pullen, M. U. Lamperth "Hybrid Vehicle Simulation for a Turbo-Generator Based Powertrain", Journal of Automobile Engineering, Proc IMechE Part D –Vol 212 Part D pp 357-368. 1998
(15) K. R. Pullen, M. R. Etemad, M. U. Lampérth, "Description of a Performance Simulation Methodology and Software for Electric and Hybrid Vehicles", Paper 97EL036, 30th ISATA Proceedings – Electric, Hybrid and Alternative Fuel Vehicles, Italy, 1997
(16) M. U. Lampérth, "Hybrid Vehicle Simulation", MSc Imperial College, University of London, 1996
(17) M. U. Lampérth, K. R. Pullen, M. R. Etemad, "Development of an Analysis Software to Allow Electromechanical Battery Modelling for Electric and Hybrid Vehicles" Paper 97EL035, 30th ISATA Electric, Hybrid and Alternative Fuel Vehicles, Florence, 1997
(18) M. U. Lampérth, K. R. Pullen, "How size and Performance of Hybrid Electric Vehicle Components are Influenced by Acceleration Patterns", SAE Technical Paper Series 1999-01-2909 Future Transport Technology, California, 1999
(19) M. U. Lampérth, K. R. Pullen "Impact of Acceleration Patterns on Size and Performance of Hybrid Electric Vehicle Components", 32nd ISATA – Electric, Hybrid and Alternative Fuel Vehicles, Florence, 1999
(20) A. Walker, S. Wilkins "Heavy Vehicle Performance on Drum Brae Cycle", internal report Imperial College, 2001

Braking Systems and Vehicle Performance

Influence of tyre properties to modern vehicle control systems

K-H HARTMANN and **H GRÜNBERG**
Truck Tyre Division, Continental AG, Hanover, Germany

SYNOPSIS

The tyre is responsible for the transmission of all forces and torques from the vehicle to the road. Only a very small contact area (for truck, the size of A4 paper) has to transmit high forces under all braking and steering conditions.

Tyres on steering axles basically have other properties than tyres on driven axles. The properties can be described in longitudinal force – slip curves and lateral forces – slip angle curves.

Moreover these properties are an input to new vehicle control systems such like ESC (Electronic Stability Control) and ABS systems. An ABS controls the tyre. The efficiency of an ABS control unit depends on tyre characteristics. Some tyres are responsive to this control but some do not respond well to ABS system.

To obtain more safety and higher comfort we have to use synergies between tyres and control systems. It is necessary to tune the tyre and the control systems to obtain the highest performance.

This report shows different truck tyre properties and the influence of these properties on modern truck braking and vehicle control systems.

1 START SITUATION

The tyre has the task to transfer all forces acting on the vehicle to the road. It can transfer either high lateral forces or high longitudinal forces. If both are required, the vector sum of both forces is the maximum possible (fig. 1).

In the past, the brakes caused the tyres to lock. From this point on, the transfer of forces was dependent entirely on the mechanisms between the road and tyre. The vehicle could no longer be steered.

Since the introduction of modern braking systems the interaction of the tyre with the road is continuously monitored and regulated. The electronics prevent the tyre from locking with the result that the vehicle remains manoeuvrable and controllable.

The lateral forces acting on the vehicle are also monitored and regulated by modern vehicle control systems (ESC – Electronic Stability Control). The tyre plays an important role in this.

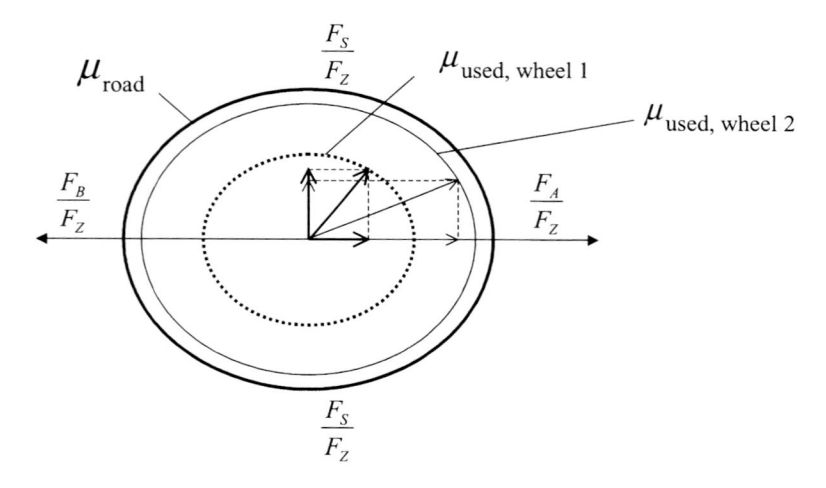

fig. 1: Kamms cycle

The increase in safety with such systems is considerable. Despite this, however, truck accidents show considerably higher consequential damage than car accidents (see fig. 2). 6% of all accidents are fatal and approximately 32% result in serious injury. The main reasons for truck accidents are speed and an insufficient safety distance.

The Continental AG started a project "Safe Driving". The target of this project is to increase the safety of trucks. The concept "Safe Driving" included the braking distance of modern trucks. The goal is to identify important influencing factors and to develop a new tyre with reduced braking distance.

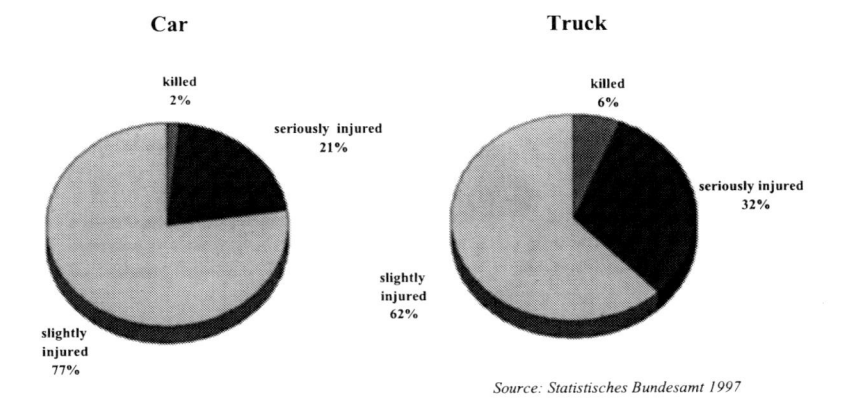

fig. 2: Accidents caused by trucks, Reasons for 33367 injury accidents

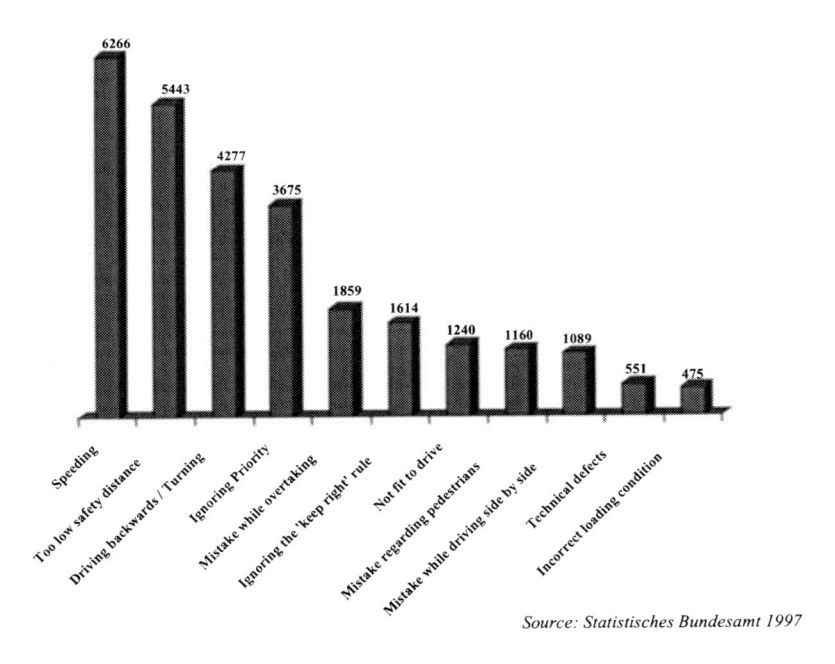

fig. 3: Accidents caused by trucks, Reasons for 33367 injury accidents

2 INFLUENCE OF LONGITUDINAL CHARACTERISTICS OF THE TYRE ON BRAKING DISTANCE

2.1 Functionality of ABS-Systems

Modern ABS-systems register the wheel revolutions with sensors at the wheels. The wheel speed can then be calculated from this information. A reference speed for the whole vehicle is than calculated from the wheel speed. Based on this reference speed, wheel slip can be calculated for each wheel.

An ABS-system monitors the wheel speed. A considerable change in wheel speed (high wheel deceleration) is thus an indicator for the initial stages of wheel locking. A further indicator for the initial stages of wheel locking is the slip of the wheel.

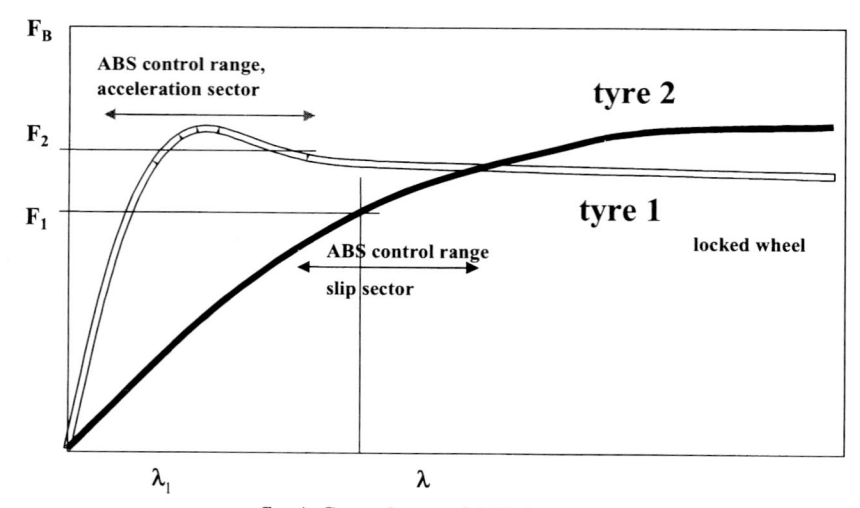

fig. 4: Control area of ABS-Systems

Fig. 4 demonstrates the control philosophy: Tyre 1 shows a pronounce maximum in the low slip range. As soon as this tyre reaches its adhesion coefficient it will lock. The wheel speed is abruptly reduced, the ABS-system will register this and reduce the brake pressure sufficiently to allow the wheel to turn again.

Tyre 2 develops its maximum adhesion coefficient very late. This maximum is not very pronounced, so the slip increases continuously without suddenly locking. The ABS-system registers the high slip and correspondingly decreases the brake pressure. Tyre 2 still has adhesion reserves, which can not be used by the braking system.

The two curves in fig. 4 clearly demonstrate that tyre characteristics and braking system can be and must be tuned to each other. An increase of the adhesion potential of the tyre alone does not automatically result in the shortest possible braking distance.

2.2 Results from braking tests

Figures 5, 6, and 7 show braking results on different road surfaces with test tyres having different tread pattern and compounds. The difference in results corresponds to results with tyres from the market.

The results are shown in comparison to a reference tyre. A value larger than 100% indicates a shorter stopping distance and a value of less than 100% indicates a longer stopping distance. Fig. 5 shows ABS stopping distance on dry asphalt. As can be seen, the difference between the best and the worst tyres in stopping distance is approx. 35%! On wet asphalt as shown in Fig. 6, this difference is approx. 28%.

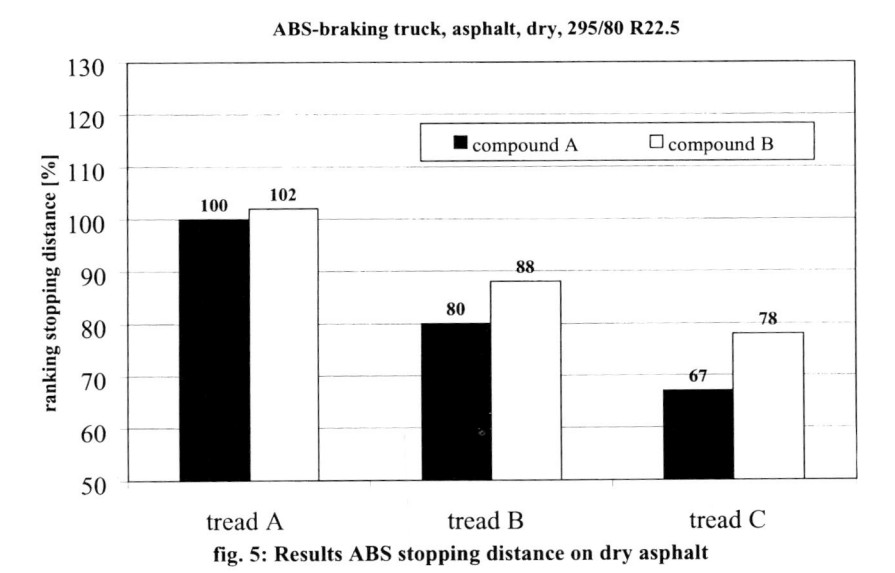

ABS-braking truck, asphalt, dry, 295/80 R22.5

tread	compound A	compound B
tread A	100	102
tread B	80	88
tread C	67	78

fig. 5: Results ABS stopping distance on dry asphalt

This means that from a speed of 80 km/h on a dry road, a vehicle equipped with best tyres requires a stopping distance of approx. 40 m. A vehicle with worst tyres still has a speed of 22 km/h at this distance.

Fig. 7 shows the ranking for locked-wheel braking on a wet road. Here the results are reversed: the tyre with poor results in ABS-braking shows a considerably better result in locked-wheel braking. The corresponding longitudinal force – slip curves (fig. 8) provide an explanation for the differing results.

Tyre A transmits high forces at low slip. For high slip values on the other hand the forces transmitted are relative low. The tyres with tread pattern C, however, show different characteristics. At low slip values the transmitted forces increase relatively slowly and at high slip values the transmitted forces stay relative constant.

ABS-braking truck, asphalt wet, 295/80 R22.5

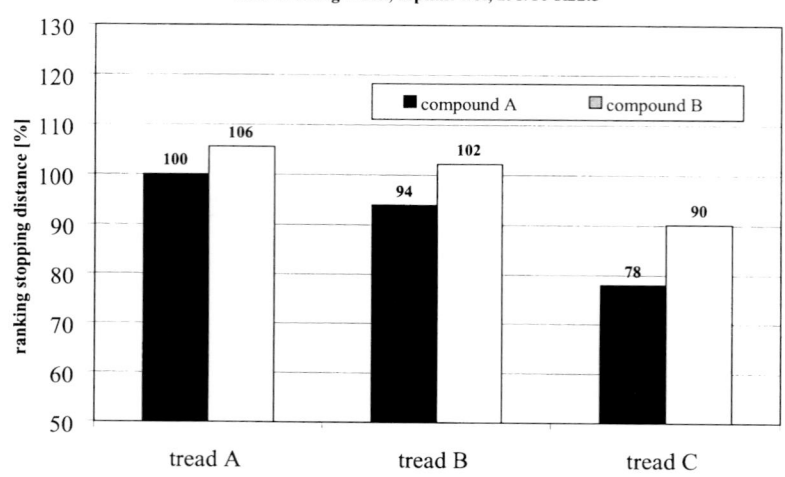

fig. 6: Results ABS stopping distance on wet asphalt

locked wheels truck, asphalt wet, 295/80 R22.5

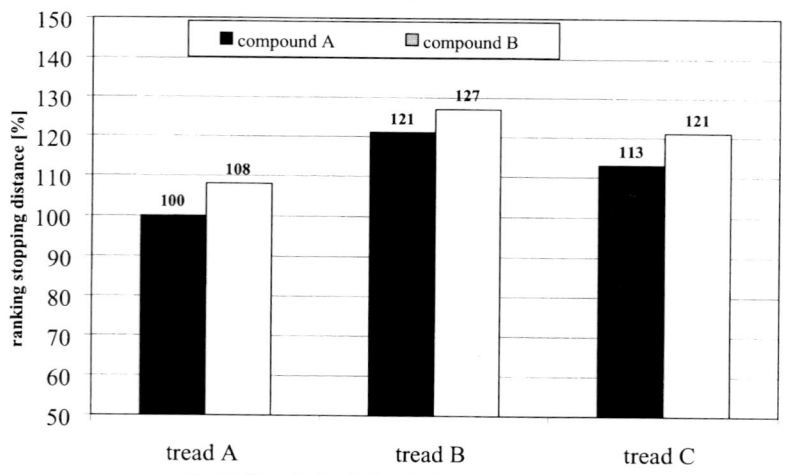

fig. 7: Results locked wheels on wet asphalt

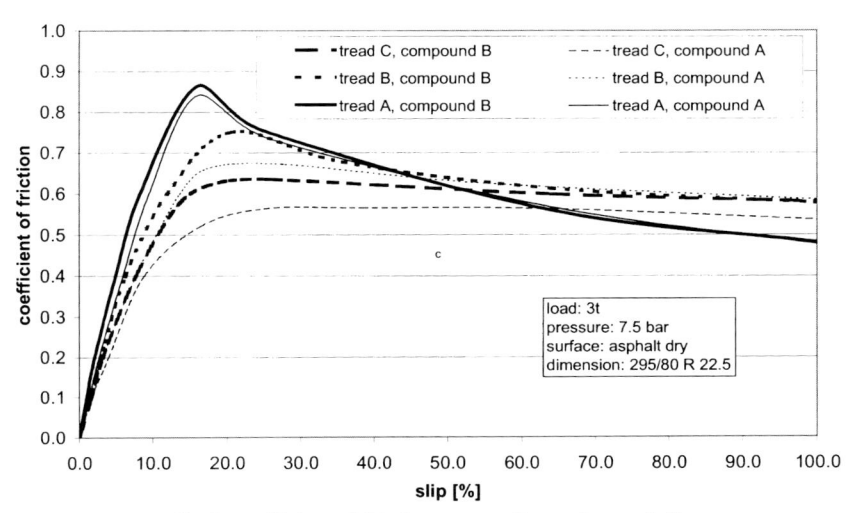

fig. 8: coefficient of friction versus slip on dry asphalt

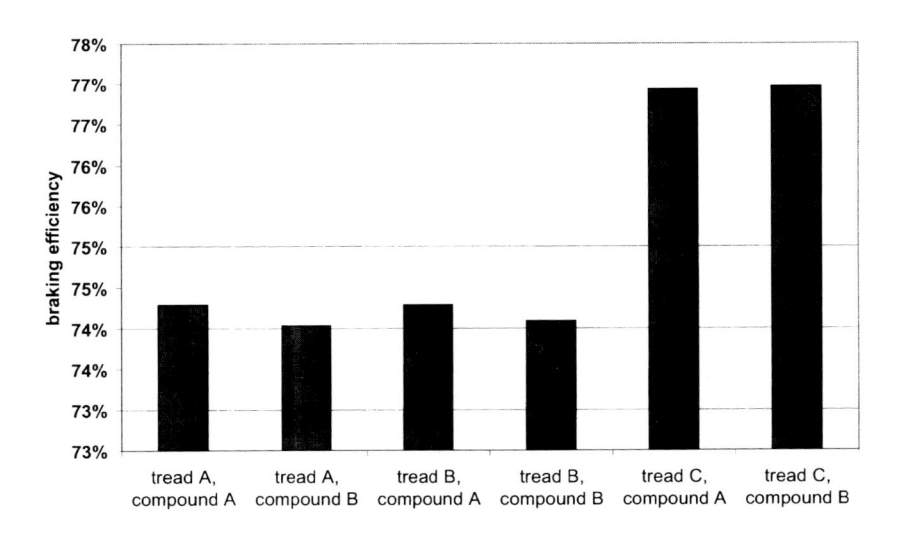

fig. 9: Braking efficiency

Thus the efficiency of ABS-braking can be calculated as follows:

$$\eta = \frac{\mu_{braking}}{\mu_{slip-curve}}$$

$\mu_{braking}$ is the coefficient of friction calculated from the stopping distance and $\mu_{slip-curve}$ is the maximum friction coefficient from the measured µ-slip curves. The efficiency of the ABS braking (fig. 9) with tread patterns A and B amount to approximately 74%. The efficiency of the ABS-braking with pattern C, however, is higher – approximately 78%.

On average, approx. 75% maximum friction coefficient is used for braking. There is certainly further potential for improvement in the area of ABS-braking control.

Furthermore the shape of the µ-slip curve appears to have an influence on ABS-braking.

3 INFLUENCE OF LATERAL CHARACTERISTICS OF THE TYRE ON LATERAL VEHICLE DYNAMICS

It is well known that tyre characteristics have an influence on the lateral dynamics of the vehicle. This is particularly significant in the case of trucks.

The first vehicle control systems (ESC) for trucks in Europe were introduced in tractors at the end of 2001. These systems control vehicle movement at the stability limit and thus increase the safety of the vehicle. The input for an ESC is the yaw rate, steering angle and the lateral acceleration.

The yaw rate and the lateral acceleration are dependent among other things on tyre characteristics. The tyre influences the yaw rate particularly during dynamic steering manoeuvres. The lateral acceleration is influenced by the tyre mainly under constant steering angles.

In contrast to passenger cars, trucks have different tyres on the front and rear axles. Furthermore, trucks usually have twin tyres on the rear axle. As a result of these factors the tyre characteristics on the front and rear axles are different.

Fig. 10 shows two different lateral forces – slip angle curves. Tyre 1 is on the front axle and tyre 2 is on the rear axle. A heavy vehicle drives around a corner so that the lateral forces on both axles are equal. The curves show that the front axle requires a smaller slip angle than the rear axle. This has a noticeable effect on the yaw movement of the vehicle. It is thus possible that an unsuitable tyre combination could have a considerable effect on the yaw movement of the vehicle. Consequentially, for vehicles with ESC, this could lead to a premature activation of the vehicle control system although the vehicle is still far away from its stability limit.

Fig. 11 shows coefficient of friction – slip angle curves of tyres as mentioned in the first part of the paper. Fig. 12 shows the corresponding ranking for constant wet cornering. Here the influence of tyre characteristics on the maximum lateral acceleration can be clearly seen.

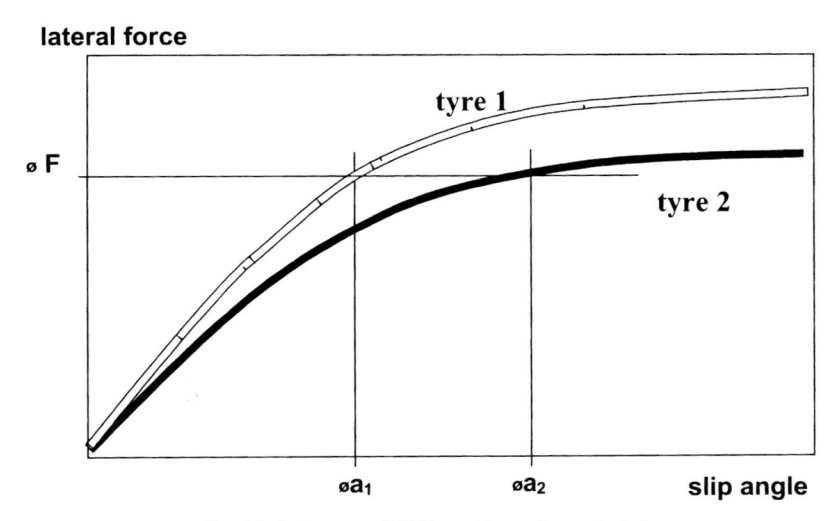

fig. 10: Influence of different tyre characteristics

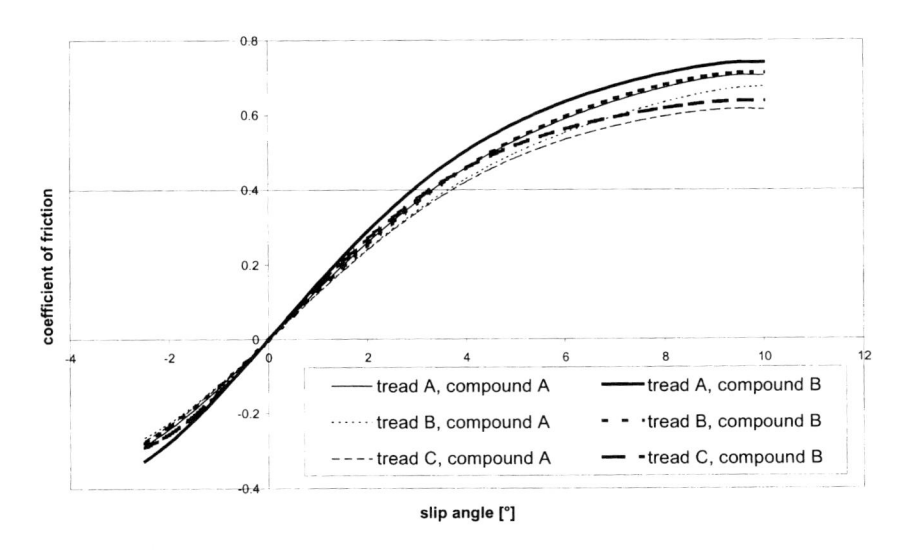

fig. 11: Tyre characteristics of different tyres on dry asphalt

fig. 12: **Ranking of the lateral acceleration during wet cornering**

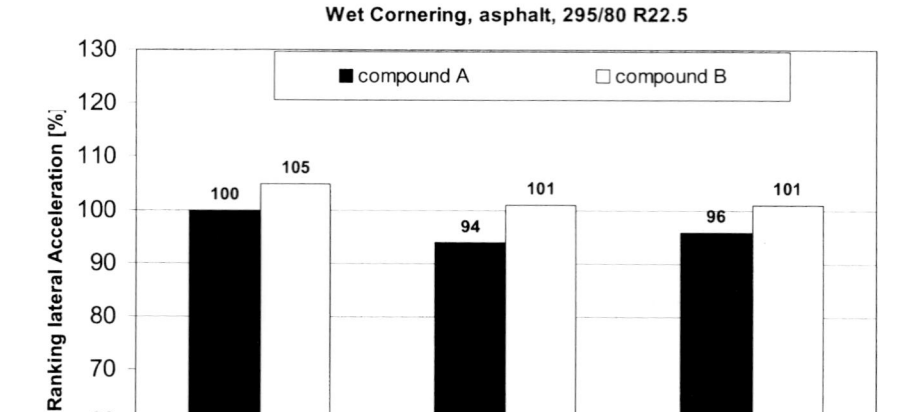

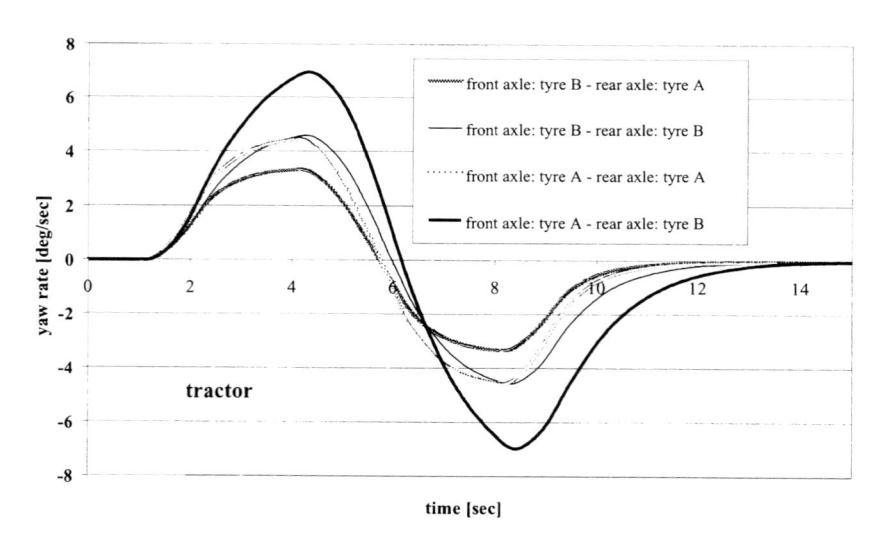

fig. 13: **Yaw rate versus time for different tyres and combinations**

fig. 14: tyre curves used in the simulation

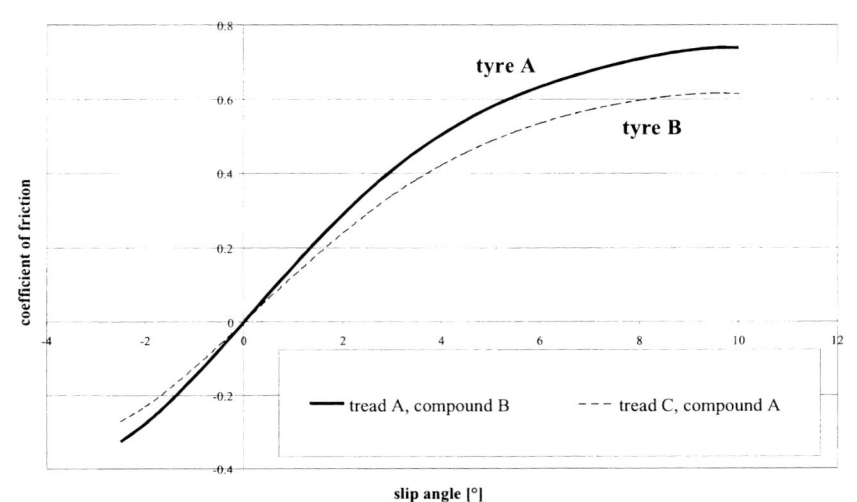

Fig. 13 shows the vehicle behaviour during a double lane change. The calculations were carried out for the two tyres that showed the largest difference in lateral characteristics (fig. 14). The yaw rate calculations were made with the following tyre combinations: front axle tyre B -rear axle tyre A; front axle tyre B -rear axle tyre B; front axle tyre A -rear axle tyre A; front axle tyre A -rear axle tyre B.

The difference between the various yaw rates can be clearly seen -the yaw rate of the vehicle with an unsuitable choice of tyres can be twice as high as with the best yaw rate combination. This means an earlier or later regulation of the vehicle control system.

REFERENCES

1 **Schroeder, C., Koehne, K. U., Kueppers, Th.** (1997) Automobiltechnische Zeitschrift, ATZ, Band 99, Heft 6, Seite 322-328

2 **Hahn, W.-D., Weber, R.** (1989) Messung der Führungseigenschaften von Nutzfahrzeugreifen, VDI-Berichte Nr. 741, S. 25

3 **Schröder, C., Köhne, K.-U., Küppers, Th.** (1996) Heavy Truck Modelling and Validation Concerning Handling. AVEC'96, International Symposium on Advanced Vehicle Control at the Institut für Kraftfahrwesen Aachen, 24th-28th of June

4 **Burckhardt, M.** (1993) Fahrwerktechnik: Radschlupf-Regelsysteme, Vogel Buchverlag Würzburg, ISBN 3-8023-0477-2

5 **Hoepke, E., Appel, W., Brähler, H., Dahlhaus, U., Esch, Th., Gräfenstein, J.** (2000) Nutzfahrzeugtechnik, Grundlagen, Systeme, Komponenten, Vieweg Verlag Braunscheig, ISBN 3-528-03898-5
6 **Grünberg, H.** (1999) Untersuchungen des Radradienverhältnisses und des Schlupfaufkommens als Beitrag zur Fahrsicherheit, VDI Fortschritt-Berichte, Reihe 12, Nr. 378, VDI-Verlag Düsseldorf, ISBN 3-18-337812-4

Brake performance monitoring for commercial vehicles using estimated tire–road friction information

T DIECKMANN
WABCO Vehicle Control Systems, Hanover, Germany
A STENMAN
NIRA Dynamics AB, Linköping, Sweden

SYNOPSIS

The idea of identifying slippery road conditions to improve safety is rather old, but still no commercial system is available. The *brake performance monitor* (BPM) is a new approach for tire/road friction detection for commercial vehicles.

The BPM uses only information from known wheel speed sensors used by ABS or EBS (Electronically controlled Braking System) and other information available on the vehicle's CAN bus. No additional sensors are necessary. It is the target of a common project between WABCO [9] and NIRA Dynamics [8] to prove the reliability of this function. Afterwards an integration of the algorithms into the EBS is intended.

A basic approach from NIRA Dynamics is adapted to commercial vehicles together with WABCO along with the European research project CHAUFFEUR II [7]. The output of the BPM shall allow the trucks in automated driving mode to react like a human driver and e.g. increase following distances on slippery road surfaces or reduce vehicle speed in dangerous situations. However, the system can contribute to ACC systems (adaptive cruise control) as well or deliver input to an internet-based map with dynamic data on road conditions, to mention a few other applications of the technology.

1 INTRODUCTION

Future driver assistance systems like *ACC* (adaptive cruise control), *Collision Warning*, *Vehicle Stability Control* or *Platooning* will be able to perform better, if they have information about the available maximum friction between the vehicle and the road surface. Knowing the current braking capability is crucial when selecting speeds, choosing distances or automatically applying brakes.

WABCO as a supplier of vehicle control systems for commercial vehicles has initiated a common project with NIRA Dynamics, an R&D company focusing on signal processing and control for vehicle dynamics applications. The project goal is to create a Brake Performance Monitor (BPM) that, even during normal driving, allows the prediction of the current deceleration capability. The approach taken here is to adapt and extend an existing friction estimation algorithm for cars to commercial vehicles.

The system does not require any additional sensors than what is already available on a modern truck equipped with ABS/EBS brakes. The tire-road friction is estimated using standard CAN bus signals (such as wheel velocities, engine torque, engine speed) and advanced sensor fusion methods. By monitoring quantities related to the so-called slip curve of the tire, relevant friction information is derived efficiently and accurately on-line.

The friction estimation algorithm is active during all driving modes including ABS/EBS braking, spin/stability control and normal driving. It also features automatic tire calibration, so no driver intervention is needed.

Generating a μ-level estimation or even a warning signal is one part of the task, making the system robust is another. NIRA Dynamics already has know-how in μ-detection for passenger cars. The co-operation with WABCO will guarantee quickest progress in adapting BPM to truck applications.

Prototypes of the brake performance monitoring system are currently under evaluation at WABCO and other partners of the CHAUFFEUR II consortium. Within the project frame it is intended to find out about the abilities and limits of this technological approach. The basic feasibility of the concept has been verified, current test results show the ability to detect slippery road conditions. Future work therefore has to identify robustness.

The outline of this paper is as follows: Section 2 describes the technical solution, while Section 3 describes the basic principles behind the friction estimation algorithm. Section 4 provides experimental results. In Section 5, finally, the conclusions are drawn.

2 TECHNICAL SOLUTION

The prototype device of BPM is a stand-alone unit connected to the vehicle CAN bus. However, the final solution for a BPM would, of course, be an integration into ABS or EBS with no additional hardware. To support a later transfer into an existing control system, the use of an integer based processor is kept in mind during the development phase.

The output of BPM indicates low-µ conditions. Furthermore, a relative BPM-value is given, that adjusts individual braking capabilities according to load and vehicle configuration. Test runs were carried out, where several parameters have been varied to study their influence on deceleration capabilities. The results were then integrated into the BPM.

The function is based on monitoring a tire property called "longitudinal stiffness" that describes the amount of force the tire transmits at a given slip value. However, the slip values investigated here are very small in a region of only 0.01% (!). The tire "stiffness" decreases on slippery surfaces. This effect has been proven for passenger cars several year ago [2]. For commercial vehicles it exists as well and the magnitude of this effect will be investigated in relation to disturbing influences from various other parameters.

Fig. 1: BPM hardware prototype

2.1 BPM Input
The BPM utilises information from existing ABS/EBS wheel speed sensors and engine data already available on the CAN bus. This implies that no additional sensors are needed.

Wheel velocities
Wheel velocities of all four wheels are taken into account. At each vehicle side (left and right), the slip of the driven wheel (rear axle) is calculated by taking the non-driven front wheel velocity as reference (zero slip assumption). The standard message according to SAE J1939 "high-resolution wheel-speed" is used to transmit wheel-speed information for calculating slip values with a resolution in the range of 0.01%.

Engine torque and speed
Measurements of engine torque and speed are needed to calculate traction forces at the driven wheels. The information is taken out of existing CAN bus messages. For use within CHAUFFEUR II there will be two different versions of the BPM: One version will be used on generally standardised commercial vehicle CAN bus following SAE J1939 specification, the other version will be able to work on the DaimlerChrysler-specific IES bus.

Rear axle load
One main influence on the tire stiffness is the vertical load on the driven axle. The tire stiffness changes as a function of the length of contact patch between tire and road. Load can influence tire stiffness in a similar way and magnitude as the µ-level between road and tire. So it is necessary to measure the amount of load and to compensate for that influence to obtain a reliable friction estimation.

Therefore a WABCO ECAS (Electronically Controlled Air Suspension) system is installed in all BPM-equipped trucks. The ECAS unit transmits the axle load information using a standard SAE J1939 message.

Other
A few additional messages and vehicle status flags are also incorporated to improve BPM performance. State information from ABS/EBS is used to block algorithm execution during braking (since there is no reference for slip calculation available then) or when the clutch is engaged. Furthermore it is checked whether or not ABS becomes active during braking.
If available, ambient temperature will be taken into account to add an additional plausibility check for low-µ warnings, thereby making the output signal more robust.

2.2 BPM Output
The BPM outputs the following information on the CAN bus:

Relative Brake Performance
This value describes the deceleration capabilities of the vehicle relative to a "normal standard". The decrease in braking capabilities due to load is taken into account here. Thus for full load of the truck, the output value will be in the region of 90%, compared to 100% when empty. The solo tractor will also only have about 90% due to limited brake pressure.

In a future step more detailed influences can be taken into account here as well, for example tire properties (if known or sensed), the location of the center of gravity on the trailer (if known or sensed) and similar inputs.

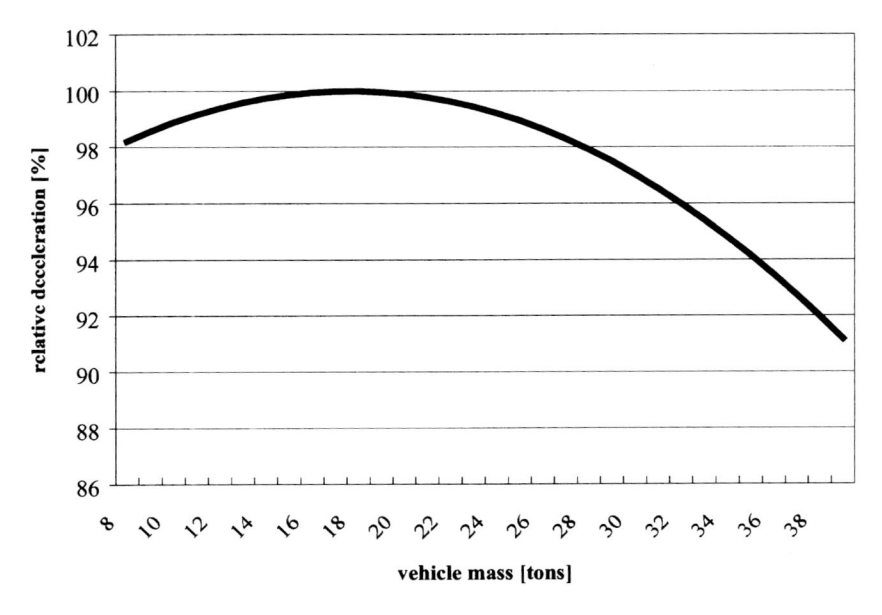

Fig. 2: Relative deceleration capability as function of load

Road friction
This is the estimated value for maximum friction available between tire and road. It does not describe the currently applied friction, but the maximum value available if needed (e.g., for emergency braking).

Currently there is a low-µ warning signal, that is activated when the slip slope falls below a threshold value. Friction levels below roughly 0.4 can be detected and a warning can be generated even during steady state driving conditions. The target is to achieve a resolution of about three classes for µ, high, medium and low. A continuous resolution of the available µ-level will probably not be possible due to real world "noise" conditions (different roads, tires and so on).

Road Friction Confidence
The confidence value given here indicates if the road friction is based on a good variety of input data, delivering a good estimation, or if rather fuzzy input data contributes to a high degree of uncertainty for the friction estimation. If this value is low, the road friction value should be used with care in the vehicle control algorithms.

2.3 Use of outputs

Distance and Speed
The output information of the BPM can be used in various ways: The distances between the trucks within a platoon – independent of whether it is CHAUFFEUR or ACC – can be adapted to the friction level. Any human driver would do it naturally and so for technical systems this is

important from a liability point of view as well. With BPM the vehicle-based system can "learn" to incorporate environmental conditions into its control strategy.

Another idea is to adapt the speed of the truck to environmental conditions, especially for a "leading truck" of a CHAUFFEUR II platoon. The leading truck has a special responsibility for all the other trucks linked to it. As a consequence e.g. the platoon speed could be reduced with reduced µ-levels.

Alternatively automated driving mode can be stopped and responsibility can be given back to the driver when slippery road conditions are detected. Another possibility is to distribute the information on low-µ conditions to other vehicles via radio link or cell phone network. Moreover, ABS and vehicle stability control systems can be conditioned to current friction levels thus reducing stopping distances and increasing safety. These are just some ideas and probably there are much more applications that can make use of this feature.

3 PRINCIPLE OF µ-ESTIMATION

As mentioned earlier the BPM system estimates the available tire/road friction on-line by monitoring quantities related to the slip curve. The slip curve describes how the wheel slip s (the relative difference in speed between the driven and non-driven wheel) relates to the utilised traction force, see for instance [1]. It is well known that this relation changes with the road surface, see Fig. 3 which shows (artificial) examples of slip curves for asphalt, gravel and ice.

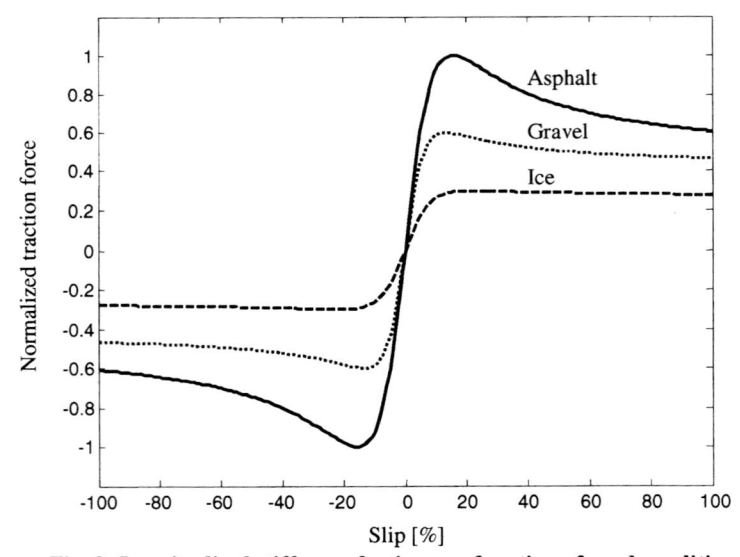

Fig. 3: Longitudinal stiffness of a tire as a function of road condition

In the BPM the slip curve is monitored in three distinct operating modes; *normal driving*, *ABS braking* and *stability control*.

3.1 Estimation of Tire Stiffness

During normal driving the wheel slip is small, say $|s| < 3\%$, so there is approximately a linear relation between normalised traction force μ and slip, i.e.,

$$\mu = k \cdot s$$

In this mode the estimation routine works by continuously estimating the so-called *slip slope* or *longitudinal tire stiffness* k (see, e.g., [3]) and translating this estimate to a friction value; the smaller the stiffness, the lower is the friction. A problem however, is that the tire stiffness k is highly tire dependent. Therefore the system is complemented with an automatic tire calibration routine that adapts to the currently used tires, see Section 3.4 below.

3.2 ABS and stability control modes

ABS braking and stability control actions gives slip and traction force measurements near the peak of the slip curve. Since the peak values reflect the maximum available friction for the surface in question, this yields that it is rather straightforward to estimate the maximum available friction in these control modes.

3.3 Influences (load, tire, tire pressure)

Unfortunately not only the µ-level between tire and road influences the tire stiffness, but also other factors. For example any parameter that changes the length of the contact patch between tire and road changes the stiffness value and, of course, the tire has a very large influence.

The load influence is compensated by use of the ECAS axle load information as described above. A pressure sensor is added to the standard rear levelling air suspension. With the pressure information the load can be calculated. The correction was derived empirically.

Tire pressure is constantly at 8 bar in commercial vehicles in Europe. Unlike in passenger cars there is no load depending adaptation of the tire pressure. So effects due to pressure variation are not existing (unless of course a tire is damaged and loses pressure). Such an event could, in fact, disturb BPM. However, there will soon be systems available on the market to monitor tire pressure, e.g. like WABCO's IVTM (Integrated Vehicle Tire Monitor).

One remaining influence comes from the tire itself. Tread depth and geometry have a large influence on longitudinal stiffness. For example there are large differences between new and worn tires, and summer and winter tires. The BPM will automatically detect the current tire properties and adapt to it.

3.4 Self learning and calibration

One of the most important tasks to be solved within the project is self-learning and calibration of the system to allow adaptation to changing properties (like tread wear) while on the other hand reliably detecting dangerous road conditions. The basic principle adopted here is to use a slowly time-varying threshold that is continuously compared to the estimated tire stiffness. From this comparison it is possible to get an estimate of the available friction. The adaptation of the threshold is tuned so that it would adapt to a new set of tires within half an hour of normal driving.

3.5 Relative BPM value

A relative BPM value is calculated by utilising total vehicle mass. Contrary to the axle load on the driven wheels, which is measured by ECAS, the total vehicle mass is estimated. Here existing bus data is used as well. The basic idea is to compare measured vehicle acceleration to utilised engine torque. Following a function empirically derived from measurements, the relative braking performance will be calculated as a function of current total vehicle mass (both truck and trailer).

4 TEST RESULTS

Several test runs and experiments have been performed in order to investigate the feasibility of the BPM concept.

4.1 Slip slope estimation

The first consideration is the tire stiffness (slip slope) estimation. Figure 4 shows the result from a test run on asphalt and wet basalt stones (for simulating slippery road conditions). An empty (unloaded) truck was driven three laps on an oval test track passing over a braking area consisting of wet basalt stones. The triangles located at the time axis indicates the transitions from asphalt to blue basalt.

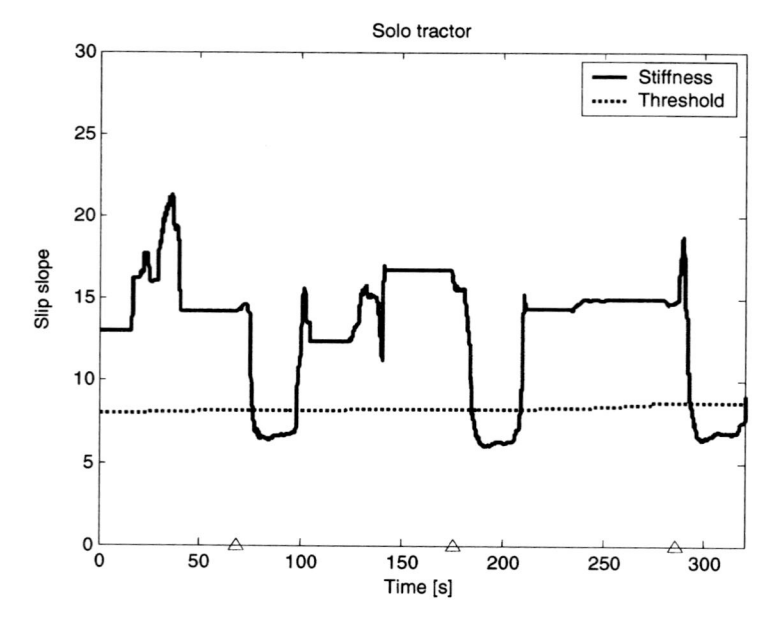

Fig. 4: Estimated tire stiffness when driving on dry asphalt and wet basalt stones.

The solid line is the tire stiffness estimated by the BPM. One can clearly see significant changes in stiffness between the two surfaces. The dotted line is the adaptive threshold mentioned in Section 3.4. The BPM computes its friction estimate by comparing these two quantities.

4.2 General braking capabilities

Although significant work has been carried out to improve ABS over the years, general figures on braking capabilities for different road surfaces are rare. Therefore a measurement program was started to collect these figures and some interesting effects were detected:

The following figures represent an overview of observed steady-state deceleration with various heavy trucks and tractor/semi-trailer combinations:

Surface	Max. deceleration
• Dry	7 – 8,5 m/s^2
• Wet	5 – 6,5 m/s^2
• Wet cobblestone	1 – 2 m/s^2
• Ice	< 1 m/s^2

Reproducibility proved to be good on a short term base, however especially on wet surfaces there is hardly a long term reproducibility. Differences up to 30% in maximum deceleration occurred on wet surfaces when measurements were repeated after 4 months (same truck, track & tires).

Load was identified as another influence. In measurements the maximum deceleration level dropped by 10% for a fully loaded vehicle equipped with EBS. As expected, best decelerations were obtained without load. However, solo tractors are handicapped due to a limited brake pressure that prevents a frontal roll-over. On one hand it stabilises the truck, on the other it decreases deceleration level. These influences are taken into account in the relative braking performance value.

Tire pressure, although definitely important for driving safety, did not prove to be relevant for braking capabilities. A reduction to 6 bar delivered similar maximum decelerations as the original 8 bar.

Apart from the road surface condition the tires are the most important influence on braking capabilities. It is inherent part of the approach described here, that the tires are part of the measurement system itself. The slip-slope is part of the tire/road interaction and it contains information on the friction level. So it is expected that the BPM will automatically take into account, that for example winter tires deliver higher forces on snow than other tires. A road surface condition sensor would never be able to do so.

4.3 Influence of Load

As mentioned in Section 2.1 the load is taken into account via an external axle load signal provided by the ECAS. It is interesting to see whether or not this load signals carries enough information to compensate for a change in load. Figure 5 shows a same kind of test run as in Figure 4, but with a trailer of weight 40 tons attached to the truck.

We see that the stiffness curves in the two figures have roughly the same shape and are located approximately at the same levels, which indicates that the load compensation works quite satisfactory in this case.

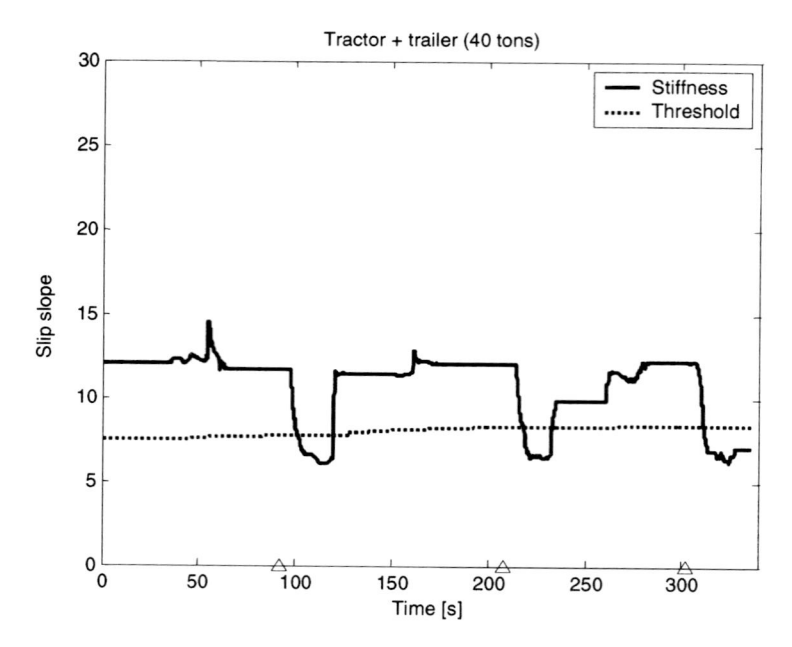

Fig. 5: **Same test run as in Figure 4 but with load.**

4.4 Results of winter tests

This section contains results from winter tests performed at real low-μ conditions in northern Finland. The test vehicle was a Mercedes-Benz ACTROS truck equipped with Continental HDW winter tires. The testing procedure was similar to those described in sections 4.1 and 4.3: The vehicle was driven over three laps on a circular route passing over a braking area consisting of five parallel lanes with different surfaces (packed snow, ice, asphalt, ice and packed snow). Outside the braking area the road was covered with snow.

Figure 6 shows the result from a test drive with a truck and a medium loaded trailer making transitions between packed snow and ice. As before the small triangles at the bottom of the plot act as markers for the transitions between the two different surface types.

We see that the slope estimator is able to track the changes in friction and that a correct binary friction information is produced when the slip slope is rated using the threshold as borderline between high and low friction. The increase of the threshold value at time t=180 is caused by a stability control regulation that provides the system with more knowledge about the actual friction level.

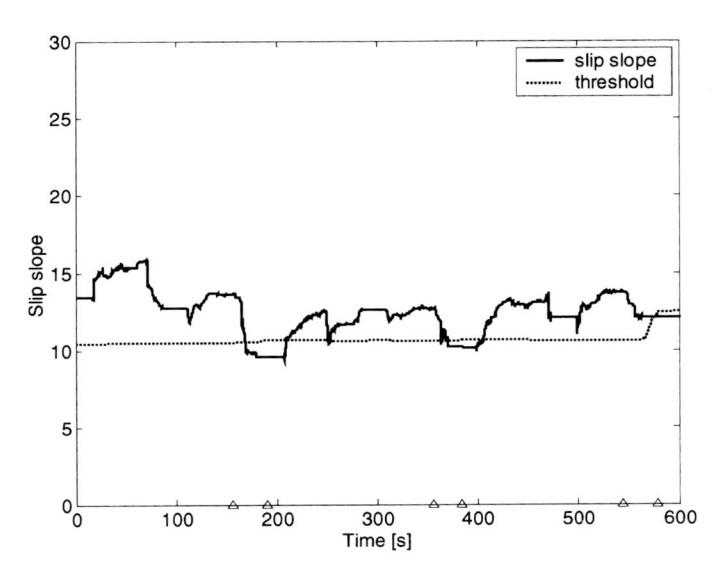

Fig. 6: Test drive with a medium loaded trailer and transitions snow/ice.

Figure 7 shows test results from a similar experiment but with transitions between snow, asphalt and ice (in that order).

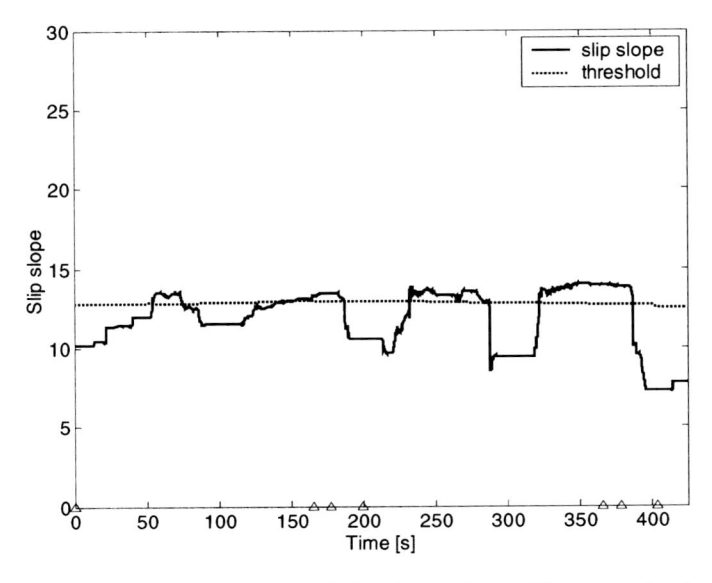

Fig. 7: Test drive with medium loaded trailer and transitions snow/asphalt/ice.

As in the previous experiment we see that the slip slope estimator is able to correctly track changes in surface type but that the slip slope level is changing rather much up and down on the snow road outside the braking area.

5 CONCLUSION

The BPM is an attempt to obtain an estimator for the µ-level between road and tire with no extra sensors. The layout of this prototype system offers the potential to include the software algorithms at a later stage directly into the ABS/EBS.

The algorithms will use wheel slip together with traction forces to estimate µ-level. The use of NIRA Dynamics' signal processing expertise together with WABCO's experience in vehicle control systems for commercial vehicles shall allow for best results in a given cost and time frame.

The general possibility of detecting low-µ conditions with the approach described above is proven. Since the exact µ level depends on several external factors such as tire type, tire wear or tire temperature it is clear that the method is best suited for tracking changes in friction conditions. Remaining future tasks comprise improved self-calibration, and an increased overall robustness. For this purpose more tests are currently prepared.

6 BIBLIOGRAPHY

[1] E. Bakker, L. Nyborg, and H.B. Pacejka, Tyre modelling for use in vehicle dynamic studies, SAE paper 870421, 1987

[2] T. Dieckmann, Assessment of road-grip by way of measured wheel-variables, FISITA, London, 1992

[3] F. Gustafsson, Slip-based estimation of tire-road friction, Automatica, 33(6): 1087-1099, 1997

[4] H. Grünberg, Untersuchung des Radradienverhaeltnisses und des Schlupfaufkommens als Beitrag zur Fahrsicherheit, Fortschritt-Berichte VDI, Reihe 12, Nr. 378, 1999

[5] W. Pasterkamp, The tire as sensor to estimate friction, Delft University Press, ISBN/CIP 90-407-1538-6, 1997

[6] B. Witte, Stabilisierung der Gierbewegung eines Kraftfahrzeugs in kritischen Fahrsituationen, Fortschritt-Berichte VDI, Reihe 12, Nr. 254, 1995

[7] www.chauffeur2.net

[8] www.niradynamics.se

[9] www.wabco-auto.com

From tyre contact patch to satellite

A R WILLIAMS
Consultant, Solihull, UK, formally of Dunlop Tyres Limited, UK

1. INTRODUCTION, AIMS AND OBJECTIVES.

From a vehicle manufacturer's point of view, the fact that the Highway Engineer has greater control over the performance of the vehicle in many of its aspects of performance than the Vehicle Designer and certainly more than the Tyre Engineer may be a surprise to many. Such performance factors as friction in the wet and dry conditions, handling and stability, comfort, noise generation and even rolling resistance/fuel economy can be very much controlled by the condition and texture of the road surface.

Once this is realised, there is the challenge to isolate the various textural levels of a road surface and their relationship to both the tyre and vehicle performance. Once this has been achieved, then it is possible to consider ways in which the specification for a road texture within its various texture levels may be related to tyre properties and to even suggest how the tyre may be redesigned to anticipate the ideal road surface specification. Further, it may be possible to consider the whole system of tyre suspension and electronic systems associated with vehicle control and comfort such that they also respond appropriately to the basic information that exists between the tyre and road surface.

Indeed the emphasis on approval testing of tyres and legal requirements could more appropriately be applied to road surface textures in order to more effectively compliment the tyre and vehicle, which should be considered as a system.

This paper attempts to summarise the known information, reviews the known research work that is at present underway and tries to foresee how the application of such knowledge may be applied in the future, together with the consequences for the "traditional" tyre manufacturers approach.

2. THE HISTORY OF TYRE/ROAD STUDIES.

It cannot be acceptable that a developed nation's road transport system becomes, due to lack of applied expertise in the highway industry, unsafe from a driving point of view, particularly under wet conditions. Indeed the World Health Organisation when considering road accidents as a whole has indicated that deaths from road traffic accidents could become the second greatest single cause of premature death in the early years of this century. Many of these accidents will be in developing countries where the mix of traffic and rate of change may require actions different to areas with a fully developed road transport system.

The European situation highlights the current level of variability. Portugal for example has Europe's most dangerous roads with a death rate of four times that of Britain, which is considered to be the safest in the EU.

Britain recorded 64 fatalities per million population in 1995, whereas this figure for Germany is 116, 145 for France and 217 for Portugal

A total number of 44,195 people were killed in road accidents in the EU during 1995. While this figure is a 14% reduction on the 1989 figure despite a sharp rise in traffic density, the cost to the community at large is a sad reflection on the overall technology of the road transport industry.

In many ways the history of the road is not dissimilar to that of the tyre in that it was first created to improve comfort and give load support. However, the history of roads goes back to the times of the ancient Egyptians where evidence of grooving in paved surfaces to improve traction has been found. Today we expect roads not only to provide support for the vehicle, which is an increasing challenge with the evolution of heavy trucks and their tyre/wheel configurations, but also to provide skid resistance, noise reduction, and spray reduction. The use of natural tar, and over the last generation's bitumen, to bond the prescribed grading of aggregate, sand and filler has been partly in commercial competition with cement bound surfaces. Only more recently have steps been taken to enhance the performance of the binder through the addition of appropriate additives. In many ways this approach parallels the compounding of base polymers in the tyre industry.

While the growth of the worldwide road network continues, there is an increasing realisation of the cost of maintaining such infrastructure and the need to increase the efficiency of a given road network. It is therefore seen as essential that the specification for roads and road textures is appropriate to their use. Indeed the progress and application of road technology may eventually limit the growth of the road transport industry as a whole.

It is perhaps a surprise to realise that the study of how aggregates pack together and how their grading may influence the surface texture and bulk durability was only seriously undertaken some 30 years ago, and originated from earlier studies of how coal packed into railway wagons. It may also be a surprise to know that the research partially pioneered by Dunlop resulted in the first designed wearing course materials for roads that optimised the tyre to road interactions, (1).

In the late 1960s the Tyre Engineer was fully aware that the road surface macrotexture supplemented the tread pattern in terms of bulk water removal and therefore aided the high-speed wet handling of vehicles, (2). At that time the macrotexture was being promoted primarily as aiding friction through bulk hysteresis losses within the tread compound. Indeed it was not many years before this that racing tyres were designed to have the highest possible resilience and it was only the pioneering work of Dunlop Research Centre in Birmingham and Professor Tabor at Cambridge University that led to a complete reversal of this concept, (3).
At the same time, macrotexturally rough surfaces with a highly polished microtexture, such as rounded pebbles were being widely adopted on proving grounds of the tyre and vehicle industry world-wide. Today such surfaces are recognised for their ability to differentiate tyres with siped tread patterns and natural rubber and synthetic polymers in the tread compound. Certainly at that time the geometric micro, macro and megatexture was not fully understood and still today requires further communication and practical application of the knowledge gained from recent years of research.

The availability of the scanning electron microscope in the 1960s allowed the aggregate microtexture to be studied as a function of the petrographic and mineralogical characteristics, the degree of traffic simulated polishing and the effect of climatic changes. We now know that the microtexture on the micron scale has a very significant influence on wet friction . The level of aggregate microtexture also governs the rate of abrasion of the tread compound. These studies together with work on aggregate gradings led to Dunlop laying road surfaces in many parts of the world under the trademark "Delugrip" Road Surfacing Materials, (4).

3. FACTORS IN THE RELATIONSHIP BETWEEN TYRE AND ROAD SURFACE.

The work of the last 25 years has shown that the road surface can dominate performance of tyre and vehicle in terms of wet and even dry handling, comfort and stability, noise generation and more recently tyre rolling resistance and therefore fuel consumption and also handling and stability. While this fact has now been realised for a number of years, it is surprising that as yet there are no appropriate standards (nationally or internationally) for monitoring the full texture range of the road in the new and in-service state. It was some 20 years ago that Dunlop first specified the ideal road texture configuration to optimise the tyre/road interaction as seen at that time. It is interesting to note that even today such a specification remains valid, (5).

The pioneering Dunlop road research led to the appreciation of the critical nature of aggregate microtexture on the micron scale, in terms of optimising wet roadhold without excessive abrasion to the tread compound, and the combining of this with the design of a drainage efficient macrotexture that optimised the tyre to road contact area. Indeed this is the same principle behind tread pattern design. The only very significant difference is that tyre tread patterns are designed to remove the water at all vehicle speeds, yet roads generally fall into specific speed categories. It was possible to configure a drainage theory that would optimise the drainage capability of the aggregate geometry with the maximising of the contact area, (6).

However, more significant for the tyre industry was the realisation that the microtexture levels could be optimised for wet friction and the macrotexture levels are minimised to reduce rolling resistance of the tyre. This knowledge was then able to be converted into the further understanding of the dynamic characteristics of the tread polymer. Through the study of road surface textures, Dunlop were able to dispute the then widely held view that increasing the tread polymer glass transition temperature improved wet grip performance. This had the contravening affect that while improving wet grip, it inherently increased the tread component contribution to the tyre's rolling resistance.

A new wet grip/rolling philosophy was developed based on measuring the dynamic properties of tread compound polymers in the laboratory under conditions approaching those existing at the tyre/road interface, both under rolling and wet sliding conditions, (7). It was found that under rolling conditions low values of loss modulus (E*) and loss modulus divided by the complex modulus squared, measured at relatively low frequencies (up to 120Hz) and relatively low temperatures and high values of loss factor measured at relatively high frequencies (in the range of 50kHz to 1MHz) and relatively high temperatures, in the range of 100-150 degrees C. were related to tyre rolling resistance and wet braking performance. This new philosophy enabled a new polymer to be developed in conjunction with Shell, which contradicted the established practices in so far as it enabled tyre rolling resistance to be reduced, while improving wet grip performance. This concept became the forerunner of a new generation of relatively high 1,2 butadiene content solution styrene butadiene co-polymers, which are commonly used today.

This example was perhaps the first of how tyre technology can be influenced by knowledge of the tyre to road interaction. A second perhaps more significant example will be given later in this paper.

4. CURRENT HIGHWAY ENGINEERING ACTIVITY.

The World Road Association (PAIRC) has global representatives of the scientific community of Highway Engineers working together to improve our road infrastructure and dissipate knowledge gained. One of the working committees within this organisation is that on road textural characteristics. Participation in this work gives the Tyre Engineer considerable information regarding the condition of roads world wide. Attendance also allows the opportunity to influence some of the thinking of Highway Engineers to the benefit of the tyre and vehicle industry.

The increasing availability of monitoring equipment for wet road friction has been recognised in studies promoted by the PIARC Technical Committee on Road Textures in order to harmonise the values of frictional measurements for the wet road performance assessment, (8). This has promoted a globally acceptable approach to the harmonisation of road surface monitoring for wet friction relatively independent of the test facility being used. Additionally it has promoted the availability of in-service testers for road wet friction monitoring.

It is interesting to note that Britain was the first country to establish and publish limits for wet road friction based on the site severity. It is hoped that this practice will spread through out all regions with a developed road transport system.

A similar programme of work has just been completed in order to monitor the lateral and longitudinal unevenness of roads, so important in terms of vehicle handling stability characteristics. Additionally road unevenness can influence wear on the vehicle and driver fatigue, and energy losses within the system that may equate to fuel efficiency. Longitudinal unevenness equates also to rutting and increases the risk of aquaplaning. Today many sophisticated measuring systems are able to measure longitudinal and lateral unevenness within the traffic flow, however even with modern data logging and analysis equipment their results do not always correlate. The experiments which have been conducted by PIARC in Japan, the USA and Europe are targeted for the benefit of Highway Engineers. This may be a criticism of the work in so much as the Vehicle Engineers should also be aware of the parameters involved and have a strong say in what constitutes a satisfactory road surface in terms of lateral and longitudinal unevenness.

The results of this study, will allow the availability of data bases for both Highway and Vehicle Engineer uses. The results should not only create appropriate limits of megatexture and unevenness from a highway engineering point of view, but also should enable the vehicle manufacturing industry to participate in any discussion in order that appropriate limits, as viewed from the vehicle manufacturing aspects, to be implemented on a global scale.

More recent studies on the macrotexture configurations pioneered by Dunlop in conjunction with the UK Transport Research Laboratory, (9), have involved the measurement of the dynamic contact patch pressure distribution between the tyre and various forms of road macrotexture. These studies have included dynamic contact pressure distribution measurements on the new generation of road textures that abide closely to the concepts discussed previously, and also on a series of replica surfaces with simplified geometric configurations that can be understood from a mathematical modelling approach. These surfaces and the modern range of wearing course surfaces have been evaluated within a dynamic rolling road facility to measure the tyre characteristics of noise, rolling resistance and comfort. The results have been used to consider again the influence on the tread compound within the tyre contact patch of the contact pressure distribution in terms of dynamic characteristics of tread compound formulation. Further work is required to appreciate the results but it seems that additional considerations will be needed in the measurement of the dynamic characteristics of tread compounds to accommodate the high frequency and very high strain rate experienced by the tread rubber in the dynamic contact patch environment.

The availability of replica road surfaces representing simple geometric shapes has enabled a greater theoretical understanding of the pressure distribution within the contact patch and the void areas between that constitute the drainage potential of the surface. Clearly it is not possible to optimise the tread pattern design without some basic specification for the road surface texture. The application of these studies by Highway Engineers to improve the standards regarding macrotexture configurations could assist the Tyre Designer in designing more efficient tread patterns.

Both in terms of rolling resistance resistance and noise, the carefully selected theoretical/geometric surfaces appear to span the tyre range that has been evaluated. This is important in using these surfaces in order to create a theoretical understanding of both rolling resistance and noise and in being able to perform indoor tests on both surfaces which are mathematically explicable and whose geometry covers the whole spectrum of practical surfaces in use and tyre variables.

Looking further ahead, the availability of these surfaces will be a critical part in a study to mathematically predict the external noise generation of a rolling tyre on real road surfaces.

The conclusions from these studies have enabled pilot scale road trials of new forms of road texture to be laid for further investigations including wet friction tests. The conclusions from this study, which will have an influence on future highway engineering, is that densely packed gap graded, angular aggregates produce good wet friction by improving the drainage within the tyre/road interface. Additionally the amount of contact area between a tyre and road surface can be predicted from numerical methods obtained from profile measurements, and this has allowed a numerical model to be divided into contact and none contact parts. The roughness of the contact surface determines the level of contact and the pressure distribution between the tyre and road surface and indirectly determines the potential volume for water drainage beneath the contact interface. In this way the contact roughness and level combine with the microtexture to control wet friction. Ideally the tread compound properties need to be measured at high strain rates during development of tread compound formulations and tyre properties need to be measured on surfaces that simulate real roads. It was also noted that high contact roughness, texture depth and unevenness produce higher rolling resistance values, the latter two needing to be controlled. Reducing the maximum stone aggregate size and texture, particularly in the higher texture ranges, will reduce noise levels significantly.

5. THE APPLICATION OF MODELLING TO TYRE AND ROAD.

The modelling of the tyre/road interface may be considered the current frontier of tyre modelling. The complexity of introducing real tread patterns and real road surfaces, with water present, in a rolling, braking and driving situation is indeed complex. The introduction of real textured surfaces or indeed those of a geometrically reproducible theoretical configuration adds another order of complexity to the models. However, if true rolling resistance of the tyre is to be predicted, if true external noise generation is to be predicted, and indeed if a more accurate prediction of vehicle internal noise through tyre/road, suspension and body is to be made, then this problem will need to be solved. A further degree of complexity arises when introducing water and trying to predict wet friction characteristics over the speed range. Here the elastohydrodynamic lubrication phenomenon dominates and finite element software packages that successfully handle this double complexity of tread pattern, road texture and water are only just available.

Success has recently been achieved in creating the complex model to predict driver monitored internal noise within the vehicle originating from the road through the tyre, suspension and body. This has used a FEM approach, but difficulties connecting the road, tyre, wheel and suspension with the body remain. Laser scanning vibrometry has been used to investigate the modal behaviour of the tyre. As the excitation frequency increases, the tyre can no longer be represented as a collection of springs and dash pots. The dynamic responses become like a composite toroidal shell structure. To compare the theoretically derived modal behaviour of a tyre with it operating in a rolling condition, being excited by such inputs as road texture has been progressed through the use of laser scanning vibrometry, (10). It has been stated, (11), that the first radial natural frequency for the standing wave is dominant in noise transmission from the road and that the torsional resonance's are important to vibration transmitterability, and also in noise transmitterabilty above 60Hz.

Studies of the complete modal behaviour of the whole tyre, i.e. two sides, together with the tread band, their relationship and their understanding to tyre design characteristics remain to be further investigated. Finite element methods are not applicable for the full external noise prediction theory and wave propagation methods are beginning to show success.

Energy loss models related only to each component of the tyre and therefore predicting the energy losses in the complete tyre are just beginning to utilise the surface inputs,

The relationship of tyre forces and moments to the conditions placed on the tyre by rate, temperature, conditioning and road texture have been studies within the European "Time" project. This industry/customer joint project has already indicated that the influence on maximum lateral force for the same tyre on different surfaces under dry conditions can be 10% and that the influence on cornering stiffness and aligning stiffness of the same tyre can be up to 13% for cornering stiffness and 21% for aligning stiffness. The research has also indicated that the track surface can influence the characteristic curves, but it does depend strongly on tyre design variables, (12). These observations are of course critical when considering the mathematical modelling of a tyre in such work software packages as Adams etc. to predict whole vehicle performance.

The remaining area of tyre/road modelling prediction recently worked on is that of comfort. Here it has been possible to investigate the impact of longitudinal force variation on the rolling tyre/wheel system through the suspension while running on surfaces of dissimilar macrotexture within the laboratory. It has been possible to demonstrate real time results and mathematically predictable results for the longitudinal response characteristics of the tyre under various input conditions, (13).

6. HOW TYRE DESIGN MAY RESPOND TO IMPROVED ROADS.

There are three approaches to answering this challenge. The first is the example of the French Road Authorities who in conjunction with their national tyre and vehicle industry established in the early 1990s a joint research programme. This approach may truly be termed vertical engineering and is producing significant advances in French roads and also has the valuable spin-off in terms of sharing knowledge between the road, vehicle and tyre making industries, (14).

The second approach is exampled by one company in Europe in extending it's own product, the tyre, through to electronic braking systems and suspension manufacture Clearly this has the advantage in terms of balancing the responsibility of each component relative to its optimum performance within the system and may have commercial advantages. It maybe that some vehicle manufacturers see simplification in terms of purchasing, specification and possibly of manufacturing through such an approach.

The third approach is one of enhancing the tyre design from the greater understanding of the tyre/road contact patch. This approach may also see the utilising of electronics within the tyre technology. Perhaps the pioneering work in the introduction of the "Warnair" deflation warning system that uses the electronics in the ABS system to monitor the change in tyre rolling radius and relates this value to tyre inflation pressure. This system, which may be considered the first step in relating tyre performance and electronics, has been successfully used in many 100,000s of Toyota vehicles in the USA and is fitted to the new BMW Mini. It is expected from experience to date and test information that the introduction of tyre deflation warning systems will not only be a major safety feature, make a significant contribution to improved fuel consumption through better pressure maintenance, but also reduce the necessity to use runflat tyres. It is expected that tyre deflation warning systems will be a standard fitment on cars in the very near future.

The tyre industry has now moved on to the concept of Smart Tyre Technology, which is initially aimed at the truck tyre market. Already systems are being successfully evaluated that can monitor and log tyre inflation pressure, tyre structural temperature, tyre mileage and identification. There is evidence that the use of smart tyres on trucks will have positive cost benefits in addition to safety and fuel conservation.

The move towards intelligent tyres is an area of intense research. It is envisaged that the monitoring of the tyre contact patch continuously in order to measure the ground plane stresses and therefore the interfacial friction will enable key inputs to be fed into the vehicle control systems. With satellite systems, where
the vehicle position within the traffic environment is already known, then if this is used with the interfacial friction then vehicle control could become absolute. Such an approach will have cost benefits in reducing the number of sensors needed on a vehicle as the tyre becomes the key sensor and in terms of traffic control and safety, (15).

The potential for smart and intelligent tyres and the concept of vertical or cornering engineering is already making significant changes in the commercial structure of the tyre industry. It may be envisaged that electronics will finally break the compromises grappled with by tyre compounders for so many years.

The second fundamental change may be to the tyre itself. The current trend to run-on and run-flat tyres which add weight to the assembly ignore what may be learnt from considering the tyre design from the contact patch upwards. A new concept is about to enter production trials, (16), This concept was derived from the understanding of the contact patch. Comparing the contact patch of this new approach with conventional tyres, one can see a significant difference.

The shape of the contact patch is such as to be adaptive to the vehicle demands of roadhold, and is capable of significant run flat capability coupled with a considerable weight reduction. The improved shape of the tyre also allows for a reduced gap between tyre and vehicle body, so reducing dynamic air drag.

7. THE MONITORING OF ROADS AND THE CONSEQUENCES FOR TYRE DESIGN.

We are aware that vehicles and tyres are subjected to considerable legal performance specifications in terms of component and whole product evaluations. It is surprising that roads are not subjected to a similar range of legislation governing their properties, particularly those that are safety related, such as wet skidding resistance and environmental related factors such as noise. It is for the tyre industry and perhaps the vehicle industry to remind our highway engineering colleagues that we are in the same business and must work together in order to promote the safety and environmental satisfaction of the road transport system. This will only be achieved through the concept of vertical engineering, the breaking down of barriers of different disciplines and the working together within various research and legislative bodies representing different sides of the same industry.

It is now possible to monitor roads with equipment that does not disrupt the traffic Flow. Such equipment can evaluate surface texture and wet skidding resistance and soon will be able to evaluate rolling tyre to road noise. Such advances, together with modern data analysis techniques, will allow road surface monitoring for friction (safety), noise (environment), macrotexture (rolling resistance), and lateral and longitudinal unevenness (handling/stability and comfort).

8. CONCLUSIONS.

This paper has described possibly a unique approach within the tyre industry, of using knowledge gained of the road texture to consider how the tyre performance may be enhanced. However, it would seem unlikely that rubber compounding alone will provide the tyre with an expectable level of friction under all conditions, no matter what the design or condition of the tread pattern may be. It can be shown that the tyre can enter an area of "unsafe" friction levels on any normal journey due to the presence of variable road conditions.

To effectively reduce accidents associated with friction levels lower than can be handled by the present tyres and braking systems, even allowing for reasonable improvements in road friction consistency, will demand a wider understanding of the "vertical engineering" approach to road transport and the application of electronics to tyre technology.

Finally it is stressed that only by the co-operation of the various technologies within the road, tyre, and vehicle industry, will significant progress be made to enhance safety and reduce the environmental intrusion of our road transport system in developed areas and so promote road transport in developing countries as a viable system.

REFERENCES.

1. G. Lees, The Rational design of aggregate gradings for asphaltic composition, Proc. Assoc. Asphalt Paving Technologists, vol. 39, 1970, pp 60-67.
2. B. J Albert, J. C. Walker, Tyre to wet road friction at high speeds, Proc. Inst. Mech. Engs. 180, (2A), 4 pp105-121.
3. D. Bulgin, D. Hubbard, M.H. Walters, Road and laboratory studies of friction of elastomers, Proc. Int. Rubber Conf., London, 1962.
4. A. R. Williams, G. Lees, Topographical and petrographical characteristics of road aggregates and the wet skidding resistance of tyres, Quart, J. Eng. Geology, vol. 2, no. 3, 1970, pp 217-236.
5. A. R. Williams, Road surface texture design, Proc. Of the fourth symp. Univ. of Salford, UK, 1975.
6. R. Bond, G. Lees, Road surface bulk water drainage, Tyre Science and Technology, vol. 6, no. 2, 1978, pp125-158.
7. R. Bond, A new tyre polymer improving fuel economy and safety, Proc. Roy. Soc. London, vol. 399, 1985.
8. P.I.A.R.C. Technical Committee on Surface Characteristics, C1, Int. P.I.A.R.C. experiment to compare and harmonise texture and skid resistance measurements, 1995.
9. A. R. Parry, Macrotexture and road safety (MARS), final report, Project report PR/CE/56/58,Co67L/94, 1998.
10. A. P. Bridgewater, Modelling of tyre models for the prediction of vibration inputs at vehicle wheels, Doctor of Philosophy Thesis, Univ. of Birmingham, UK, 1997.
11. M. G. Pottinger, K. D. Marshall, J. M. Lawther, D. B. Thresher, A review of tyre/pavement interaction induced noise and vibration, The tyre and pavement interface, A. S. T. M., S. T. P., 929, 1986, pp 183-287.
12. J. van Oosten, M Augustin, R. Gnadler, H. J. Unrau, Tyre measurements, forces and moments, E. C. Research Project Report "Time".
13. D. Allison, R. S. Sharp, In plane vibration of tyres and their dependence on wheel mounting conditions, Vehicle dyn. Conf., Univ. Delft, 1998.
14. T. Christophe, Y. Dalanne, J. P. Serfass, Road/vehicle interaction, comparison of two surfacing in terms of rolling resistance, vibration, comfort and noise. P. I. A. R. C. Technical Committee on Surface Texture Characteristics, Montreal Congress, 1995.
15. J. Jessop, Electronics in tyres, Inst. Of Materials, Int. Rubber Conf., Birmingham, UK, 2001.
16. M. Tanaka, K Asano, D. Major, Development of the light weight run-flat tyre. Inst.of Materials, Int. Rubber Conf., Birmingham, UK, 2001.

Intelligent braking management for commercial vehicles

E-C VON GLASNER, H MARWITZ, R POVEL, and C KOHRS
DaimlerChrysler AG, Stuttgart, Germany

ABSTRACT

The development of electronic intelligence and the increasing knowledge of driving dynamics make intelligent vehicle systems available for series production, which substantially enhance the active safety of commercial vehicles.
Through the implementation of advanced subsystems, which can be integrated into the basic electronic braking system, it will be possible to expand the possibilities of introducing assistance systems, which help and relieve the driver from stress in critical situations.
The driver will be relieved of all duties which could divert his attention or cause severe stress. As a consequence, the active safety of commercial vehicles will be considerably increased.

1   BACKGROUND

Currently, about 70% of the total volume of goods transported in Germany is carried out by commercial vehicles (Fig. 1). According to the latest forecasts, this figure is expected to increase dramatically in years up to 2010. The commercial vehicle will therefore continue to play a dominating role as the prime mover of merchandise in Europe [1].
The reasons for this development are the well-known system-related advantages of road transport that is:
- door-to-door deliveries,
- rapid servicing of the territory,
- application flexibility and

- relatively low costs

non of which will not even come close to being matched by any other transport system. This trend will continue to experience even greater intensification with increasing European integration, which will invariably open more and more to the East.

2 INTRODUCTION

In many analyses during the past 20 years, it has been shown that operational factors affecting the driver and the vehicle such as
- traffic density,
- climatic conditions and
- road conditions

must be accepted as invariable constants. Simultaneously, criticism has been voiced in these analyses that the burdening flow of information to the driver is permanently increasing.

For safety reasons, it has therefore become necessary to offer the driver consequent relief from additional burdens while driving the vehicle. One such relief can be attained by applying intelligent systems, which substantially raise the level of active safety.

Above all, systems for increasing the active safety have been of primary importance over the years and will remain the main domain of vehicle development in the future. Such systems are equipped with high electronic intelligence and assist the driver by eliminating the decision-making process in critical situations, which exceed driver-response capabilities.

On the basis of the many analyses over the past 20 years, it has become clearly evident, that an increase in vehicle braking performance represents an important aspect in the efforts towards more active safety on our roads.

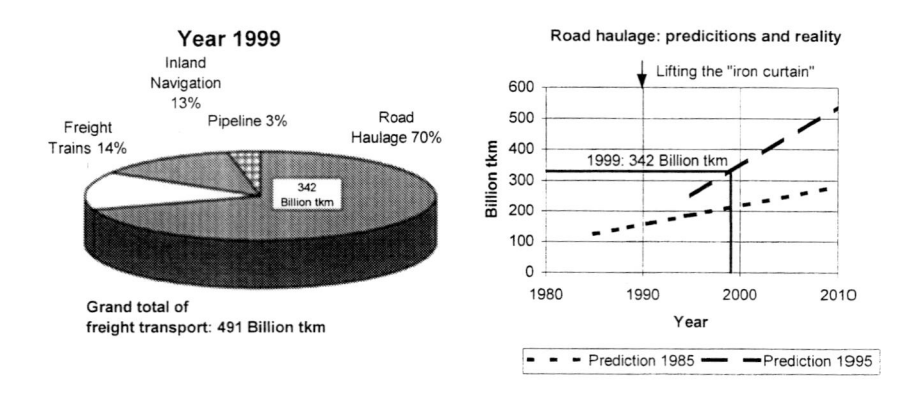

Fig. 1 Development of road haulage in Germany (predictions and reality)

In consideration of this fact, electronic braking management systems were already conceived and proven feasible in prototypes in the late 1980s.

Today's basic electronic braking system, which can be augmented with so-called "driver assist system-add-ons", represents the exemplary consequence derived from this development of the 1980s.

3 INTELLIGENT BASIC BRAKING SYSTEM WITH ELECTRONIC BRAKING FORCE DISTRIBUTION AND INTEGRATED ANTILOCK- AND TRACTION CONTROL SYSTEMS

The basic braking system consists of a two-circuit, ECE R 13-conform service braking system superimposed with higher-level electronic activation.

The brakes on the front and rear axles are assigned to pressure control loops, by means of which the pressure setpoint value determined by the controller in the electronic braking system can be converted to actual braking pressure.
The antilock- and traction control systems no longer function as stand-alone systems, but are integrated into the basic braking system logic.
Such full-electronic braking systems make it possible to continually and optimally adjust the braking force distribution of the vehicle to driving situations.

As a consequence of the rapid response of the electronic control system in comparison with conventional braking systems, shorter stopping distances can also be attained.
With the assistance of an intelligent differential control, it is guaranteed that optimum load-dependent braking force distribution is available.
This equates to
- high driving stability,
- good steering capabilities and
- short deceleration times.

Moreover, a permanent monitoring of the brake performance reserves is guaranteed.

Fig. 2 shows the configuration of an electronic braking system on a truck [1, 2]. In comparison with the conventional service braking system
- conventional brake valve,
- load-sensing valve and
- antilock system- and traction control valves

are replaced by new components.
What remains unaltered with respect to conventional braking systems are
- energy generation,
- energy conditioning,
- energy storage as well as
- the parking braking system.

The braking desires of the driver are transmitted to a central electronic unit via the brake pedal and an electric brake sensor.

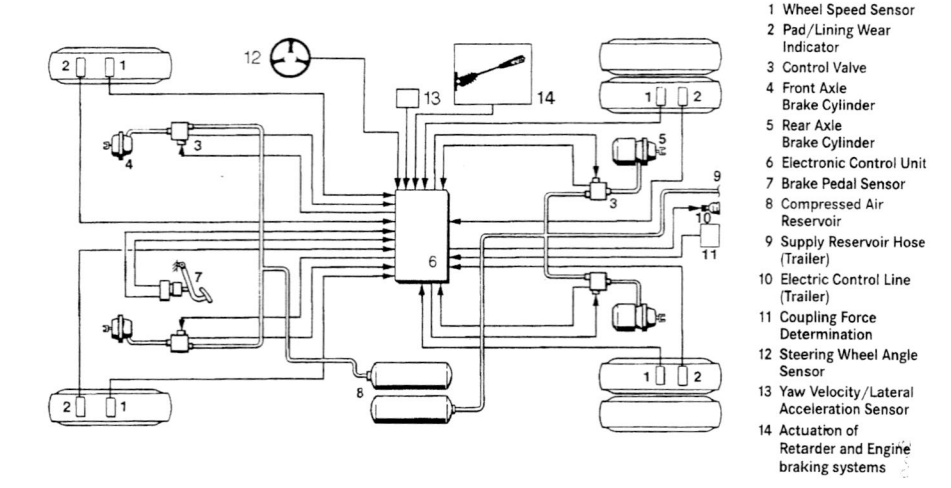

1 Wheel Speed Sensor
2 Pad/Lining Wear Indicator
3 Control Valve
4 Front Axle Brake Cylinder
5 Rear Axle Brake Cylinder
6 Electronic Control Unit
7 Brake Pedal Sensor
8 Compressed Air Reservoir
9 Supply Reservoir Hose (Trailer)
10 Electric Control Line (Trailer)
11 Coupling Force Determination
12 Steering Wheel Angle Sensor
13 Yaw Velocity/Lateral Acceleration Sensor
14 Actuation of Retarder and Engine braking systems

Fig. 2 Working principle of an electronic braking system

Wheel speeds are recorded by the customary antilock sensors.

With the help of a differential slip control, the front-wheel- and rear-wheel braking pressures are optimally controlled in correspondence with the driving situation via a pressure control module.
The differential slip control is based on the measurement and evaluation of wheel speeds and braking pressures.
With differential slip control, optimal braking force distribution in conjunction with an outstanding utilisation of tyre-to-road adhesion and a well stepped braking system can be attained.
After recognition of an overbraking tendency on one axle, a redistribution of the braking pressure follows on the front and rear axle in order to minimise the speed differential of the wheels at the individual axles and, again, to make optimum use of the tyre-to-road adhesion.

In order to obtain a good sensitivity of brake response, a delay control is implemented in the overall system. Thus, a certain pedal travel always corresponds to a certain delay, independent of vehicle load or the road inclination.

The antilock- and traction control system operations represent in the most cases only additional software components in the overall scope of control.
Optimal starting traction is guaranteed by engine and differential braking control.

On failure of the electronic basic braking system, e.g. by a total electric power failure, all axles of the tractor and trailer still can be braked redundantly in a conventional manner.

4 INTEGRATION OF ENDURANCE BRAKING SYSTEMS

Engine braking systems and retarders are endurance braking systems not subject to wear, which relieve the wheel brakes of the service braking system.

Retarder action is generally on the rear axle(s) of a vehicle. This additional braking power alters the installed braking force distribution.

Endurance braking systems have, above all, an effect on the braking stability of the vehicle at low tyre-to-road adhesion levels (rain, snow, ice, etc.). Therefore, the interaction between the endurance- and service braking systems needs to be monitored and optimally controlled by an intelligent braking management system.

All basic braking system operations always have priority over all endurance braking system operations. If necessary, the basic braking system admits only limited amounts of the additional braking action from the endurance braking systems. This is generally referred to as "blending".

The software algorithm in the basic braking system acts as a watchdog over the braking forces acting on the rear axles(s), so that excessive wheel slip on the rear axle(s) is eliminated, which would otherwise destabilise the vehicle owing to the lack of side forces.

The electronic basic braking system, therefore, offers active control for all endurance braking systems installed in the vehicle as a function of the tyre-to-road adhesion levels.

The incorporation of commercial vehicle endurance braking systems into the basic braking system also increases the efficiency of the latter. In addition to the higher braking power, this has a wear-inhibiting effect on the brakes and contributes to an increased brake pad life [3].

5 STABILITY CONTROL SYSTEMS WITH ROLL-OVER PROTECTION

Meanwhile yaw compensation systems in form of a vehicle stability control, like the well-known systems in cars, are standard equipment in some commercial vehicles (Fig.3).
The stability control is incorporated as a subsystem into the basic braking system and will become a self-evident feature in commercial vehicles in the 21st century.

The primary areas of application for such stability control systems in commercial vehicles are substantially different from those used on cars.
On bends with a low adhesion coefficient between the tyres and the road surface, the adhesion limit of the tyres is reached even at relatively low lateral accelerations. By means of a stability control system, this risk can be reduced by braking the vehicle to lateral accelerations, which are appropriate to the driving situation.
The danger of jackknifing is especially high, when the tractor oversteers, in which case the trailer continues to be understeered and moves in a straight direction. By appropriate activation of the brakes on the individual wheels, the tractor can be stabilised at any point.

To record the desired handling behaviour of the driver,
- steering wheel angle

- wheel speed as well as
- lateral acceleration

are measured. Derived from these values, the desired handling behaviour is defined.

The actual handling behaviour is recorded via an additional measurement of the yaw speed. Deviations resulting from a comparison between the "desired handling behaviour" and the "actual handling behaviour" result in the immediate intervention of the stability control.

There are individual intervention possibilities of the stability control for the front- and rear-axle brakes. The basic braking system offers separate control for each individual brake.
When the vehicle understeers, braking intervention is necessary on the rear axle; if the vehicle oversteers, braking is necessary on the front axle.

An additional important safety-relevant feature of the stability control is the prevention of vehicle tilting in curves, when driving with a high lateral acceleration on dry roads.
With the assistance of a roll-over-protection subsystem, each wheel is checked for excessive slip, when exceeding a certain lateral acceleration by a corresponding short braking pressure build-up [4].

Should it be necessary for the braking pressure to be reduced to zero, when a wheel looses contact with the ground, this indicates the initialisation of tilting behaviour. By immediate activation of the braking system tilting is prevented.

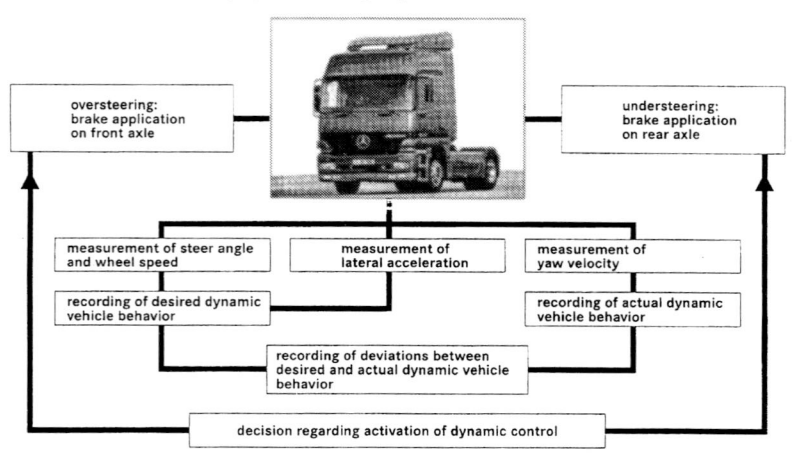

Fig. 3 Vehicle stability control

6 OPTIMISATION OF THE COMPATIBILITY BETWEEN TRACTOR AND TRAILER

The coordination of the braking effects of the individual tractor/trailer elements, especially in cases, where tractor/trailer combinations are often changed, cannot be achieved by using conventional equipment.

Poor tractor/trailer compatibility, e.g. by an insufficiently braked trailer, causes a disproportionate thermal load on the individual brakes of the tractor - shown in the top section of Fig. 4 - and may result in excessive brake lining / pad wear on the tractor.
In addition, the poor mating compatibility may have a negative effect on the stability of the tractor/trailer combination [4].

In order to prevent disturbing coupling forces between tractor and trailer (see Fig. 5), it is necessary to establish compatibility between tractor and trailer by means of an appropriate braking force distribution using electronic intelligence.

This means, that an appropriate participation of the trailer in the braking action of the entire rig should be ensured by optimising the relevant braking force distribution on both tractor and trailer.

To the extent that this is permissible by defined legal limits (ECE R 13), coupling forces between tractor and trailer are kept to a minimum, so that negative effects on the stability of the rig can be excluded.

This is done either by an intelligent trailer control system or a compatibility control system integrated into the basic braking system.
The latter type of coupling force control can be attained without having to measure the coupling forces between tractor and trailer.
The brake pressure is controlled in accordance with the specified setpoint, this usually being in the middle of the "ECE-Braking Band".
Should the trailer deviate from legally defined values, the braking pressure is corrected accordingly, ensuring at the same time, that the legally defined limits are not violated (Fig. 6).

Operation at the optimal working level of the electronic basic braking system on the tractor guarantees not only minimal brake pad wear, but also a maximum degree of safety.

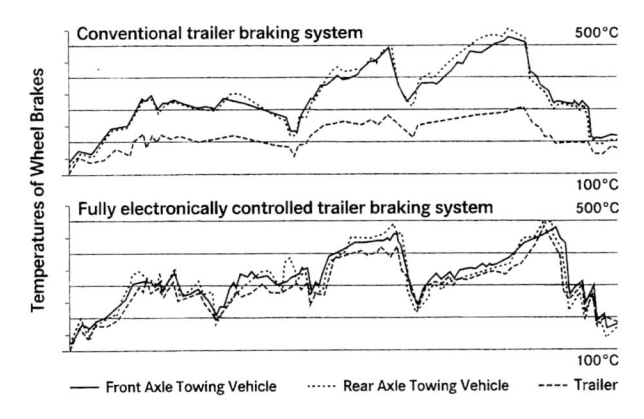

Fig. 4 Harmonisation of thermal loads of brakes, when decelerating a tractor/trailer combination

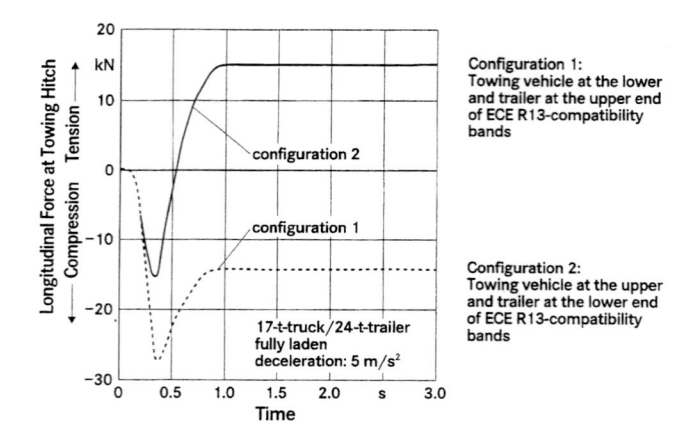

Fig. 5 Coupling forces of a tractor/trailer combination during braking

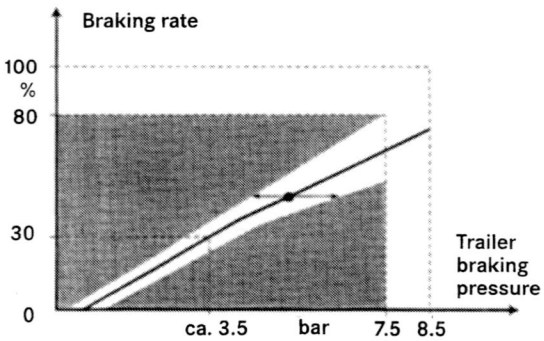

Fig. 6 Tractor/trailer compatibility control

7 ASSISTANCE SYSTEM FOR EMERGENCY BRAKING

The electronic basic braking system offers another advantage, in that a so-called "Brake Assist" corrects a possible driver error, should full braking be necessary. This can simply be done by an additional software integration.

The most important aim is the reduction of the stopping distance by early recognition of a critical situation. Thus
- head-to-tail collisions may be avoided, and
- collision speed may be drastically reduced and
- accident consequences dramatically alleviated.

The situation is recognized as a "panic stop" by measuring the speed, at which the brake pedal is activated (Fig. 7).
The required rapid increase in braking pressure follows by
- means of a pump or
- means of aerating a chamber in the brake booster or

activation of the reservoir pressure in the case of full power braking systems.
By releasing the brake pedal, the driver himself can determine the application period of the "Brake Assist".

The prerequisite for an application of this system is an antilock braking system, so that the individual wheels do not lock, when applying full braking pressure.

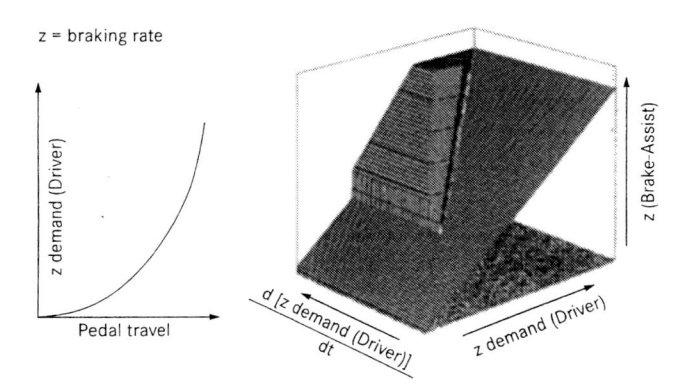

Fig. 7 **Working principle of a brake assist**

8   HILLHOLDER

The purpose of this subsystem is to relieve the inexperienced driver of the difficulties of uphill starting.

The brakes on the rear axle of the tractor are activated by "locking in the braking pressure" via the ABS- valves. Braking pressure reduction is triggered, when the clutch is engaged.

The ease associated with activation of the system can be improved to even a greater degree by a correspondingly stepped pressure build-up, if an incline- forward- and reverse-gear recognition is fitted.

Here as well, ABS is a prerequisite for the production application of this system in commercial vehicles.

9 BRAKE WEAR CONTROL

The electronic basic braking system offers a practical combination of a multitude of operations.
For example, it is possible, even while maintaining all active safety aspects, to minimise vehicle servicing costs. By separate control in accordance with specified wear criteria for the brakes of front and rear axle(s) on tractors and trailers, brake wear can be mutually harmonised.
Fig. 8 shows this harmonisation of brake wear on a tractor with a trailer.
As a result of the uniform load on all tractor brakes, overall wear on the brakes can be minimised. If the trailer as well is equipped with an electronic braking system, a further increase in the overall economy is possible.

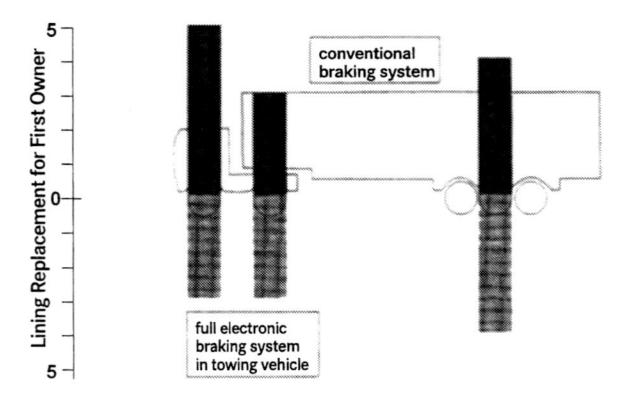

Fig. 8 Harmonisation of lining/pad wear

This results in equal service- and brake pad/lining change intervals for the entire rig. As a consequence, the downtime costs owing to maintenance work can be reduced substantially.

Using the subsystem "pad wear control", scheduled service planning is possible. Comparing only the service costs of a rig with an electronic basic braking system with those of a rig with a conventional braking system, it becomes obvious, that several thousands of EURO can be saved, based on approx. 100,000 miles / year and a 4-years service life.
In the case of non-critical partial braking at low speed differences, which represents by far the majority of all braking operations, the braking pressures on the front and rear-axle brakes are no longer exclusively applied using the criterion "adhesion", but also using the criterion "difference in brake wear". This results in a harmonisation of brake wear, i.e. an uniform brake wear on all axles is attained by activating only those brakes, which have the least wear on the pads/linings.
If an emergency braking needs to be initiated, conversion to adhesion-adaptive braking behaviour by the differential slip control follows automatically, i.e. all wheels are uniformly braked with constant slip.

The electronic basic braking system provides the driver with the additional possibility of being permanently informed of the condition of the braking system and the brakes.
The comprehensive diagnosis and monitoring operations of this new braking system are a prerequisite for effective fleet logistics.

10 PREDICTIVE EMERGENCY BRAKING SYSTEM

This safety system which is currently in the research arena could be a standard system in every commercial vehicle in the future. The purpose of this system is to initiate an automatic emergency braking under the following conditions:
- An obstacle (slower or stopped vehicle) is in front of the vehicle and
- the driver himself cannot avoid hitting the obstacle by steering commands

which means that the crash is inevitable without immediate reaction. The emergency stop will be started without the driver pressing the braking pedal. This initiation threshold must be dependent on the vehicle type and its lateral dynamics, its loading condition and its relative position to the obstacle.
Fig. 9 shows a possible scenario. If the system detects vehicle 1 as a noticeably slower or even stationary obstacle, the emergency braking is started at the threshold which is dynamically calculated [4].
Not in all cases the crash can be avoided completely, but the collision speed and therewith the collision energy can be reduced noticeably by a so called "electronic crumple zone".

Like in many others of the presented subsystems this system mainly consists of software, the additional costs and effort for hardware, in this case radar sensors are relatively low.

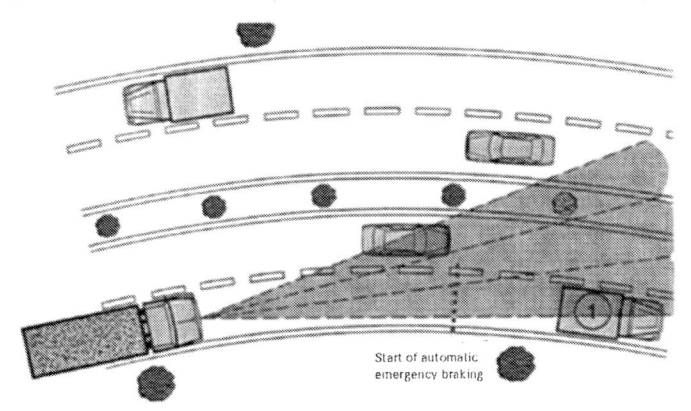

Fig. 9 Radar distance-monitoring for ACC and predictive emergency braking system

11 ADAPTIVE CRUISE CONTROL (ACC)

In contrast to the previously presented system the adaptive cruise control is not a safety system but a comfort system to unburden the driver. With radar sensors the distance and relative speed to a vehicle driving ahead is measured (Fig. 9). If the distance to this vehicle is becoming smaller and could fall below the minimum safety distance, the cruise speed is

slowed down by reducing the throttle and also by application of the service brake up to a certain deceleration. If the relative speed between the two vehicles is too high so that even with application of the brake the distance is becoming too small, that driver is warned by a signal.
Today this system already is standard equipment for a variety of commercial vehicles.

12 ELECTRO-MECHANICAL BRAKE (EMB)

As shown above the development of braking systems is characterised by the increasing employment of electronic control systems. The latest development has been the introduction of an electro-hydraulic braking systems for passenger cars.
The final end of this development could be a full-brake-by-wire-system (EMB), what means that not only the operation and control but also the actuation of the wheel brake works electrically and replaces hydraulic and pneumatic actuating systems that are in common use today.
By the future development of the foundation brake particularly the following potentials for improvement are going to be raised.
The main customer related advantages are:
- Shorter braking distances
- Optimised pedal feeling under all circumstances
- Easily adjustable pedal position and optimised force/way characteristics
- Increased passive safety by crash optimised pedal kinematic
- Increased active safety by redundancy and intelligent failure reaction
- Easy integration of advanced functions.

Manufacturer related advantages are:
- Integration of service braking and parking braking system
- Reduction of weight and volume especially of the operation facilities
- Flexible package, less interfaces, reduced assembly effort (plug and run)
- Platform independent standardised actuation and pedal modules for left and right hand drive
- Integration in vehicle command systems
- Improvement of corporate image by positive impact on traffic safety.

Regarding passenger cars and light commercial vehicles the advantages of electro-mechanical braking systems are undoubted. The complete removal of all braking hydraulics provides the so called "dry" brake.
Regarding medium and heavy duty commercial vehicles the over all benefit of EMB is considered less consistent. The conceptual advantages are smaller because today's use of compressed air already provides an external actuation force. Moreover, the compressed air facilities cannot be removed completely by EMB equipped vehicles, because it is needed for subordinated functions as air suspension. Additionally during a long period of transition compatibility with air braked trailers is necessary.
A challenge of EMB systems for passenger cars and light commercial vehicles is to achieve the dynamic characteristics provided by hydraulic systems, when activating and especially releasing the brake, i.e. for ABS control.

Regarding truck brakes on the other side, problems of layout of the mechanical and electrical concept arise from the huge clamping forces, caused by axle loads.

Considering compliance questions of by wire systems there are solutions foreseeable.

According to a proposal to modify ECE regulation 13 brake by wire systems can be certified as far as they fulfil certain demands of documentation of the development process, failure strategy and the aspects of complex electronic automotive control systems.

For EMB in specific there is still lack of specifying the energy supply because the regulations i.e. for pneumatic truck brakes cannot be transferred. This requirement could be covered together with the discussion about a 42V on-board system.

Summarising these aspects there is a challenge for the braking manufacturers to find solutions on the fields of mechanics and electronics, that raise the full potential of optimising braking systems specifically for commercial vehicles as well as they have to fulfil all future legal requirements.

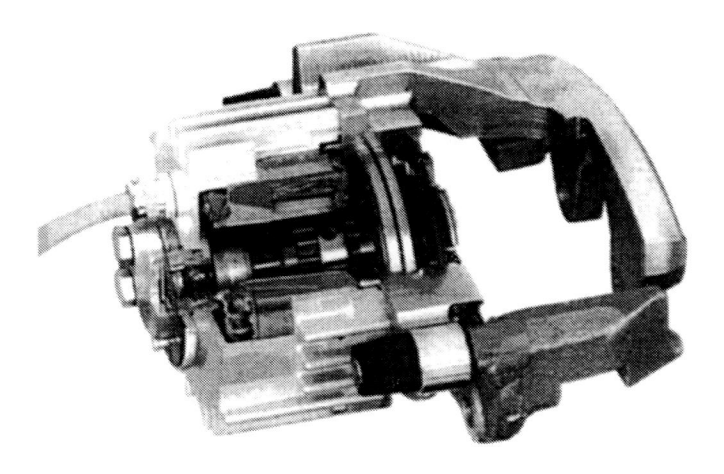

Fig. 10 Electro-mechanical brake

13 SYSTEM INTEGRATION AND ELECTRONIC NETWORKING

Future generations of vehicles will contain a large number of electronic main systems and subsystems. In such systems, all vehicle components are monitored and corresponding diagnostic evaluation is carried out. Diagnostic- and service systems realise economic aspects, functional safety and reduce maintenance times on vehicles.

Such a development strategy requires all electronic systems to be networked. The basis for such networking is a high-performance and safety-relevant computer communication system, which intelligently monitors all vehicle systems.

One example of how braking management can be integrated into this system is shown in Fig. 11. Simultaneously, all previously discussed functions of the braking management as well as the corresponding subsystems are described in this figure.

The network must operate with standard systems architecture and must be able to supply fail-safe safety-relevant information on brakes, steering, chassis and powertrain.

Through the application of an emergency circuit in breakdown situations, the vehicle must at least be capable of being driven to the nearest shop.
The networking is attained using high-speed- and low-speed CAN buses.
The proper inter-operation of the individual systems is monitored by a master system. This master system controls the individual subsystems with high artificial intelligence.

```
┌─────────────────────┐      ┌─────────────────┐      ┌─────────────────┐
│  Information and    │      │   Suspension    │      │    Drivetrain   │
│Communication Management│    │   Management    │      │   Management    │
└──────────┬──────────┘      └────────┬────────┘      └────────┬────────┘
           │                          │          CAN-Bus       │
───────────┴──────────────────────────┴────────────────────────┴──────────
```

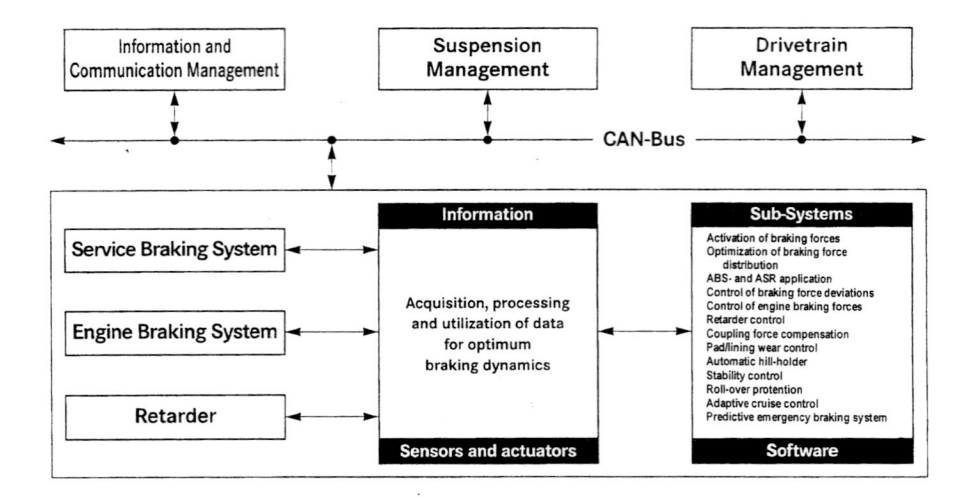

	Information	Sub-Systems
Service Braking System		Activation of braking forces
		Optimization of braking force distribution
	Acquisition, processing	ABS- and ASR application
	and utilization of data	Control of braking force deviations
Engine Braking System	for optimum	Control of engine braking forces
	braking dynamics	Retarder control
		Coupling force compensation
		Pad/lining wear control
Retarder		Automatic hill-holder
		Stability control
		Roll-over protention
		Adaptive cruise control
		Predictive emergency braking system
	Sensors and actuators	**Software**

Fig. 11 Braking system management

14 CONCLUSION AND OUTLOOK

The last decade showed an enormous increase of the usage of electric and electronic systems in automotive technology. In the examples mentioned in this paper the advantages of this trend are pointed out clearly in the field of braking systems.

Based on a basic system of electronic braking activation, there can be developed and implemented a variety of different assistance- and comfort-systems by adding only little hardware, i.e. sensors, but mainly software. These additional subsystems interact with the basic system and use most of the basic system's hardware. The costs for these subsystems are quite low compared to conventional "stand-alone"-systems with their own hardware. Furthermore the value of these systems for drivers and other road users is quite high, seen from the point of active safety.

This paper shows systems which are state of the art, but also possible developments for the next decade. Of course, the progress will not stop and more and more new electronic systems with high intelligence will be introduced, which will make a substantial contribution to relieving the driver from stress and allow him greater freedom to observe the immediate surroundings. The driver will be relieved of all duties which could divert his attention or cause severe stress. As a consequence, the active safety of commercial vehicles will be further increased.

15 REFERENCES

[1] R.Povel, E.C.von Glasner :
"Active Safety of Commercial Vehicles – The European Status",
JSAE Spring Convention 2000, Yokohama, Japan

[2] E.C. von Glasner, H. Marwitz, R. Povel:
"Active Safety of Commercial Vehicles - The European Status",
IPC11-Conference 2001, Shanghai, China

[3] M. Bergmann, E.C. von Glasner, H. Marwitz, R. Povel
"The Influence of Retarders on the Braking Behaviour of Commercial Vehicles"
IPC11-Conference 2001, Shanghai, China

[4] R. Povel, E.C. von Glasner, K. Wüst:
"Electronic Systems Designed to Improve the Active Safety of Commercial Vehicles"
SAE do Brasil-Conference 1998, Sao Paulo, Brazil

The intelligent brake system

E GERUM
Knorr-Bremse SfN, Munich, Germany

ABSTRACT

The air disc and EBS (electronic controlled brake system) became a standard in the truck in the recent past. Main challenges in the future of the brake system are the introduction of ESP (electronic stability program), further improvement of brake performance, reduction of complexity of the brake system and establishing an intelligent, standardised interface for the foundation brake. Knorr-Bremse SfN reports a solution of these issues. An EBS with a system architecture of distributed intelligence with wheel individual control fulfils the above mentioned demands and is the best platform for ESP. To meet the requirements of less complexity and of an intelligent wheel brake, we show prototypes of the integration of EBS wheel module and caliper of air disc. A report is given on test results from driving tests.

This concept is a consequent and intermediate step towards an electronic actuated foundation brake. It is shown that the KB EBS concept of today meets the requirements for future system architecture of an electric actuated foundation brake. The Knorr-Bremse SfN concept of an electric actuated foundation brake is demonstrated as well as a report about driving tests and a comparison of today's brake system, the future intelligent pneumatic actuated brake system and the electric actuated brake system.

1. INTRODUCTION

The commercial vehicle brake industry must prepare a long-term strategy regarding the entire brake system. The truck brake system has not changed for several decades. The components became smaller and smarter, with slightly improved performance, but the basic architecture and the layout of the brakes system remained unchanged. The previous system consisted of the following components:
compressor, governor or unloader valve, air-dryer, air-dryer control, multi-circuit protection valve, reservoirs, footbrake valve, relay valves, load proportioning valve, brake chamber, slack adjuster and foundation brake.

A first change occurred in the early 80s with the introduction of ABS; however this was an add-on system where the pneumatic control system remained unchanged.

In the mid nineties, EBS (electronic brake system) was introduced which caused a major change in the system layout because data processing was now realized by means of electronics and not by pneumatics. This system is today's state of the art. Due to the fact that there will be continuous significant progress regarding performance and costs of microcontrollers as well as in the field of sensors and high speed data transfer, a large variety of additional features can be and will be realized to the user's benefit. These additional features cause another challenge for the truck OEM (original equipment manufacturer) and supplier due to the significantly increased complexity of the future system. These two facts result in the assumption that there will be very fast development or evolution in the truck brake system, which may even become a revolution due to the high speed of innovation. It is very possible that none of the brake components of the most advanced trucks of nowadays will survive in a new truck model launched in 2010 or 2012. And the same will apply again for a truck launched in 2020. Therefore, it is an important issue in the industry to prepare a common product strategy for future brake systems in order to prevent the industry from detours or even dead ends in product development.

2. DRIVING FACTORS FOR THE FUTURE BRAKE SYSTEM

Costs are the key-driving factor for the future development in brake systems. Business features like fun or prestige are not market factors as in the case of, for example. the passenger car. In the past, the supplier's main target was to reduce the costs for the OEM. This will be still a key factor in the future. But there are two other cost factors, which will have to be considered more and more.

First of all, the life cycle costs or costs of ownership become a more important issue. In consequence, only those features can be sold which are able to show a reasonable payback in a short period of time. Another cost block will become a key factor in the future: the cost for the society. This means the impact to the environment due to pollution, wasting of energy and production of carbon dioxide as well as the loss of human lives or injuries and the economic loss due to traffic jams and destruction of infrastructure such as roads.

3. ISSUES REGARDING THE FUTURE BRAKE SYSTEM

Issues regarding the driving factors for future brake systems are as follows:
- Reduced component costs
- Reduced system costs
- Increased reliability
- Easier service
- Easier handling
- Enhanced safety
- Increased transportation capacity
- Saving of resources
- Reduced emission

All these factors contribute to the reduction of life cycle costs at least then when society requires that the costs are covered by the source of responsibility.

4. CONSEQUENCES FOR THE BRAKE SYSTEM

In the recent past, the brake system had very few functions which could be shown on a control flowchart such as: load proportioning, predominance and automatic slack adjuster. With the help of ECBS (Electronically Controlled Brake System) and EAC (Electronic Air Control), a huge variety of functions can be realized today such as: load proportioning without sensors, ABS/ASR, ESP, ROP, electronic slack adjuster, ACC (Adoptive cruise control), hill start aid, coupling force control, intelligent compressor control, wear control, self diagnosis brake system, 'brake assistant'. It is foreseeable that this increase of functions will also continue in the future. This means that the system will become much more complex. This is explained as follows:

A former simple pneumatic valve like a relay valve had one single function and consisted of ten parts only. A more sophisticated valve like a load proportioning valve had several functions and about 100 parts – see Figure 1.

A simple electronic system like basic ABS has approximately 50 functions and 1000 parts. In this case, both the electronic components as well as the software are counted. One line of code of the functional algorithm in high level language is counted as one part because the work to design, document, test, release and maintain one line of code can be compared with the work required for one mechanical part.

A modern most advanced EBS with integrated ESP (Electronic Stability Program) has several hundred functions and its parts amount to more than 10,000.

This proves that over a time period of several decades, an exponential growth of functionality and complexity took place. This can be verified with the R&D efforts made by the OEMs and in paricular by the suppliers.

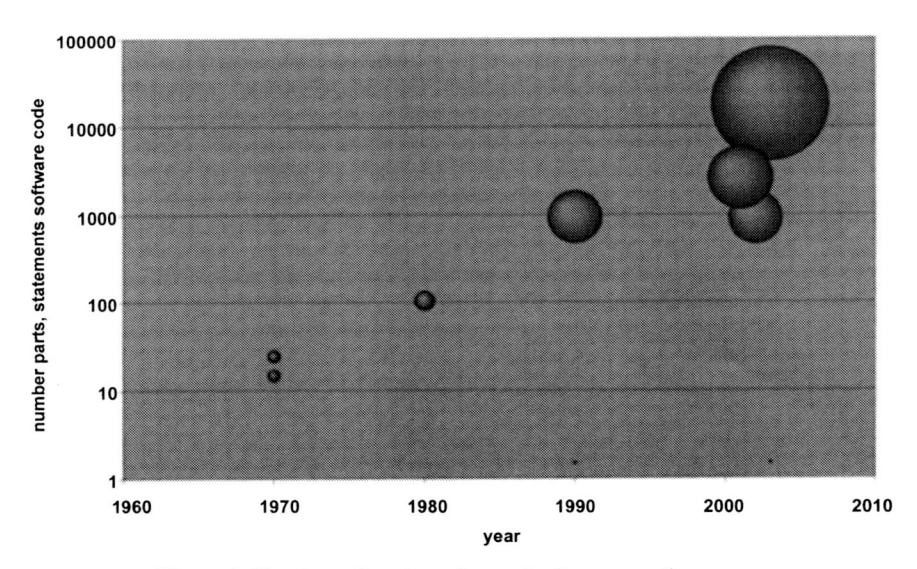

Figure 1: Number of parts and complexity versus time

Due to the fact that a significant change in this development is not foreseeable, it becomes obvious that measures must be taken in order to reduce this complexity. Otherwise, the whole system can no longer be handled.

This will be explained in the example as follows. Let's assume that a truck OEM wants to introduce a new foundation brake and a brake control system. Considering today's sharing of tasks in the case of state of the art systems, the truck OEM has to deal with the following topics:

Friction materials
Stiffness caliper
Slack adjuster
Installation conditions
Electric connectors and ports
Type brake chamber
Force amplification caliper
Bearing and hysteresis force transfer
Type EP converter
EBS functionality
EBS ECU
Data bus interface
Interface between:
EBS, brake chamber, caliper, friction materials

There is at least one person with specific know-how for each line. This shows that the truck OEM has to make enormous efforts in order to create a brake system without knowledge of the brake components. However what is really necessary from the truck maker's point of view in order to realize a brake system is only the following:

Brake force in kN or % braking ratio per wheel or axle
Geometry conditions like envelope and fixing points
Type of connectors and fittings
Data bus

It will be necessary in the future to reduce the number of interfaces and to make the interfaces as simple as possible which means that the interface has to be reduced to the basic physical factors such as brake force and to eliminate any details regarding a specific design which is today's state of the art. In consequence, this signifies a plug and play philosophy which is very well known in the PC industry, but without any doubt a brake system must be much more reliable. This basic philosophy is not absolutely new in the truck industry. Engines and transmissions do already have such an interface nowadays. A state of the art engine has an integrated electronic control, the actuators for injection, a data bus for the desired torque or engine speed and well defined fixing points and interfaces for the infrastructure. The brake system must be developed to the same level.

5. CONSEQUENCES FOR THE TRUCK OEM

In the past, the truck OEM managed the evolution of the components for the entire brake system. By means of EBS, a part of this task has been transferred to the supplier providing a complete brake control system.

In future, even more responsibility will be passed on to the supplier so that eventually the entire brake system will possibly be designed and supplied by a system supplier. This will enable the truck OEM to reduce the effort required for system integration. The forementioned transfer of tasks from the truck OEM to the system supplier does not mean reduced R&D efforts for the truck OEM. The contrary is true. However by means of this strategy the increase in R&D efforts of the truck OEM might stay manageable.

6. THE BRAKE SYSTEM OF THE FUTURE

A today's state of the art brake system with EBS and air disc consists of 15 part numbers and 25 parts for a 4x2 truck not taking into consideration fittings, cables, consoles and pipes – see Figure 2.

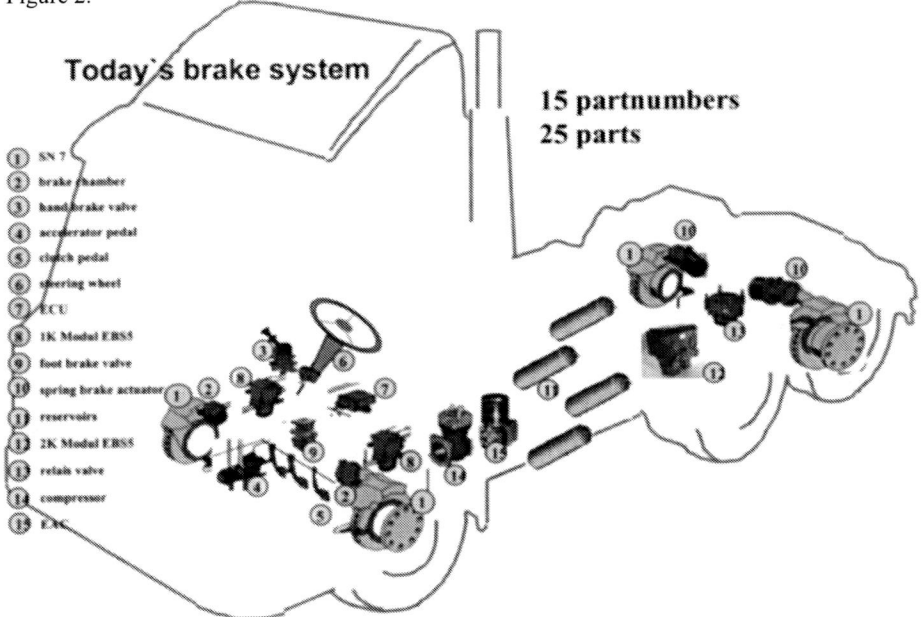

Figure 2: Today's brake system

The vision for the far future is that the entire brake system can be created by using only 3 basic components – see Figure 3. Each component is a mechatronical component with micro-controller and data bus interface in order to communicate with the rest of the vehicle. The plug and play feature is fulfilled.

These three components are:
Man –machine interface
Air supply module
ADIC (Air disc with Integrated Control)

The number of parts and part numbers are reduced to a third with the help of this concept, which causes a significant reduction in logistic costs and installation efforts. Testing at the truck OEM assembly lines can be reduced due to the fact that all these components are intelligent and are supplied 100% tested. This results also in a cost reduction for the truck OEM and consequently in a cost reduction for the fleet owner. The reduction of costs for society can be achieved by the reduction of pollution, which will be explained later.

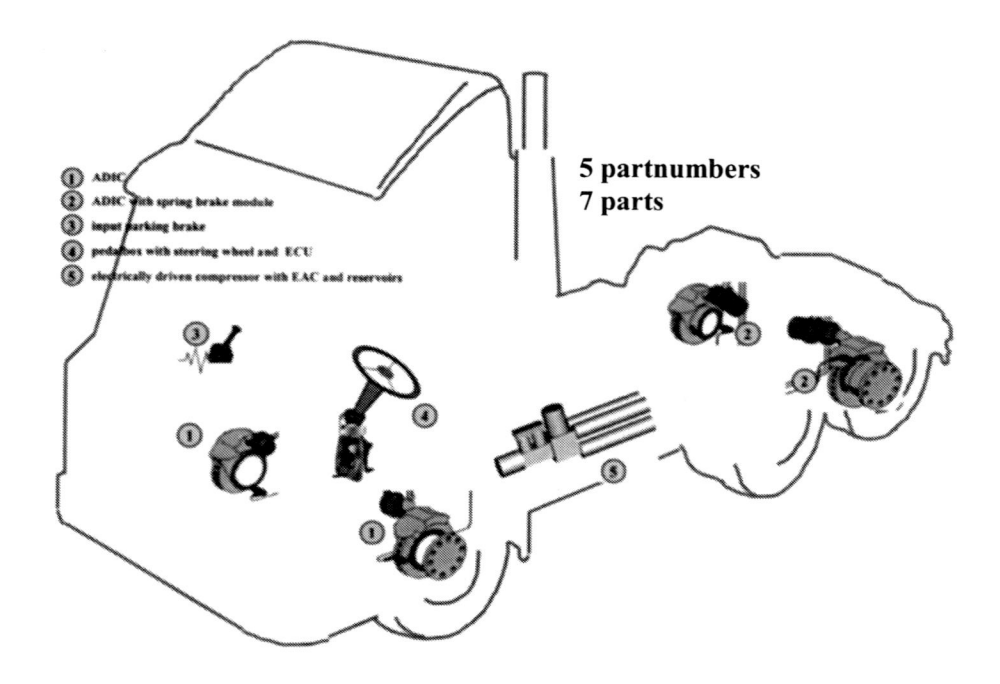

Figure 3: Vision of future brake system

7. MAN-MACHINE INTERFACE

The man-machine interface in cars and trucks has not changed for more than 100 years. There is still a steering wheel, accelerator pedal, brake pedal and clutch pedal. The coordination of the input signal is managed by the driver. For more than 10 years, it has been state of the art in civil aircrafts to fly by wire and to have a side stick as a man-machine interface. But also in control devices for locomotives or commuter trains, ships, agricultural tractors or construction vehicles a side stick is state of the art. An electronic input device will be the future even for the truck. First of all, the mechanical components such as steering column, pedals with sensors, converters and electronics will be integrated into one mechatronical unit.

The challenge of such a system does not reside in the mechanical design, but in the logistics and the flexible assembly because a large number of variants has to be handled and just in time delivery is required.

Step by step, this system will be developed into an electronic man-machine interface. This interface will operate in parallel to another input interface, which will receive the target trajectory for the vehicle from external navigation systems, traffic control systems or from fleet management.

8. AIR SUPPLY MODULE

The main driving factors in this field are as follows.

Pollution with oil: The contamination of oil due to the oil-lubricated compressor has been a problem since air brakes exist. The technology is now available to design a dry running compressor which meets the requirements of the truck industry. Key measures are new materials for piston rings and sliding materials for the cylinder or piston. Another big advantage of a dry running compressor is the increased lifetime of the air dryer. Nowadays, the air-dryer cartridge must be replaced due to the oil contamination of the desiccant. If there is no oil contamination, the lifetime of the desiccant can be the truck's lifetime. This allows also for a new freedom in design of the air-dryer because easy exchange of the desiccant is no longer necessary.

Energy saving: A pending issue for a dry running compressor is its lifetime and any measure to reduce the number of revolutions during the lifetime is desirable. Therefore, such a compressor is first of all electrically driven. This means that the compressor is only running during the supply phase whereas it will be stationary during the rest of the time, unlike today where the compressor runs at engine speed all the time and consumes a significant amount of energy. During an intermediate phase, the dry running compressor will still be engine driven, but with a clutch or a variable transmission. Both will reduce the overall revolutions of the compressor and the energy consumption.

Saving space in the engine compartment: Future systems to reduce NOX and particles might result in an additional space demand in the engine compartment which is not available. Therefore, it is desirable to relocate the compressor to another place in the chassis. This problem is solved by the electrically driven compressor mounted to the chassis. However, this solution requires a much more powerful electric power supply. This will be realized gradually when more and more electrically driven servos will be available and when the 42 Volt system will be introduced.

Modular design: As explained above, it is necessary to form mechatronic, integrated modules. Therefore, a link between the electrically driven compressor, air-dryer, control of compressor, air-dryer and the reservoirs is the future. Perhaps a unit with 4 reservoirs would create problems with respect to installation. However there is the solution where small electrically driven compressors are provided for each reservoir. This would mean that the function of the multi-circuit protection valve is no longer necessary and that the electronic control of the module will become easier.

9. ADIC (AIR DISC WITH INTEGRATED CONTROL)

This module integrates the today's brake caliper, brake chamber and the electronic control for one wheel, including micro-controller, solenoids, pressure sensors and relay valve. The advantage for the truck OEM is that he will be supplied with a 100% proven unit of brake chamber, caliper and electronic control.

The installation efforts and costs for wiring harness are reduced as well as the logistic costs. The test at the end of the assembly line at the truck OEM will be reduced as well. Furthermore, an improvement in function is given. Today, the length between EBS modulator and brake chamber is between 1 m and 3 m. This distance will be reduced to more or less zero with a positive effect on delay time and air consumption.
The presence of an intelligent system in the caliper opens new freedom in design of the caliper, e.g. it is easily possible to acheive an electronically controlled management of clearance between pads and rotor enabling an increased performance of the brake and a significantly reduced demand in space. Other options are a new basic design for the caliper which reduces weight and increases durability.
There might be a concern that electronic control at the caliper cannot withstand the rough condition in this environment. Firstly, a small electronic board in the caliper due to the presence of intelligence in the caplier exists already today, i.e. the signal conditioning board for the potentiometer of the analog wear sensor. This indicates that conventional electronic technology is able to operate in this area. On the other hand, it is state of the art that complex electronic systems with powerful micro-controllers are located at the engine. Temperature, stress and acceleration are on similar levels as for the caliper.

10. ELECTRO-MECHANICAL ACTUATION

It is desired by the industry that electrical actuation of the foundation brake will be realized in the long-term. The main reason is to eliminate the air from the vehicle or to use it only for suspension and trailer. An obstacle regarding the electro-mechanical actuation is today the required electric power which exceeds significantly even the future power supply in a truck. But there are projects under investigation which might solve this problem on a long-term basis.
It is important that the architecture of the future brake system as described above is absolutely the same as required for an electro-mechanically actuated brake system. Therefore, this concept is a step-by-step evolution towards an electro-mechanically actuated brake system.

11. BRAKE BY WIRE

Today's EBS is not understood as a brake by wire system because there is a pneumatic back-up line in case of a failure. It is of no relevance how the brake force is created, but how the information is processed and transferred by wire and only by wire. A system without pneumatic back-up line is required due to cost reasons with or without electro-mechanical actuation. This means a very big step forward regarding R&D efforts and the truck's layout. There is a requirement for a 'safe' information line and power supply.

Even in the case of one failure, a braking performance according to the legal requirements has to be achieved. In any case, a second battery is required with monitoring of the charging condition and protection against short-circuits with the first battery with relation to the function of the multi-circuit protection valve. The size of the second battery is defined by the legal requirements regarding the brake applications and other safety systems such as steering. Today's CAN bus is neither a 'safe' bus nor does it fulfill real time requirements. Both are indispensable for a brake by wire system. A real bus has to be designed. Furthermore, automotive standard micro-controllers need to be available at a reasonable price. This will most likely not be realized by the truck industry. The truck industry must cooperate with the passenger car industry in order to modify the car solution to the needs of the truck.

Figure 4 shows a proposed pathway to this future brake system, which eventually in any case will be a system without a backup but it is still open as to whether the brake will be actuated pneumatically or electromechanically.

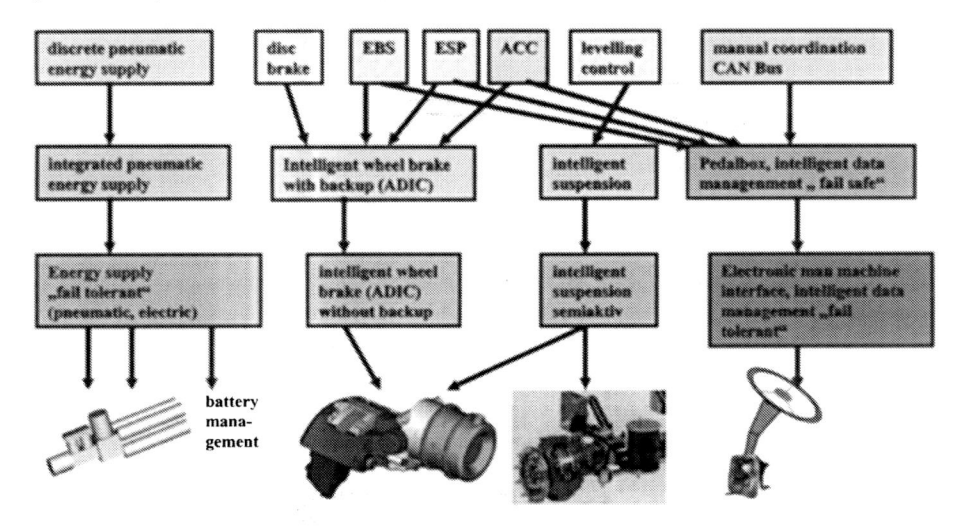

Figure 4: Road map for the evolution of the brake system

12. AUTONOMOUS DRIVING

The performance of the brake system including EBS with all its increased performance and ESP seems to reach a level where improvement in performance of the entire brake system might no longer be necessary. Safety still needs to be improved; however this is not related to the brake system. Nevertheless, it is still necessary to prepare the brake system for future systems including line guidance, driver drowsiness, ACC and in the long-term autonomous driving. This results in a brake demand next to the driver's input. This demand has to be a 'safe' message, otherwise tragic accidents could occur.

The future brake system must be equipped with a proper interface for this future external brake demand.

13. CONCLUSIONS

There is a strong need to develop today's brake system step by step towards a modular system with intelligent components. The end of this development might or might not be an electromechanically actuated brake system. In any case it should be a brake by wire system without pneumatic backup.

Current position and development trends in air-brake systems for Mercedes-Benz commercial vehicles

T MARKOVIC
Daimler Chrysler AG, Stuttgart, Germany

SYNOPSIS

With the increasing demands on commercial vehicles in the last few decades the air-brake system has been continuously improved. In addition to this the performance, operating convenience and economic efficiency of the brake system was increased by the use of the Telligent®-Brake System, which was primarily introduced by Mercedes-Benz onto the market. Compressed-air actuated wheel brakes are activated via electrical control signals and integrated pressure control circuits. Optimised ABS/ASR functions are integrated in the system.

Due to the electronic control the stopping distance is shortened and economy by harmonisation of the brake lining wear and integration of the endurance brake system is increased. In addition auxiliary functions like the brake assistant, the stability control and the adaptive cruise control are integrated in a simple manner.

1 INTRODUCTION

Intensified environmental requirements, a higher density of traffic, the increase of pay load, an enormous competition pressure in transportation traffic as well as the constantly rising of the driver's strain let the commercial vehicle requirements constantly increase over the past decades. The increase of economy and a maximum active and passive vehicle safety are the emphasis of the commercial motor vehicle development. In particular for heavy commercial motor vehicles the effect of the improvement of passive safety is limited because of the high energy content of the vehicle.

Due to its the increase of the brake efficiency of commercial motor vehicles is a central aspect for the increase of active safety on our roads. Substantial steps in the development of commercial motor vehicle brake systems were 1973 the introduction of the ECE brake assembly, 1981 the development and introduction of the anti-skid system (ABS), the increase of the brake pressure and the use of disc brakes at all axles. By introduction of the Telligent®-Brake System in the ACTROS in 1996 the capability, the control comfort and the economy of the vehicle were substantially improved.

At present for commercial motor vehicles a set of systems are developed to relieve the driver for safety reasons. By the introduction of the Telligent®-Stability Control at the end of 2001 the vehicle safety could be again substantially increased. A goal of the truck development in the next years is it to process the information of the traffic and road situation around the driver in order to recognise potential collisions in advance and to avoid the accident surely also without interference of the driver.

2 TELLIGENT®-BRAKE SYSTEM

The reduction in stopping distance is the aim whenever a braking system is modified. Irrespective of the increase in the brake-air pressure level from 8 to 10 bars or the use of the disk brake, the braking system response times can be significantly reduced with the aid of electrical signal transmission in the brake system. This has a particularly significant effect on the rear axle as here the largest distance for transmitting the control signal has to be covered. It also permits the optimum ABS control in the electro-pneumatic brake system, using the actual pressure values existing in the system at the front and rear axle to delay the ABS control intervention for as long as possible and thus once again to shorten the stopping distance.

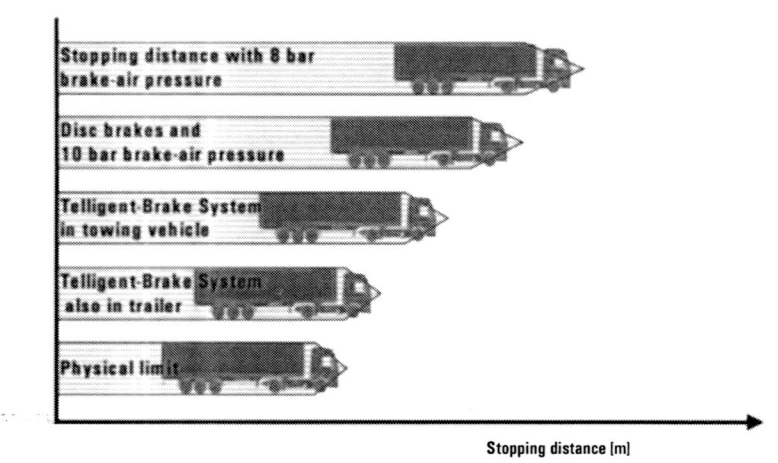

basis: 90 km/h, 40 t total weight, temperature of brakes: 300°C

Figure 1. Stopping distances

The reduction in stopping distance which can be achieved by using Telligent®-Brake System in the order of magnitude of the length of one passenger car with a fully laden vehicle combination represents a significant gain in safety. By using of electropneumatics also in the trailer the stopping distance can be reduced further, so that we could approach the physical limit significantly. Figure 1 shows the reductions in stopping distances which can be achieved by various measures to optimise the braking system.

Not only does the brake system make the remaining lining of each tractor wheel brake available, using bi-directional interfaces. Comprehensive diagnosis functions, which utilise the brake system of sensors and test the entire braking system with the aid of special test routines, are the basis for efficient trailer pool logistics. Last but not least, all adjustment work at the automatic load-sensitive brake pressure regulator is dispensed with on vehicles braked by Telligent®-Brake System, as this component is no longer required.

2.1 System configuration

Technically, the electro-pneumatic brake system (figure 2) is a system with two pneumatic and one electrical brake circuit (2p/1e system); in an intact condition the electrical circuit controls the system. The front axle brake pressure always consists of pneumatically and electrically modulated partial pressures. The rear axle brake pressure is solely electrically controlled; the pneumatic circuit is blocked. ABS and ASR functions are integrated in the brake system.

In detail the Telligent®-Brake System consists of the following components:

Two redundant brake pedal sensors and two further switches are integrated in the brake signal transmitter, which is a dual-circuit pneumatic design. The brake pedal sensors in each circuit determine the braking demand of the driver. This demand is transmitted to the central control unit as a pulse width modulated signal.

The control unit is used to control and monitor the electronically controlled braking system. It performs brake management and data communication through out the system. The control unit uses the brake system CAN-bus to exchange data with the axle modulator.

In the axle modulator two pressure control circuits are assigned to the wheels, in order to meet the requirements for service brake applications and interventions of the ABS and ASR systems. The axle modulator includes an ECU, which provides management of each of the independent pressure and monitoring functions. The redundancy valve is used in failure situations only.

The proportional relay valve modulates the brake pressure at the front wheels. It consists of a proportional solenoid valve, a relay valve and a pressure sensor. The actual front axle pressure is sensed and transmitted to the control module to complete a pressure control circuit. ABS control valves are fitted downstream of the proportional relay valve and permit an ABS control intervention at each front wheel.

The electro-pneumatic trailer control valve corresponds largely to a conventional trailer control valve and pressurises the conventional coupling heads, but the first control circuit is actuated electrically. The actual pressure at the brake coupling head is sensed to complete a pressure control circuit. For the optimum operation of trailers with an electronically controlled brake system an electrical interface in accordance with ISO - 11992 is integrated in the control unit.

The main functions as well as the communication interfaces to the axle modulator and to other ACTROS systems and the pressure control circuits of the front axle and the trailer control are integrated in the control unit of the brake system. Finally, in order to be able to implement an self-acting brake lining wear equalisation in the tractor on the basis of real wear data, the remaining lining of each wheel brake is recorded and the value supplied to the brake system.

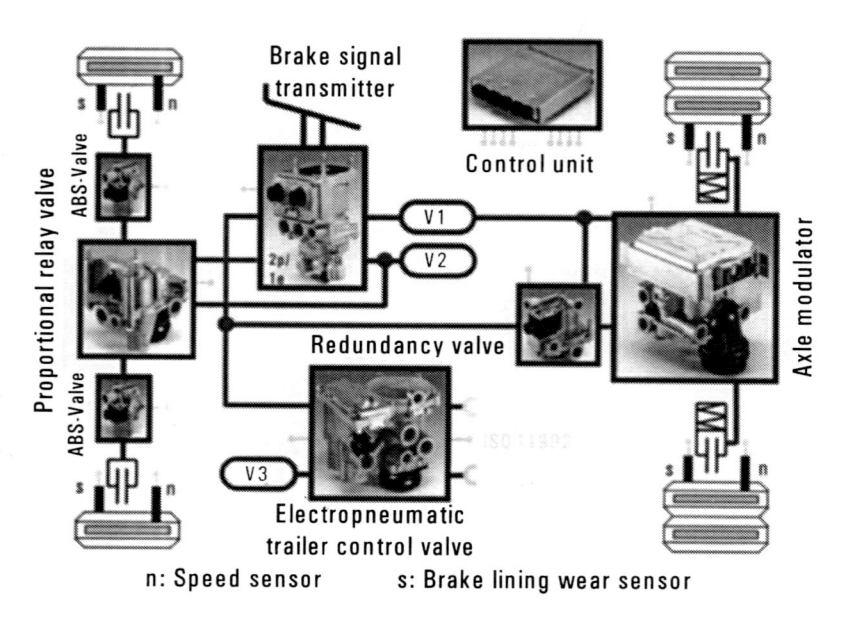

Figure 2. Telligent®-Brake System

2.2 Control Philosophy

The fundamental control in the Telligent®-Brake System is a differential slip control (DSC). The basic functions of the brake force control - adhesion, graduation, harmonisation of brake lining wear and trailer activation - are solved independently of each other and individually for each vehicle.

For this purpose, the brake system requires only the wheel speed sensors (required for ABS) already present in the vehicle together with pressure sensors for the individual brake circuits as measured parameters which are to be registered by sensors. Sensors for measuring axle loads, coupling forces between tractor and trailer or vehicle-dependent parameters such as wheelbase, tyres and similar data are not necessary. By evaluating the differential slip between the front axle and the rear axle as determined by the wheel speeds, an overbraking tendency of an axle is recognized and the brake pressures are distributed accordingly. Through simultaneous deceleration control, the mean pressure is altered. This leads to an adhesion behaviour of both axles which is close to the optimum (figure 3).

To control adhesion utilisation, the brake pressure distribution $\Phi = p_{FA}/p_{RA}$ is controlled as a function of the desired deceleration z_{demand} according to the function

$$\Phi = \frac{p_{FA}}{p_{RA}} = a + b \cdot z_{demand}. \qquad (1)$$

Φ: brake pressure distribution
p_{FA}: brake pressure at front axle in bar
p_{RA}: brake pressure at rear axle in bar
z_{demand}: desired deceleration in %

The parameters a and b essentially describe the static axle loads and their dynamic distribution. The tyre characteristics and the brake cylinder size have also been taken into account in these parameters. When starting the vehicle, these parameters are allocated start values. The criterion of intervention with regard to changing the parameters a and b is the behaviour of the differential slip

$$ds = \frac{v_{FA} - v_{RA}}{v_{FA}}. \qquad (2)$$

ds: differential slip in %
v_{FA}: velocity at front axle in m/s
v_{RA}: velocity at rear axle in m/s

Control occurs in such a manner that the value of ds is minimised or remains below a given limit. This results in an adhesion behaviour of both axles, which is sufficiently close to the optimum.

In addition to adhesion control, the deceleration control belongs to the basic elements of Telligent®-Brake System. The deceleration of conventional vehicles with a given pedal force depends strongly on the loading conditions as the brake pressure is set at the service brake valve. In order to remain independent of external parameters and to maintain optimal graduation, deceleration is controlled in the brake system by means of the mean brake pressure level

$$K = \frac{p_{demand}}{z_{demand}}. \qquad (3)$$

K: brake pressure level in bar
p_{demand}: desired pressure in bar

The driver's demand force on the brake pedal is principally interpreted as being a desired deceleration. If the actual deceleration deviates from the demanded deceleration, the value K will be corrected slightly or dynamically.

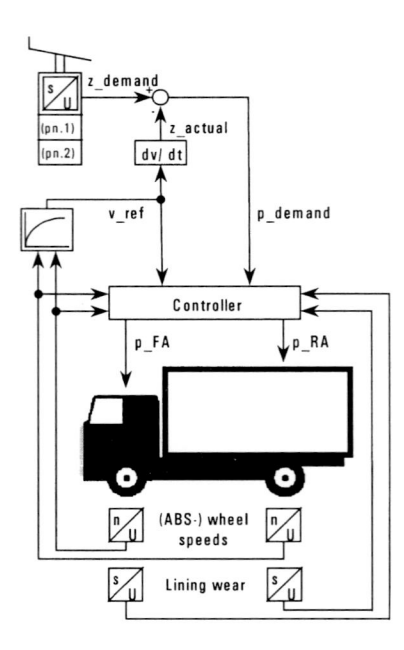

Figure 3. Differential Slip Control

The result of these calculations is the optimum physical brake pressure target values for the front and rear axle. If the deceleration demanded by the driver is now in the range of non-critical brake applications, these optimum physical braking pressure target values are corrected within the framework of the brake lining wear equalisation standards without changing the overall level of brake application.

2.3 Functions of Telligent® -Brake System

2.3.1 Trailer control

For trailer control the legal requirements of the EC/ECE Directives/Regulations must be complied with. Moreover, the spectrum of the actual existing trailer fleet has to be taken into consideration. Therefore, a coupling force control without direct sensing was developed. The pressure on the brake coupling head is set electrically to the braking rate compatibility band in accordance with the given demand deceleration on the pedal and adjusted to the desired deceleration through the deceleration control.

In the case of distinct deviations of the trailer, the control within the band limits occurs subsequently. The relation between the brake pressure level K and the static axle loads of the towing vehicle a is used as a parameter for the recognition process. The deceleration of the tractor without trailer which is established from a trip with the vehicle unladen and put into buffer storage is compared with the deceleration of the complete vehicle combination at an identical level of brake pressure. For the optimum tuning of the brake systems of the tractor and trailer, the same deceleration should be introduced for the same pressure level in the tractor - irrespective of whether a trailer is being operated at the same time or not.

The electronically controlled brake system now recognises deviations from this physically optimum behaviour of the vehicle combination and corrects the trailer braking pressure, provided the limits defined by the EC/ECE Directives/Regulations (figure 4) permit this. This means that with brake system tractor and trailer always contribute to the total braking effort of the complete rig in proportion to their weights.

In the latest version of the ECE-R13 regulation, Supplement 2 for electronic brake systems includes a definition of enhanced deceleration bands under para. 5.2.1.28 (coupling force control systems). On this basis, the deceleration band so far implemented into the control system in accordance with the EC regulations (71/320 or ECE/R13) has been enhanced moderately, i.e. without making full use of the control limits specified in Supplement 2. No warning message is generated since the excess pressure of 1.5 bar specified as a warning threshold in the regulation cannot be modulated. Figure 4 compares the enhanced deceleration band implemented in the control system with the previous system.

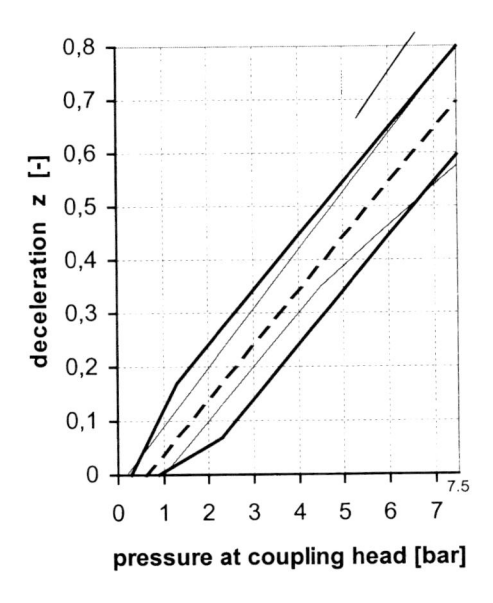

Figure 4. Deceleration band

The longer response times of the trailer brake system which are unavoidable when transmitting the control pressure via the brake coupling head often cause an uncomfortable overrun shock for the driver. The electro-pneumatic brake system avoids this by introducing a shorter, defined pressure pulse into the trailer braking system at the start of braking.

After the introduction of trailers with an electro-pneumatic braking system and data transmission via the electrical tractor/trailer interface (ISO 11992) the response times and the stopping distance of the complete vehicle combination are reduced, the optimum match of the tractor and trailer braking system is realised.

2.3.2 Brake assist

The ‚brake assist', which is well known from Mercedes-Benz passenger cars, became a standard function of the brake system. In cases of emergency this function gives support to a hesitating driver and reduces the stopping distance.

The realisation of brake assist functions with electronically controlled commercial vehicle brake assemblies is simpler than with the passenger vehicle brake systems. Passenger cars possess as well known a hydraulically power-assisted braking system, with which the foot pressure of the driver is coupled to the brakes directly. The brake booster supports the operation.

Commercial motor vehicles as standard equipment already possess a power braking system, with which the pressure on the pedal steers the brake system. In the electronically controlled brake system the driver demand is measured by brake pedal sensors and directly interpreted as retardation desire $z_{demand\ driver}$ (figure 5). No auxiliary component is required, but only a software function to realise the brake assist.

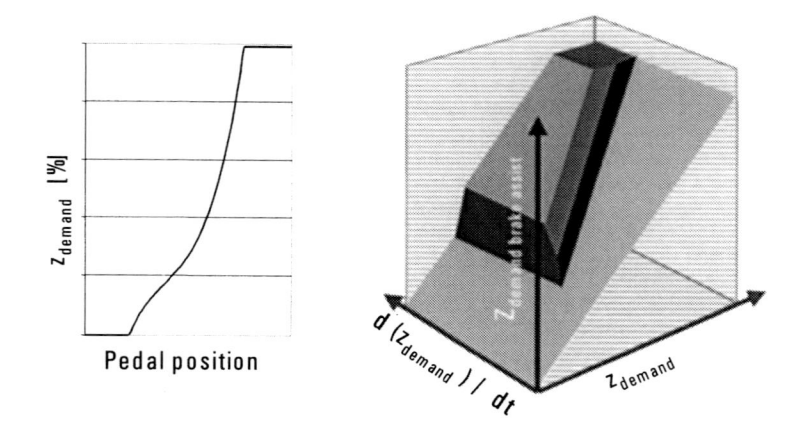

Figure 5. Deceleration rise as function of the pedal operating speed

Investigations on the driving simulator and on test areas showed it is crucial for the success of such a system that an emergency braking is recognised fast and surely, since a high percentage of drivers in emergency situations brakes too unassertively and/or insufficiently. Some drivers step either too unassertively, so that the demanded deceleration is reached only slowly, other drivers depress the pedal not fully enough, so that the maximally possible deceleration is not reached at all.

In the event of an emergency stop, the brake assist is intended to apply a higher braking force than would be requested by the driver if only the brake pedal position were considered. Since the driver actuates the brake pedal very rapidly only in an initial stage in the event of an emergency stop but subsequently hesitates in applying the increased pedal force/travel required to achieve the maximum possible braking effect, valuable stopping distance is lost. This loss of stopping distance can be avoided by using the brake assist. The fast pedal operation that is characterised by an increasing target deceleration is intended to serve as an indicator for an emergency stop. For reasons of driving safety, the ABS must be fully operational.

If the target deceleration rise $\Delta z_{demand}/\Delta t$ reaches a certain threshold, then the desired target deceleration is raised up to 100 % depending upon a number of additional conditions (figure 5). With this measure the maximum possible braking performance can be realised. If the driver decreases during an emergency braking to a lower deceleration level, the increased height is also diminished. This happens again as a function of the target deceleration.

Furthermore it must be pointed out that in general full trailer combinations with conventional brake systems and drum brakes reach a maximum level of approx. 0.6 g. Even with optimal brake equipment hardly 0.8 g deceleration can be attained, since the tyres determine then the physical limit.

2.3.3 Endurance brakes

Endurance braking systems (engine brakes and retarders) are employed in commercial vehicles to allow driving downhill at higher speeds, without permanently using the friction brakes which are inadequate for this purpose, as well as for adjusting braking phases in which case only relatively small speed differences or minor retardations are called for. Fully loaded 40 t combinations can, depending on their equipment and driving speed (> 20 km/h), achieve retardations of up to 9% with the endurance braking system which only acts on the drive axle.

Modern engine braking systems already achieve power ratings of up to 350 kW, depending on the engine speed. As a result, an adhesion of approx. 0.4 on the rear axle is necessary. Hydraulic or electrodynamic retarders can even develop much higher power ratings for a short time.

In contrast to that today brake systems, with which the endurance brake is simply added, in the Telligent®-Brake System the endurance brakes are included so far with, as they can contribute due to grip conditions safely. In order to utilise endurance brake systems optimally, they are actively controlled by the DSC. If the endurance braking system is activated alone, the differential slip limit is increased in relation to the speed so that the braking effect of powerful endurance brakes in the case of a loaded vehicle and good road conditions can be transferred without a control reversal that occurs too quickly. Through permanent calculation of the differential slip it is already possible, long before an intervention by ABS, to reduce and, depending on the subsequent situation, to reactivate the endurance brake effect.

The endurance brakes are activated with almost any deceleration procedure. This happens under permanent monitoring of the adhesion conditions as well as the brake condition. The main goal is to obtain higher lining service lives and to keep the friction brakes as cool as possible by consistent use of the endurance brakes. Communication between the brake system and the endurance braking system occurs via the vehicle CAN-bus and, by means of this, also with the engine control. As a result, the reduction concept functions independently of the type of endurance braking system installed.

The substantial problems, which had to be solved with the development of the integrated endurance brake, lay in the fact that the system must in each case learn the deceleration and slip portion of the endurance brake. Furthermore the different endurance brakes possess very different time constants to obtain their full brake power. This has to the consequence that in the short learning phase the endurance brake is not integrated.

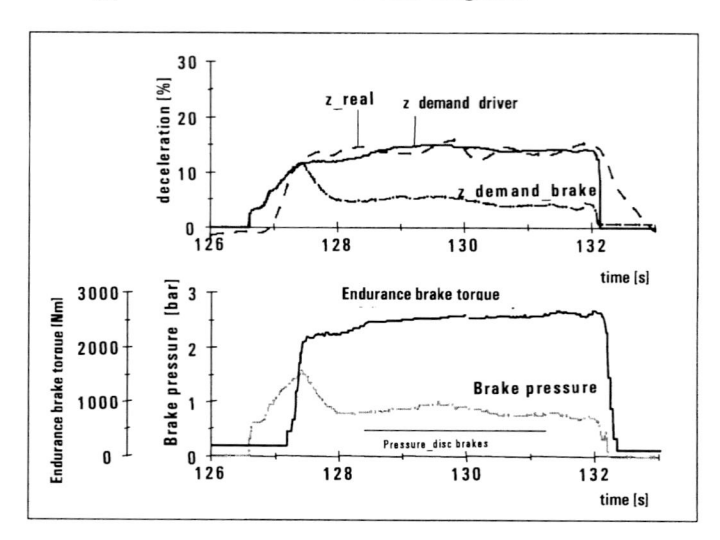

Figure 6. Endurance brake

Since the driver expects an immediate reaction of the vehicle with actuation of the brake, the service brakes must be first operated because of their shorter response times. According to the time performance of the endurance brake torque then the brake pressure at the disc-brakes can be reduced. The wheel brake pressure is however only so far reduced that a stepless and fast reaction is possible when small changes of the target deceleration are required (figure 6). This is however not possible with the today's retarder and engine brake systems.

2.3.4 Lining wear equalisation
Different brake lining wear on the axles of the tractor gives rise to frequent criticism. As the different lining wear mainly depends on the application of the vehicle and the way it is driven, the basic design of the braking system by the vehicle manufacturer can only help to avoid it in the case of average applications. However, the haulier incurs considerable cost due to the different levels of wear: either the brake linings which have worn less are replaced at the same time or the downtime of the vehicle is increased because the brake linings can only be replaced in axle sets. Here, considerable economic advantages can be achieved with the possibilities of an electronic control: The Telligent®-Brake System considers the remaining lining of all tractor wheel brakes during the operation of the vehicle. Non-critical service

brake applications or check braking which make up the great majority of all brake applications, permit a deviation from the optimal brake pressures which go unnoticed by the driver. The brake system uses this situation to counteract different levels of wear on the axles of the tractor, the specified pressure values being corrected in a way to equalise lining wear.

2.3.5 Hill-holder

Purpose of this subsystem is it to relieve the driver of a commercial vehicle when starting at a slope. If the driver starts after a deceleration, then though releasing the brake pedal a constant brake pressure is held and the vehicle cannot roll back. If the driver actuates the clutch and the engine torque increases, then the hill-holder will be disengaged automatically and the driver can start.

3 TELLIGENT®- STABILITY CONTROL

With driving along curves, obstacle-avoidance manoeuvres or rapid lane changes it can come to critical driving situations, which are sometimes no longer controllable by the driver. Also an unfavourable loading situation or a changing roadway condition influence the vehicle stability. Tilting of a commercial motor vehicle occurs almost without preliminary warning for the driver and is usually only then recognisable if it is already too late and manoeuvres do not help any longer. The Telligent®-Stability Control evaluates the actual vehicle dynamics and detects critical deviations like a breaking out, a jackknifing or a tilting and prevents these tendencies, so far physically possible. The driving dynamics of a commercial motor vehicle can vary strongly in contrast to the passenger car. This is due to different axle load, varying loading conditions, different semitrailers and finally by the air brake system.

The drive stability control completes the longitudinal control of the electro-pneumatic brake system with a control of the vehicle's lateral dynamics. The direction of the tractor, which is desired by the driver, will be determined from a steering angle sensor and a suitable reference model. The actual dynamic behaviour of the vehicle will be determined by a combined lateral acceleration and yaw rate sensor. In case of a critical difference of the demanded direction to the actual direction, the drive stability control activates the brakes of one or more wheels of the tractor and - via the brake coupling head - the brakes of the trailer and influences the engine's torque.

The basis of the Telligent®-Stability Control is the Telligent®-Brake System with electro-pneumatic brake pressure regulation. Additionally an ECU with a combined lateral acceleration and yaw rate sensor is needed (figure 7). Since for the stability control functions the lateral acceleration in the centre of gravity is needed, the ECU is frame-mounted and close to the centre of gravity. The substantial functions of the regulation are integrated in the ECU of the stability control, as well as the communication interface to the ECU of the brake system. The data exchange via brake system CAN data bus is similarly to data exchange of the axle modulator. Between steering wheel and steering column a steering angle sensor is integrated. The steering wheel angle sensor is connected to the brake CAN data bus of the brake system.

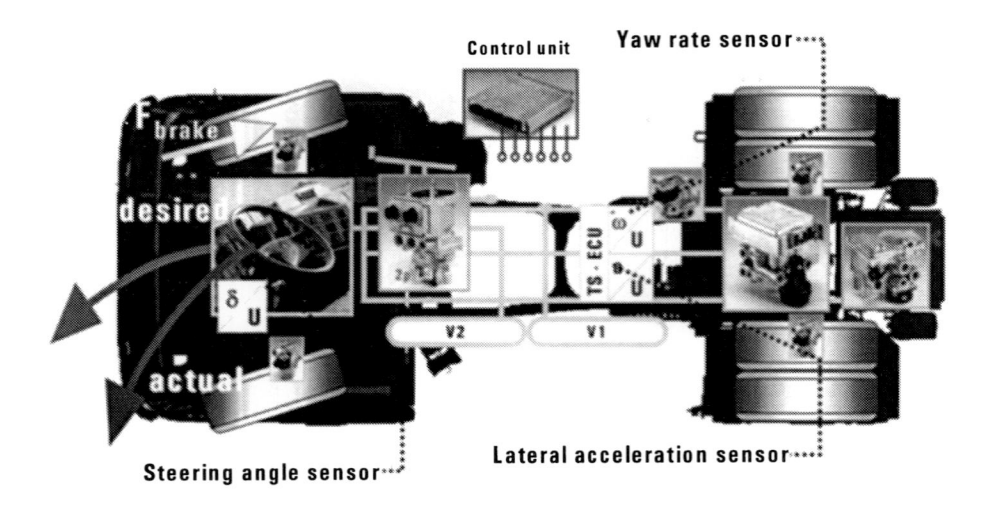

Figure 7. Telligent®-Stability Control

4 TELLIGENT®-ADAPTIVE CRUISE CONTROL

Similarly to the Mercedes-Benz passenger cars for commercial motor vehicles an adaptive cruise control was developed. With a radar detector system ahead-driving vehicles at a range up to 150 m before the own vehicle are identified by the adaptive cruise control and the distance between vehicles, the difference speeds and their changes are evaluated. For the adjustment of the chosen cruise control-speed of the vehicle to the changing traffic conditions vehicle electronics with control mechanism is necessary.

The adaptive cruise control is realised via external controlling of the endurance brakes. If a higher deceleration and also larger changes of deceleration are demanded the service brake with its high brake performance and fast responding will be used.

The electronically controlled brake system activates the friction brakes with the goal of deactivating these again as soon as the adequate endurance brake portion, which is controlled by the differential slip control, is developed.

5 OUTLOOK

5.1 Telligent®-Brake-system 2. Generation

The number of safety systems, which activate externally in the brake system, will strongly increase in the next years. Future systems will recognise the driving conditions by sensors and will automatically prevent accidents. The maximum electronic availability of electrical circuits is therefore a substantial goal of future brake systems.

As already realised with conventional brake systems, in the future the interfaces between brake components of electro-pneumatic brake systems are to be standardised. On this basis brake components of different suppliers can be used alternatively. Besides the basis for a modular brake system is created, with which the different market requirements for brake systems can be fulfilled worldwide. Thus development expenditures can be reduced and the numbers of items will be increased. The Telligent®-Brake System 2. Generation equipped vehicles will still require the air supply system (compressor, air dryer, reservoirs, etc.) since standard technology has not yet been developed to provide sufficient power to apply the commercial vehicle brakes by electrical means. By a standardisation of the interfaces electromechanical brake systems without pneumatic back-up can be integrated in a simple manner. This enables the complete elimination of air from the from the commercial vehicle brake system.

5.2 The next stage in safety

Approximately 50 % of accidents with commercial motor vehicles > 8 t are rear end collision, head-on collision and cross road accident. To reduce these kinds of accidents substantially, systems are necessary, which recognise a potential collision in advance.

The present state of the art of systems which recognise the surroundings of the vehicle are the adaptive cruise control and the lane control system. The lane control system is a safety system without interference into the driving function. The position of the vehicle and the course deviation are recognised optically and an acoustic warning will be activated if the driver is unintentionally straying from this lane.

Future active safety systems, which reduces the probability of accident, must fulfil the following functions (figure 8):

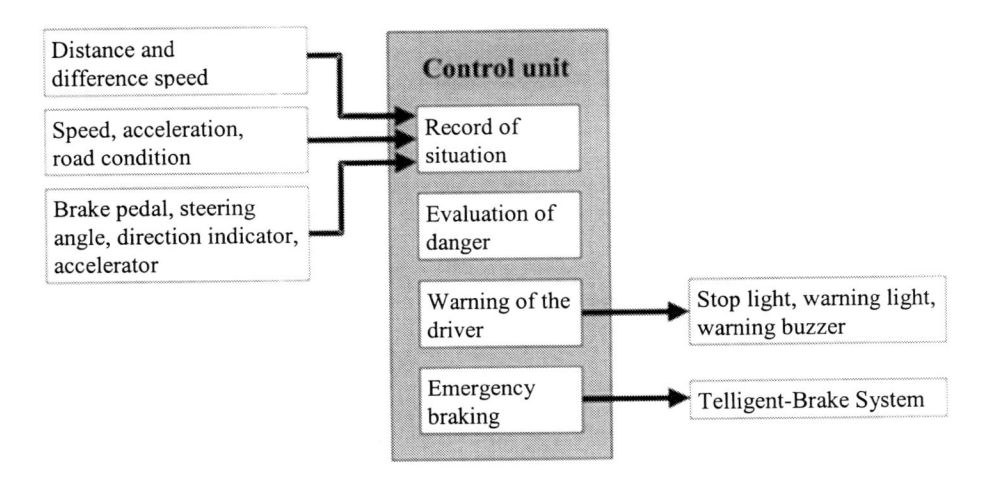

Figure 8. Active safety system

- **Record of situation**
  With the help of the distance sensor distances and difference speeds to moved and standing targets are identified and built up to a traffic scenario.

- **Evaluation of the danger of a rear end collision**
  On the basis of the distance and the difference speed to ahead-driving vehicles is evaluated whether the traffic conditions can lead to a rear end collision.

- **Warning of the driver**
  When the traffic conditions lead to a critical situation and a no driver reaction can be observed an escalating driver warning does take place.

- **Emergency braking**
  If a rear-end collision situation is so critically that the collision can only be prevented by drastic brake, the automatic emergency braking via an active interference into the vehicle brake system takes place, until the driving conditions become uncritical.

LITERATURE

1. **Pressel, J.; Reiner, M.** (1996) Treatment of Endurance Braking and Road Slopes in Electronic Braking Systems, AVAC'96.
2. **Pressel, J.; Reiner, M.** (1995) Die Basisstrategie der elektropneumatischen Bremsanlage (EPB) von Mercedes-Benz, VDI-Berichte 1224
3. **Marwitz, H.; Fischer, J.** (1998) Braking Systems on Heavy Duty Trucks, Truck & Commercial Vehicle International 1998, pp. 13-14.

Wheel movement during braking

J KLAPS
Ford Motor Company, UK
A J DAY
Department of Mechanical and Medical Engineers, University of Bradford, UK

ABSTRACT

An experimental study of wheel movement arising from compliance in the front suspension and steering system of a passenger car during braking is presented. Using a Kinematic and Compliance (K&C) test rig, movement of the front wheels and the suspension sub-frame, together with corresponding changes in suspension / steering geometry under simulated braking conditions, were measured and compared with dynamic measurements of the centre points of the front wheels. The resulting knowledge of front wheel deflections has enabled the causes and effects of steering drift during braking to be better understood in the design of front suspension systems for vehicle stability.

1. INTRODUCTION

The friction brake on each wheel of a motor vehicle generates a braking torque and a resultant braking force which are reacted by the suspension components and the subframe or chassis system [1]. Although the suspension components may be the same side-to-side of the vehicle, the subframe and/or chassis system is generally not symmetrical from side to side. The suspension, subframe and chassis systems are compliant to a greater or lesser extent, and deflections resulting from the braking forces and torques can therefore be responsible for different wheel movements on each side of the car. The kinematic effect of this can be to create dynamic changes in wheel alignment and steering geometry during braking, and on the front wheels, where braking loads are highest, such changes can be a major contributory factor to steering "pull" or "drift" during braking.

Compliance in the suspension system is necessary to achieve a good ride characteristic, but an undesirable side-effect is compliance steer, which results from the application of lateral or longitudinal forces at the tyre contact patch, and is considered to be one of the biggest contributors to straight-line stability during braking [2]. Compliance steer is affected by the design of rubber components in suspensions and can be introduced into suspension systems by elastomeric (rubber) suspension bushes, and rubber mounts for cross-members and steering racks.

Steering "drift" during braking usually refers to a relatively minor deviation from straight-line braking, although even minor deviation remains unacceptable by today's standards of vehicle driveability. Previous work by the authors [3] used vehicle tests to investigate 4 parameters associated with steering geometry, viz. toe-steer, camber, caster, and scrub radius which affected steering drift, and found that compliance in the bushes of the lower wishbone rear bush of the front suspension of the particular vehicle studied had a significant effect on steering drift during braking. Brake "pull" associated with unequal side-to-side braking forces interacting with steering geometry was not found to occur.

The vehicle tests provided an indication of the practical significance of the identified parameters in the generation of steering drift during braking on an actual vehicle. The results of the tests showed clearly that the steered wheels did change their orientation during braking, as measured by the toe steer angle. The results also demonstrated that the most effective means of controlling any tendency towards steering drift during braking was to ensure minimum side-to-side variation in suspension deflection and body deformation both statically and dynamically.

This paper presents a further study of wheel movement and suspension deflection under forces which are representative of those generated during actual vehicle braking. Using a Kinematic and Compliance (K&C) test rig, movement of the front wheels and the suspension sub-frame, together with corresponding changes in suspension / steering geometry under simulated braking conditions, were measured at different levels of suspension movement. These measurements were then compared with dynamic measurements taken from a test car. The result is a better understanding of the causes and effects of steering drift during braking which will assist in the better design of passenger car front suspension systems for vehicle stability during braking.

2. STATIC MEASUREMENTS OF FRONT SUSPENSION DEFLECTIONS UNDER BRAKING FORCES

The test car used for the investigation presented in this paper was a front wheel drive family saloon. The front suspension was a McPherson strut design, with the lower wishbone (also known as the "A-arm") pivoted to a subframe via rubber bushes, while the subframe was mounted to the vehicle body via rubber mounts. The top of the strut was mounted directly to the vehicle body via rubber bushing at the "suspension turrets".

Static measurements were carried out under one author's instruction by IKA (Aachen University) on their Kinematic and Compliance (K&C) test rig facility. The toe-steer and camber angles, caster angle, and kingpin inclination angle were measured by a standard wheel alignment test device. A 3-D coordinate measuring device was used to measure the actual position of the wheel centre points, tyre contact patch centre, strut rotation (top), lower ball joint, and the front and rear mounting point of the sub-frame to the body. The measurement accuracy was estimated to be ± 0.05 mm.

Vertical and longitudinal forces were applied at the positions of the tyre patch centres; the wheels were not included to avoid tyre deflection effects. Full details can be found in [4]. The measurements from the K&C rig are summarized below.

Steering offset
The measured steering offset varied from –6.5 mm at the nominal condition (static load/deflection) to approximately –8.5 mm at 25 mm suspension vertical compression (jounce), as shown in figure 1. The Right side steering offset was slightly greater than the Left side by approximately 1 mm at 25 mm compression.

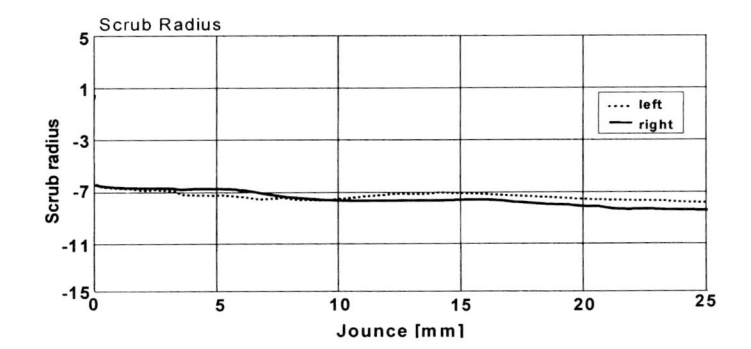

Figure 1: Scrub radius dependent on the jounce

Tyre contact patch centre position
Calculations showed that at maximum measured vehicle deceleration (9.7 ms^{-2}) the brake force at each wheel was 2800 N (front) and 1500 N (rear), so longitudinal forces of these values were applied to each tyre contact patch position on the K&C rig, while the front suspension compression was increased from 0 to 25 mm in 5 mm increments. The results are summarized in figures 2 and 3.

As the suspension compressed, the track increased, but the Right wheel showed a bigger lateral deflection than the Left wheel. As expected, the longitudinal brake forces moved the contact patch backwards; both wheels were moved approximately the same amount. These results confirmed that the steering offset change was different side-to-side, but this difference was small, insufficient to change the steering offset between positive and negative values.

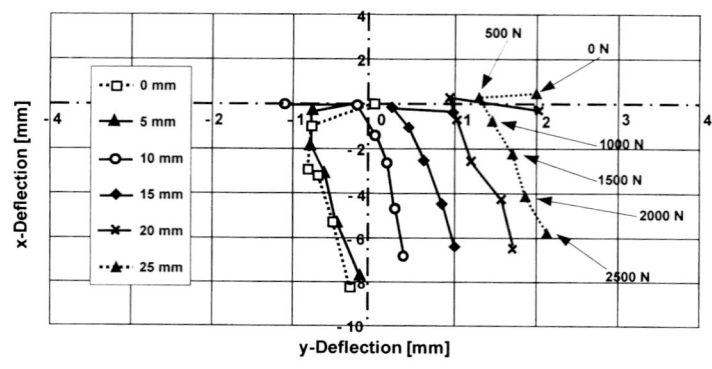

Figure 2: Horizontal deflection of the left wheel depending on compression and brake force

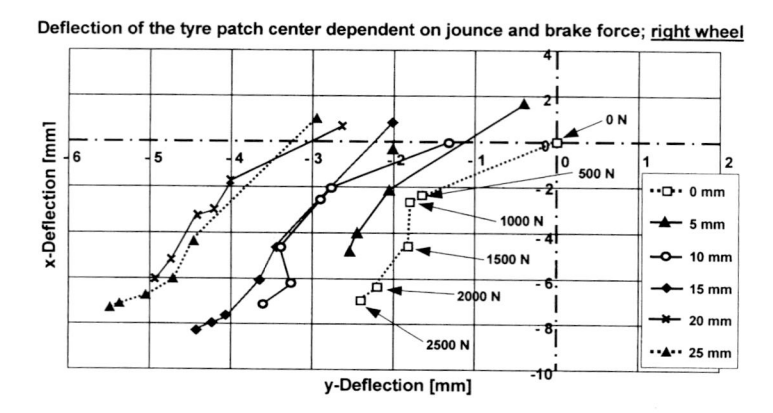

Figure 3: Horizontal deflection of the right wheel depending on compression and brake force

3. DYNAMIC MEASUREMENTS OF FRONT SUSPENSION DEFLECTIONS UNDER BRAKING FORCES

The same test vehicle was used for the dynamic tests as for the K&C tests. Deflections and movements to be measured dynamically fell into two: large movements up to 50 mm (e.g. the suspension vertical movement) and small deflections up to 5 mm (e.g. the deflection of bushes).

The instrumentation had to be tolerant of temperature, vibration and shock, and also as compact and lightweight as possible.

A "Rope Potentiometer" was selected for measuring both cases of movements and deflections. The principle of the rope potentiometer was that one end of an inextensible cord was attached to the point whose movement was to be measured, and the other end was coiled tightly around a drum attached to a rotary potentiometer. As the cord was drawn out, the potentiometer was rotated, and gave a signal proportional to the extension of the cord. This technique was accurate, robust, and convenient for use on the vehicle. Three such potentiometers were required to define precisely the movement of the point of interest in 3-D space, and as an example, the arrangement for measuring the wheel centre position is shown in figure 4. Two of the potentiometers were aligned in the X-Y plane, and the third was aligned in the Z direction. A portable computer with A/D-converter and measuring acquisition software (DIA/DAGO®) was used to log the data [16].

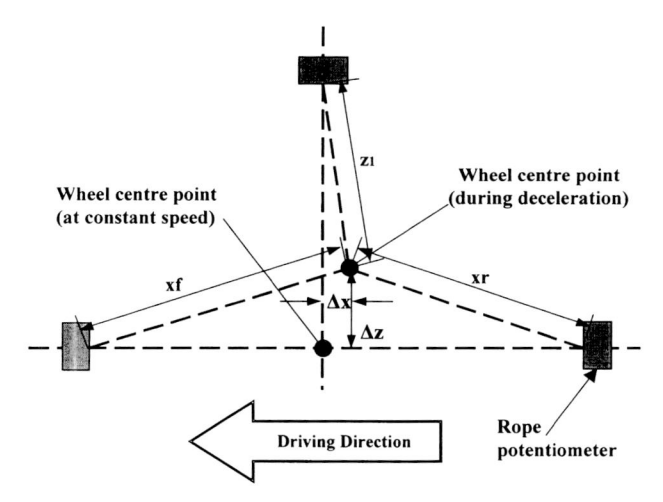

Figure 4: Assembly at the wheel centre point of three rope potentiometers.

Movements and deflections were measured as follows:
1. Subframe relative to vehicle body: 4 points; 2 in X and Y, 2 in X, Y, and Z (X, Y, and Z represent longitudinal, transverse, and vertical respectively),
2. Lower suspension arm deflection (Z),
3. Wheel centre (X, Y, Z),
4. Strut top (X),

The measurement positions are summarized in figure 5. Deceleration and other parameters were also recorded as previously described by the authors [3] [4].

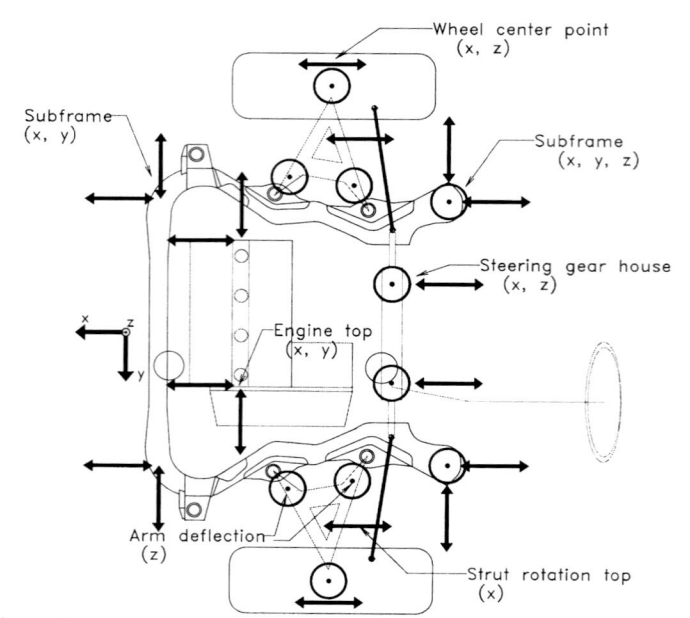

Figure 5: *Measurement positions at the sub-frame, A-arms, strut rotation top, engine and steering gear housing*

Left and Right X deflection of the subframe is shown in figure 6; the subframe moved backwards by approximately 1.55 mm during the test. There was no noticeable difference between "fixed" and "free" control (hands on or off the steering wheel).

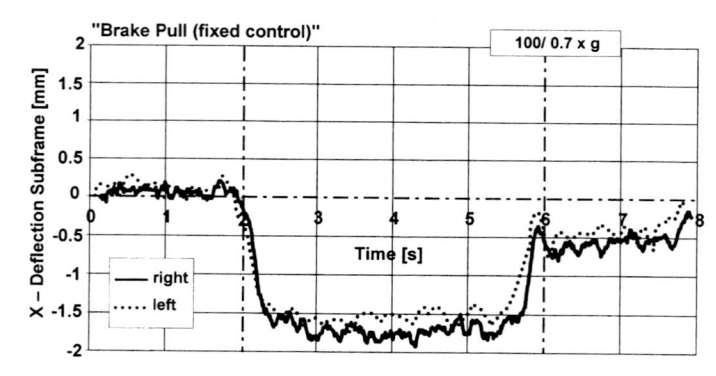

Figure 6: *X-deflection of the sub-frame; fixed control*

At the rear subframe mounting, the measured vertical deflection (Z) was approximately 1.2 mm upwards as shown in figure 7.

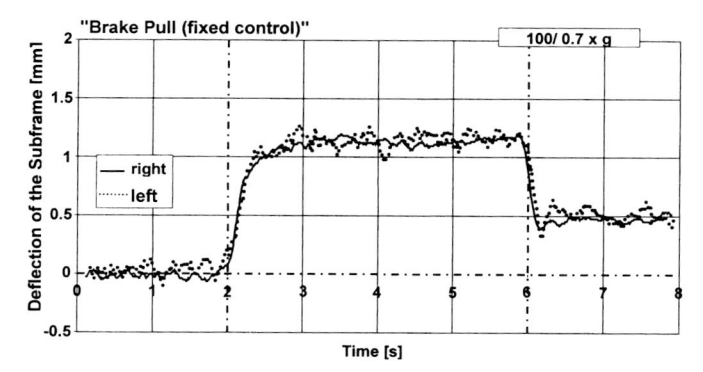

Figure 7: *Z-deflection of the sub-frame, fixed control*

Further analysis of the subframe deflection showed that there was some small "internal" deflection of the subframe (less than 1 mm), in that the front left corner and the rear right corner of the subframe moved closer together.
Because some suspension components are attached to the subframe, and some are attached to the body, these movements and deflections affect the steering geometry.

The vertical deflections of the front and rear bush positions of the lower suspension "A-arm" are shown in figures 8 and 9, which indicate approximately 2.5 mm upwards at the front position and approximately 4.5 mm at the rear position.

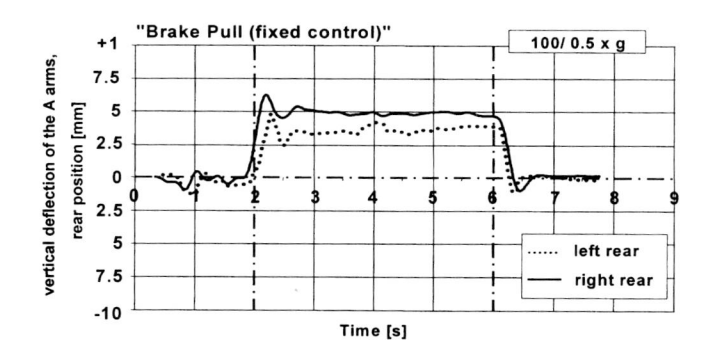

Figure 8: *Vertical deflection 'Z' of the A-arm rear position, fixed control*

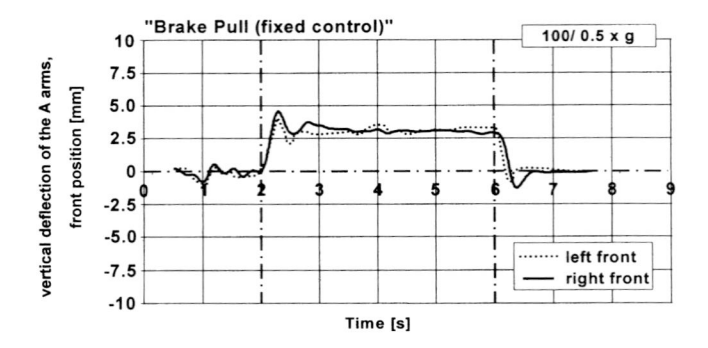

Figure 9: Vertical deflection ‚Z' at the front position at the A-arms, fixed control

The wheel centre movement is summarized in figures 10 and 11 in the vertical and longitudinal directions respectively. The peak vertical movement recorded was approximately 45 mm on the Right wheel and 38 mm on the Left wheel. The longitudinal measurement showed a movement of −10mm (backwards) for the Right wheel, compared with −8 mm for the Left wheel at the start of the test, while towards the end of the test the two sides converged to a value of 9 mm, with a definite indication of greater movement at the Left wheel.

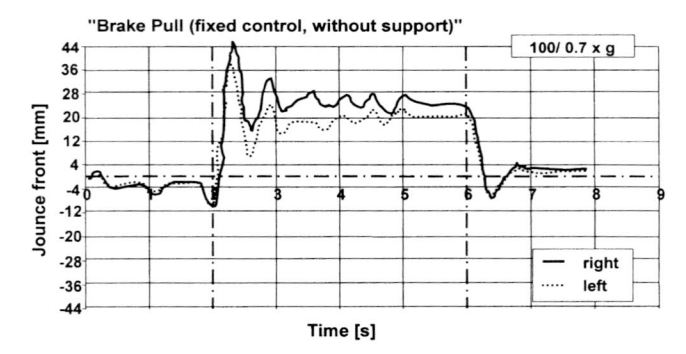

Figure 10: Jounce at the front axle

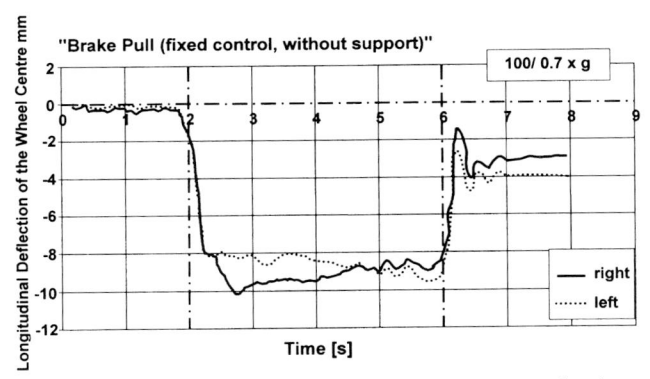

Figure 11: Longitudinal deflection of the wheel centre points, fixed control

The strut top position moved forward by up to 0.75 mm during the test (figure 12).

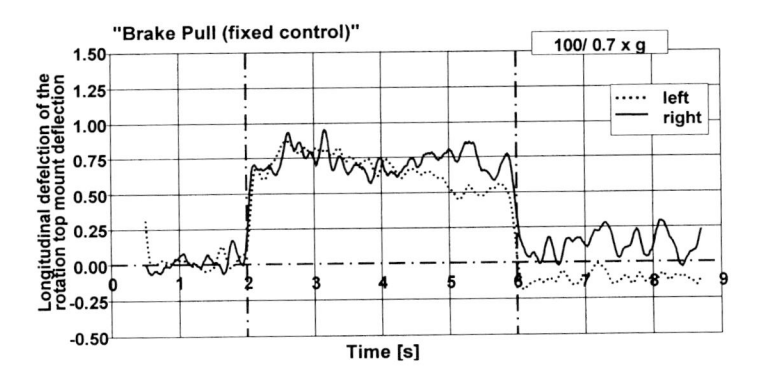

Figure 12: Longitudinal deflection of the strut rotation top, fixed control

4. DISCUSSION OF RESULTS

Both the static tests (K&C) and the dynamic measurements have shown how the suspension geometry can change during braking. The measurements have enabled changes in steering and suspension design parameters to be calculated and their effect analysed. Of particular interest was the change of steering offset and the wheel centre position during braking, which has been measured under static conditions of longitudinal braking force for different amounts of suspension compression. These measurements confirmed that not only was there a side-to-side difference but also this difference depended upon suspension compression (jounce).

Dynamic caster angle was calculated from the measured wheel centre deflection data and is considered in three parts: caster angle, caster trail (at the wheel centre) and caster offset (at the road surface). These are illustrated in figure 13. The reaction force at the tyre contact patch generates a steering force when the caster is non-zero, the magnitude of which depends upon the caster angle and the kingpin inclination. The caster angle is normally designed to be positive to give a self-aligning torque, but if the caster angle reaches a negative value, then the torque works in the opposite way. The results from the dynamic tests indicated that the caster angle did in fact change from positive to negative: this was a compound effect which included a difference of nearly 1½° between nominal and actual (+3° to +1.6° approximately), a non-zero caster trail at the wheel centre, a vehicle pitch angle of up to 1.5°, and longitudinal deflection of the wheel centre relative to the strut top. The net result was that the Right wheel in this case reached a negative caster angle during braking before the Left wheel early on in the brake application. Towards the end of the brake application, both wheels had switched from positive to negative camber, with a consequential loss of self-aligning torque.

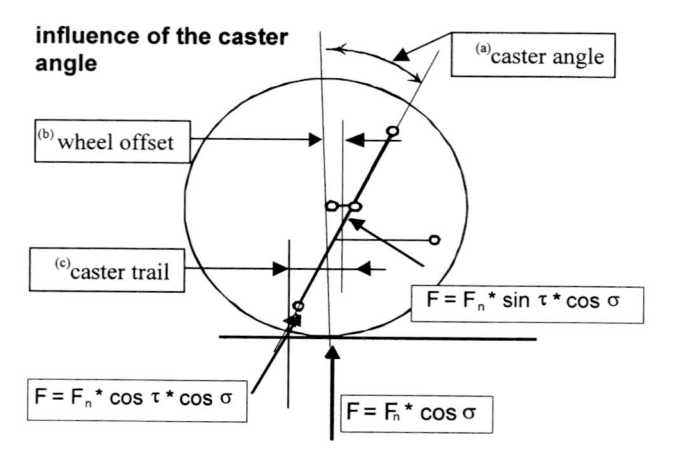

Figure 13: *Caster forces caused by the wheel load*
(kingpin inclination angle $\equiv \sigma$; caster angle $\equiv \tau$)

The maximum values of dynamic caster angle and caster trail are shown in Table 1.

	Nominal value	Maximum dynamic value; Left	Maximum dynamic value; Right
Caster angle (°)	3.00	-0.45	-0.80
Caster trail (mm)	14.64	-1.5	-3.8

Table 1: **Dynamic caster angle and caster trail**

The self-aligning torque arising from the caster is only one of several sources of self-aligning torque, e.g. the pneumatic trail of the tyre, so the change from positive to negative caster angle would not destroy the vehicle stability. However, reduction in self-aligning torque is likely to allow other causes of steering drift to be more clearly felt. This was confirmed in a further test when the suspension was modified to be able to adjust the caster angle. When the settings were adjusted to give the same static caster angle each side, no effect of different caster angles was perceived (subjectively) by the driver. When the static caster angles were adjusted to be different from side to the other, the driver noticed a greater tendency to drift to one side during braking.

5. CONCLUSIONS

The major cause of steering drift during braking has previously [3] been found to be side-to-side dynamic variation in the deformation and deflection of suspension and steering components, and not side-to-side variation in brake performance. The research results presented here confirm that finding, and give more insight into this complicated phenomenon, emphasizing that steering drift during braking is an issue at the system level and not merely component level. The phenomenon cannot be addressed in terms of any single design characteristic of the vehicle suspension or brake system design. It can be concluded that a fully integrated dynamic model of the vehicle chassis will be a most valuable tool in chassis system design for stability.

The accuracy of the measurements made depended upon the transducer accuracy, and then the computational error in the derivation of parameter values. The accuracy was estimated to be no worse than 0.5 – 1%. Therefore it can be concluded that any experimental error is unlikely to affect the results so that their interpretation is invalid.

It is again concluded that control of the compliance at each side of the vehicle is critically important in minimizing steering drift during braking. In addition, though, it is concluded that it is equally important to ensure that the compliance and resulting deflections at both sides of the vehicle are as near the same as possible. Minimizing the compliance overall is helpful in achieving this aim, but this represents a compromise in terms of ride harshness and shock transmission.

Compressing the suspension increased the track width of the test vehicle, and altered the steering offset. The change in steering offset was found to be small in absolute terms (a few mm), and could be different from side-to-side. However, it is also important to note that every change in the steering offset each side will create an imbalance from side to side because of the difference in the steering arm forces. K&C tests are a useful way of measuring static deflection characteristics in a vehicle suspension, although an integrated computer model (as mentioned above) must be seen as the way forward.

ACKNOWLEDGEMENTS

This paper presents research carried out as part of an MPhil. study with the University of Bradford, U.K.. The authors are grateful to all who contributed to the research, including staff in the Ford Motor Company, IKA (Aachen), and supplier companies. Thanks also go to the Directors of the Ford Motor Company for permission to publish this paper.

REFERENCES

1. Holdman, P., Kohn, P., Moller, B, Willems, R., "Suspension kinematics and compliance – measuring and simulation", SAE 980897, 1998.
2. Momoiyama, F., Miyazaki, K., "Compliance steer and road holding of rigid rear axle for enhancing the running straightness of large sized vehicles", SAE 933009, 1993.
3. Klaps, J., Day, A.J., "Steering drift during braking", Brakes 2000 conference, IMechE, 2000.
4. Klaps, J., "Investigation of the effects of the longitudinal stiffness of the engine subframe and suspension system during straight-line braking in passenger cars", MPhil. thesis, University of Bradford, U.K., 1999.

Materials and Modelling

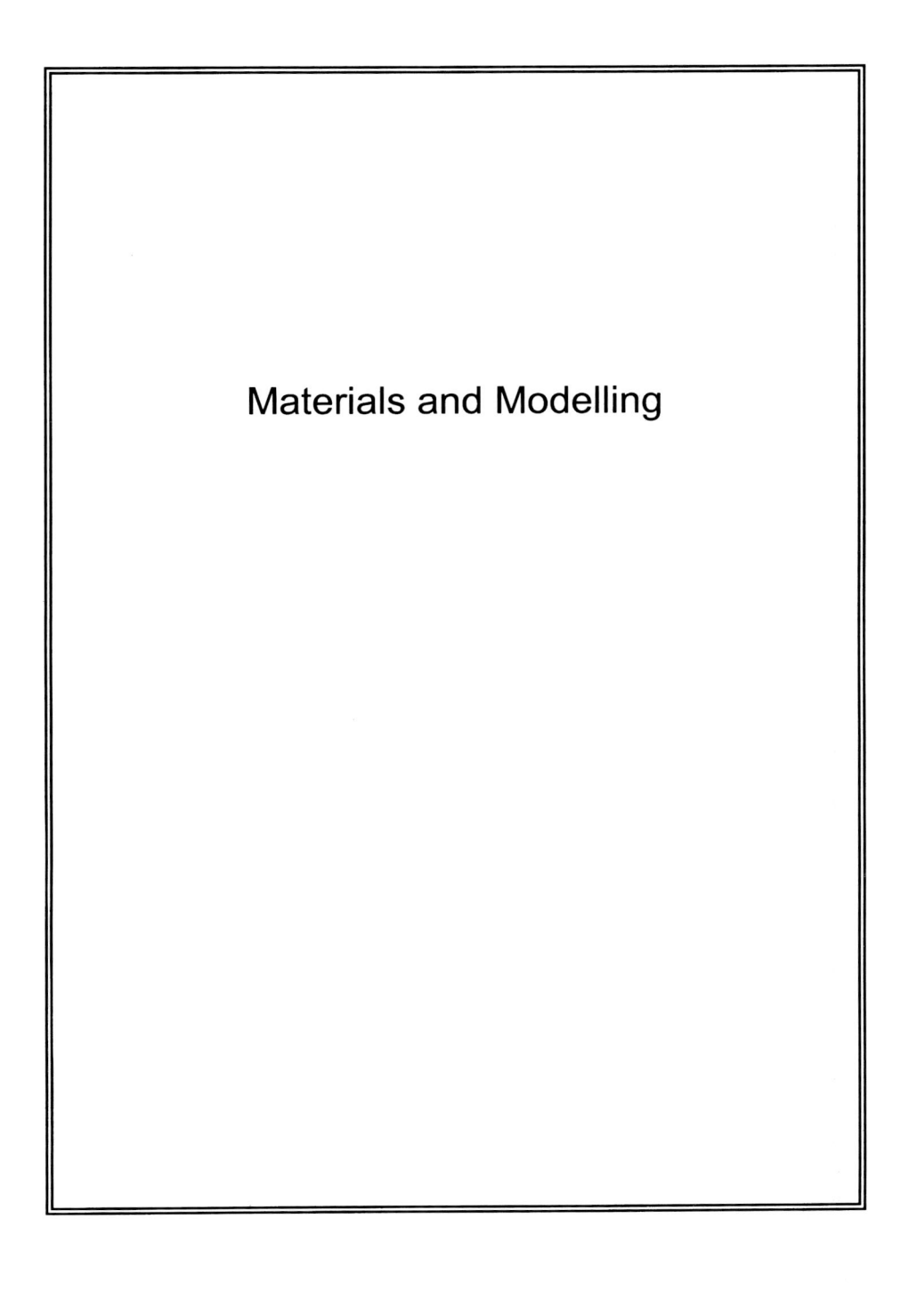

Simulation of brake sequences and corresponding brake wear

R HARJU
Brake and Suspension Systems, Scania CV AB, Södertälje, Sweden

ABSTRACT

In order to theoretically predict the wear of brake discs and brake pads on commercial vehicles a simulation structure has been built up. The structure consists of several independent parts.

The simulation structure works with road measurements. The model uses a mapping of the road topography together with the recorded speed of the vehicle that was used when the road was recorded. These data are used as the foundation of this brake wear simulation.

The results of these simulations give better possibilities to adapt the brakes to the needs of certain trucks or buses. The time for testing is shortened and the designer can get a great help while evaluating different design ideas.

1 INTRODUCTION

Testing a new brake configuration is a very complex undertaking. Consuming a lot of time and recourses with comprehensive hardware testing might seem frightening for designers within product development processes today. The result of a hardware test might also be that the product does not fulfil the requirement it is supposed to.

Using a mathematical simulation can in many cases be a very cost-effective way of developing new products. It shortens the time for testing and it gives the designer big possibilities to test different types of designs. Test methods for disc and pad wear are often very time consuming and therefore costly. However a simulation tool is not complete without a solid foundation of practical knowledge and experience.

To be able to optimise both design and usage of service brakes it is important to understand how the wear mechanisms work. Accelerated test methods in a brake dynamometer can give good estimations of the pad life on a truck in actual use. Difficulties might however occur if you would like pad-life estimations for usage in different types of vehicles. It is for example very hard to estimate the pad life of a city bus using the same pad as on a long haulage truck. The connections are difficult to extract and the influencing parameters are many. In this case the usage of a simulation model can be a great help.

This paper describes a structure of interconnected simulation models designed at Scania CV AB. Many parts of the structure have their roots in thesis-works carried out in co-operation with universities in Sweden. Three different models are used for simulating the wear of brake pads and discs. One model is simulating a truck travelling on a road. The road is built up of measured data from actual roads all over the world. Connected to this model is a model of the driver. The driver "manoeuvres" the service brakes according to a pre set pattern. The result of the interaction of these two models is then distributed to a third model, the wear model. The wear model uses several parameters from the first simulations and calculates the temperature of the pad and the disc. The wear is calculated using parameters calculated from dynamometer test data.

The result of the model can be used in many ways. Not only can different types of pads be evaluated on different road routes but also different types of design solutions for different brake configurations can be tested. The knowledge extracted from work with simulation models gives the designer good possibilities to plan further testing in the most cost-effective way. The usage of simulation models is expected to shorten the time for hardware testing and thus lowering the costs.

Further development of the simulation models is in progress, the work is expected to enhance the accuracy of simulation result and to make the model even more adaptable to different types of changes in the structure.

2 TRUCK SIMULATION(1)

The base of the simulation structure is an advanced model of a truck driving a recorded road. The initial purpose of this model was not to analyse the usage of brakes. The model was primarily design to simulate the fuel consumption of a truck. The model gives the user possibilities to test different types of gearboxes, axles etc in different driving conditions.

The model uses a mapping of the road topography together with the recorded speed of the vehicle that was used when the road was recorded. During the simulation the simulated truck tries to follow the recorded velocity.

The truck in the simulation uses the same sources of brake torque as a real truck. The engine drag torque is dependant on what gear is engaged and thus the engine speed. Also the choice of engine affects the drag torque. As a choice in the program also an exhaust brake can be used enlarging the engine drag torque. In the simulation model a retarder can be used applying a brake torque to the prop-shaft. If these brakes are not sufficient to maintain the correct speed of the truck the service brake must be applied. The driver model determines the application time, duration and magnitude of the service brake application.

3 DRIVER MODEL (2)

Designing a good driver model is perhaps the most complicated task of all. The drivers come in different varieties and shapes. Even if only one driver is used to create the model he cannot be described as simply another equation. In the driver model several limitations must be made to the driver's behaviour.

The driver model is designed to apply the brakes at a certain deviation of the simulated truck velocity from the recorded velocity. When the simulated speed is lowered the driver is designed to release the brake. The magnitude of the brake application is set to achieve the mean brake power in an average brake application.

However measurements have shown that the same limits in the speed deviation are not valid for all roads. Using one set of conditions in simulations on one road can correspond well to actual braking cycles. Using the same set of conditions on another might however correspond with less accuracy. The reason for this difference is not fully clarified but the research continues. The simulated brake signal can be compared with a measured brake signal in Fig 1.

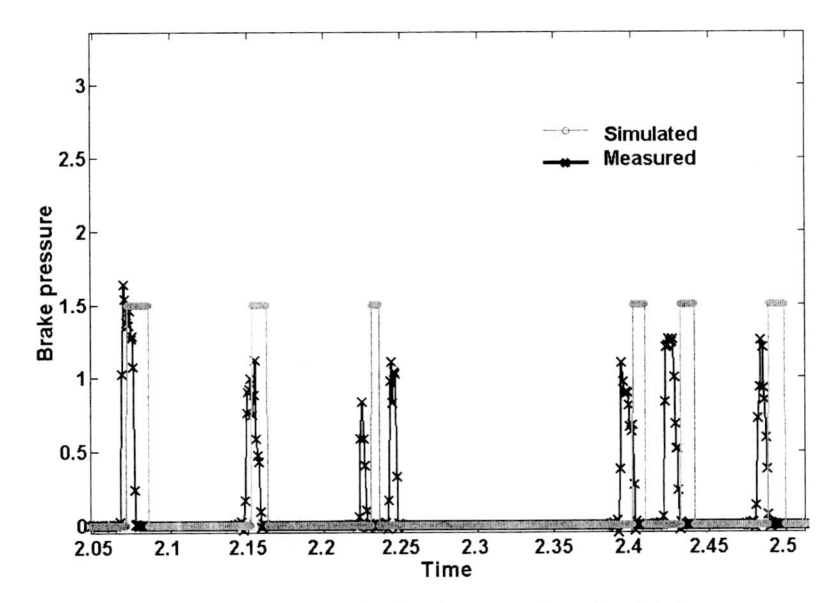

Fig 1 Measured brake signal compared to simulated.

4 WEAR SIMULATION [3]

The wear simulation can be divided into two parts: one part calculating the temperature of the brake pad and the brake disc and another part calculating the wear.

4.1 Temperature model
The temperature of the pad and the disc is calculated using a simple one-dimensional FE model. The heat increase in the disc and the pad is generated by the energy transferred into pad and disc during the brake sequence.

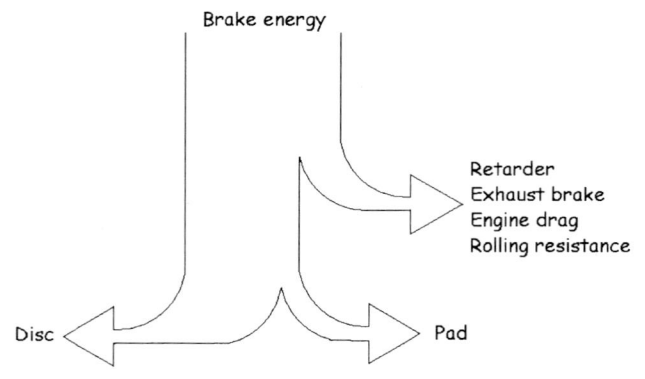

Fig 2. Brake energy distribution.

The disc is modelled using five elements and the pad is modelled using twenty-five elements. In this simulation only one side of a disc is used assuming that the same amount of energy is transferred into each side of the disc. During the brake sequence the brake power absorbed by one half of a disc is transferred into the pad and the disc. The relation between the energy put into the pad and the energy put into the disc is described by Formula (1) below [4]:

$$\frac{Q_{disc}}{Q_{pad}} = \frac{A_{disc}}{A_{pad}} \cdot \frac{\sqrt{\lambda_{disc} C_{v,disc}}}{\sqrt{\lambda_{pad} C_{v,pad}}} \qquad (1)$$

By using characteristic parameters for the pad and disc this formula shows that approximately forty times more energy is dissipated into the disc than into the pad. Due to its size, mass and heat conduction the pad however showed much higher temperature during the brake sequence, see Fig 3.

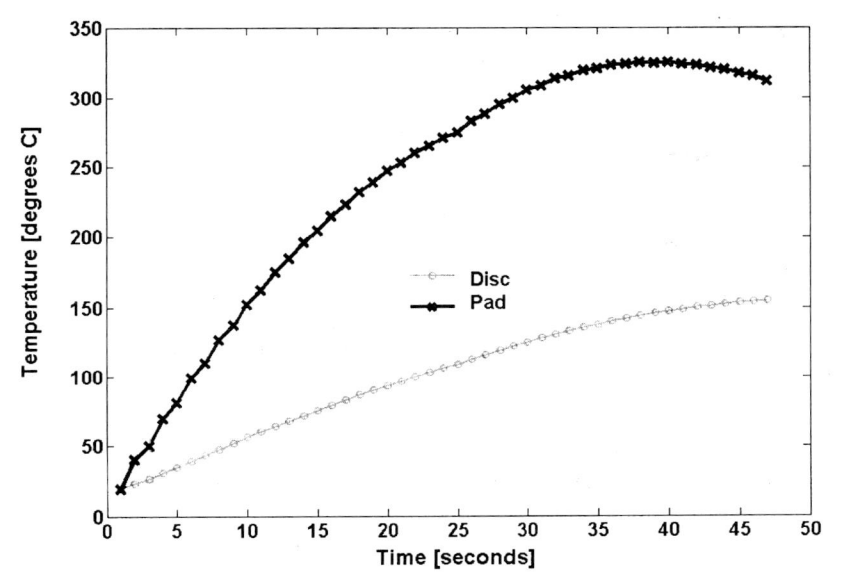

Fig 3 Simulated pad and disc temperature during braking sequence

The temperature within the pad and disc however turned out to vary in another way. Due to very low heat conductivity of the pad the temperature gradient in the pad was very large. In the disc the temperature was more evenly distributed.

4.1.1 During cooling sequence

The dissipation of accumulated heat from the constituent parts of the brake is a very complex process depending on many different factors. In order to make a reliable but yet simple model some assumptions were made. The heat radiation from brake parts was not calculated, nor was the conduction of heat into surrounding parts, such as rims and axles. Instead the convection was used to compensate for these factors.

The convection of heat varies with the velocity of the air passing above the surface of the heat emitting body. Due to the complex physical structure of a brake this velocity cannot be easily calculated. Therefore an assumption has been used also in this case. The velocity of the air passing over the surface of the brake parts is set in constant relation with the speed of the truck.

Using the measured data from proving ground test-drives a good appreciation of the temperature process during cooling sequence at different vehicle speeds was obtained. The convection could thus be adjusted accordingly. Different types of vehicles could though result in different airflow patterns and the consequence will be different cooling-rates.

Measurements from the Scania proving ground test-drive were used validating the mathematical temperature model. The validation showed that braking sequences were proportionately easy to simulate, cooling sequences though often resulted in larger errors. The simulated temperature was within 20 degrees Celsius of the measured signal during 1800 seconds of driving.

5 THE WEAR MODEL

The wear process in vehicle brakes is very complex depending on many different parameters. There are also difficulties when measuring the wear of pads and discs because the wear is hardly measurable after just a few braking sequences. Assuming the wear is correctly measured another problem occurs: parameters on which the wear surely depends can vary in a wide range during a braking sequence, for example the pad surface temperature can rise from 100°C to 700°C during braking sequence. Another problem is the non-homogeneity of pad material, composed of several different components whose individual wear characteristics are not known.

5.1 Archard's law of wear
A widely used mathematical definition of wear is Archard's law of wear [5]:

$$w_z = \frac{K}{H} \cdot F \cdot s \qquad (2)$$

The parameters K and H are material parameters and can not easily be determined. The quotient (K/H) can be assumed to vary with the temperature. According to Archard's law of wear the most significant parameters are the distance travelled and the force pressing the bodies together. Disregarding the variation of friction between bodies due to temperature changes this force is in direct relation with the braking force and thus the deceleration. In short the wear can be related to the braking energy.

It can be assumed that the wear of rubbing elements can be described similar to:

$$w_z = f_1(T) * f_2(W_{Brake}) \qquad (3)$$

Elaborating with different sets of parameter combinations in line with Formula (3) resulted in the usage of a polynomial on the form shown in Formula (4):

$$w_z = c_1 \cdot T^4 + c_2 \cdot F \cdot Q_b + c_3 \cdot F + c_4 \cdot Q_b \cdot T^4 \qquad (4)$$

Formula (4) was implemented in the wear model and a standardised wear test also performed in brake dynamometers was simulated. A set of four linearly independent equations was extracted. The set of equations was solved using actual wear measurements made in the brake dynamometer. The set of equations was solved with back-substitution and the constants c_1-c_4 were determined. The function simulating wear of a specific brake pad and disc combination is compared to real measurements of wear in Fig 4.

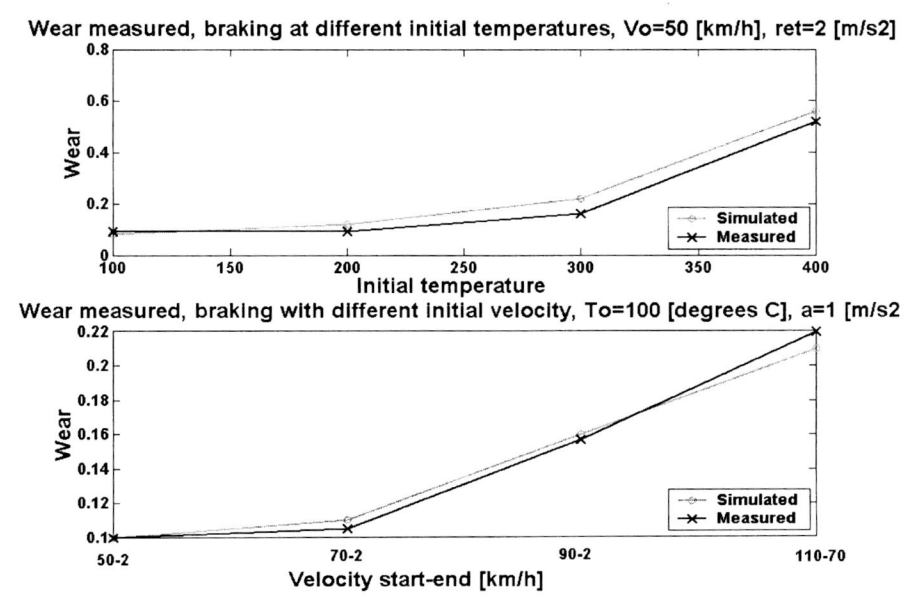

Wear measured, braking at different initial temperatures, Vo=50 [km/h], ret=2 [m/s2]

Wear measured, braking with different initial velocity, To=100 [degrees C], a=1 [m/s2]

Fig 4 Simulated and measured wear in dynamometer tests.

6 CONCLUSIONS

The simulation tools described in this paper are very simple but yet useful in the product development process. There are of course more advanced models available taking into account other factors like the influence of for example dust and road pollution. These parameters are however very dependant of specific roads. Using only the temperature can give a good picture of the temperature resistance of different pads. The usage of simulation models is expected to shorten the time in the development process and thus lower the costs.

7 REFERENCES

[1] Sandberg T. (2000) *STARS Model, Vehicle model for Scania Truck and Road Simulation*, internal report, Scania.
[2] Spagnoli A (2001) Modularization of heavy vehicle service brakes, Scania/Royal Institute of Technology
[3] Harju R, Gutgesell R (2001) Simulation of brake wear in heavy vehicle applications, Scania/Royal Institute of Technology
[4] Hohmann C. (1999) Simulation von Versleiss an Scheibenbremsbelägen, Shaker Verlag.
[5] Archard J F (1953) Contact and rubbing of flat surfaces. J Appl. Phys. 24: 981-985

Analysis of automotive disc brake cooling characteristics

G VOLLER and **M TIROVIC**
Department of Mechanical Engineering, Brunel University, Uxbridge, UK
R MORRIS and **P GIBBENS**
ArvinMeritor Heavy Vehicle Braking Systems (UK) Limited, Cwmbran, UK

SYNOPSIS

Studies of automotive disc brake cooling characteristics have been performed experimentally using a specially developed Spin Rig, as well as numerical modelling methods, FE and CFD. All modes of heat dissipation have been studied in detail and a drag brake application modelled to determine the contribution of individual modes to the disc cooling.

The Spin Rig experiments enabled the determination of the thermal contact resistance between the disc and wheel carrier. The analyses demonstrated the sensitivity of this mode to clamping pressure. For convective cooling, very similar results were obtained using Spin Rig experiments and CFD analyses. The nature of radiative heat dissipation implies substantial effects at high temperatures. The results indicate substantial change of emissivity throughout the brake application.

1 INTRODUCTION

To achieve efficient brake operation, acceptable temperatures must be maintained during all service conditions. Throughout a heavy-duty single brake application virtually all the thermal energy is absorbed by the brake disc and pad, the short braking time does not allow significant heat dissipation. Brake disc thermal capacity must be sufficient to ensure acceptable temperature rise, keeping brake temperatures within the safe operating range. For repeated brake applications or drag braking, the brake disc must be able to dissipate the thermal energy that is generated. If temperatures are allowed to become too high, deterioration of the brake structural integrity and friction performance will take place. Excessive thermal loading can result in surface cracking, judder and high wear of the rubbing surfaces. High temperatures can also lead to overheating of brake fluid, seals and other system components.

The problems of adequate brake cooling are associated with all brake types and heat dissipation has attracted much investigation and research ([1] to [12]). Convection is considered to be the most important mode of heat transfer, dissipating the highest proportion of heat to surrounding air in most vehicle braking conditions. Conduction is the least studied mode of heat transfer, whereas radiation significantly contributes to heat dissipation at high temperatures. In this paper, all three modes have been studied experimentally and their effects analysed theoretically using CFD and FE modelling. The brake assembly studied in this paper is a front ventilated commercial vehicle (CV) disc brake as shown in Figure 1 and detailed in Table 1.

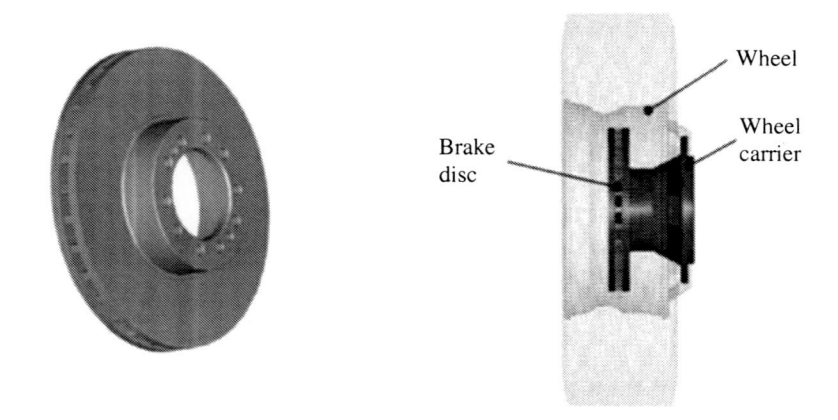

Fig. 1 Commercial vehicle front ventilated brake disc and wheel assembly

Table 1 Brake and wheel assembly data

Disc cheek OD	434 mm
Disc cheek ID	234 mm
Disc thickness	45 mm
Number of vanes	30
Disc mass - grey cast iron	33 kg
Wheel carrier mass - SG cast iron	21 kg
Wheel mass - steel	34 kg
Wheel diameter	22.5"

2 TEST FACILITY

The experimental part of the study has been conducted using a specially developed Spin Rig, see Figure 2 and Table 2. The Rig has a simple, in-line arrangement of the disc, torque transducer, speed sensor and electric motor. The Spin Rig is used for measuring disc cooling and airflow characteristics.

Cooling tests are performed on electrically or flame heated discs, enabling high test repeatability, uniform temperature distributions and accurate study of the main mechanisms of heat dissipation at different disc rotational speeds. Temperatures are measured throughout the tests using rubbing and imbedded thermocouples, as well as infrared sensors. Figure 2 shows the Spin Rig with the CV disc, wheel carrier and wheel mounted (see also Figure 3). It was considered that the tyre does not influence disc cooling in still air conditions and has not been used in tests.

The Spin Rig also enables the measurement of the torque required to spin the brake disc. This can be done at elevated or room temperatures to verify CFD results and determine the power required to rotate the disc (pumping losses).

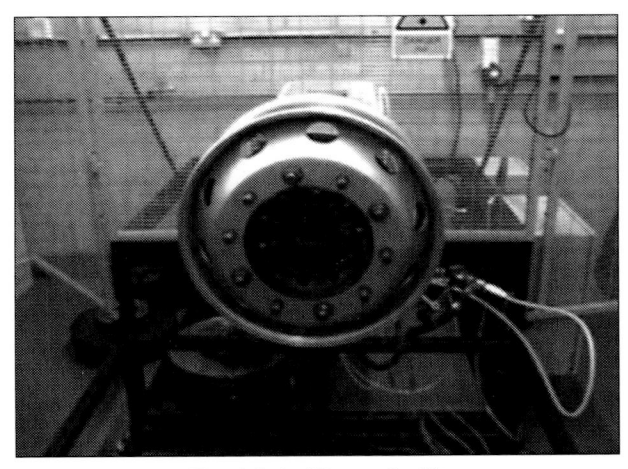

Fig. 2 Spin Rig test facility

Table 2 Spin Rig specification

Motor speed	0 – 2000 rpm
Motor power	7.5 kW
Torque transducer	0-50 Nm
Maximum rotating mass	250 kg
Maximum rotating diameter	1.05 m
Data logging	Speed, torque, temperatures
Electric heaters	8 kW

The Spin Rig tests are less expensive to prepare and run than brake dynamometer tests, due to much lower cost and complexity of the equipment, lower energy consumption and manpower required. The Spin Rig usage is aimed at experiments related to brake disc heat dissipation and airflow characteristics and not at studies of friction couple or brake structural integrity. Compared to dynamometers, achieving high disc temperatures is somewhat difficult on the Spin Rig. In addition, the heat is not generated by rubbing the friction pads against the disc surface, which results in a different condition of friction surfaces compared with discs used on dynamometers and vehicles. However, the above effects can be taken into consideration by adequate measurements and calculations. As a result, the prediction of 'real friction brake' temperatures can be very accurate.

3 EXPERIMENTAL STUDY OF CONDUCTIVE HEAT DISSIPATION

The cross section through the CV wheel assembly, as mounted on the Spin Rig, is shown in Figure 3. The wheel assembly provides two areas of conductive heat dissipation from the disc, one through the bearing assembly, the other through the wheel carrier. Heat transfer from the brake disc to the bearing must be avoided and bearing temperatures kept low. However, the wheel carrier has substantial mass and provides a conduction path from the brake disc to the wheel of the vehicle. The outer faces of the wheel carrier and wheel have direct contact with cool, fast flowing and turbulent air and these boundary conditions suggest the wheel carrier and wheel can offer substantial potential for heat dissipation. It is important at this stage to consider the tyre, overheating of this component can lead to extremely dangerous conditions and must be avoided.

Experiments have been carried out to determine the effects of clamping force and surface conditions on the thermal contact resistance (TCR) at the disc flange/wheel carrier interface. Figure 3 shows the experimental set up, the brake disc and wheel carrier are bolted to the Spin Rig shaft adapter. The shaft adapter is insulated to prevent heat conducting to the shaft. The brake disc is surrounded by an insulation box (not shown in Figure 3) and heated. The disc and wheel carrier are insulated (see Figure 3) to prevent convective and radiative heat dissipation. The heat is conducted from the disc to the wheel carrier. Embedded thermocouples have been used to determine the temperature gradients at the disc flange/wheel carrier interface. The measured temperatures enabled the calculation of the TCR.

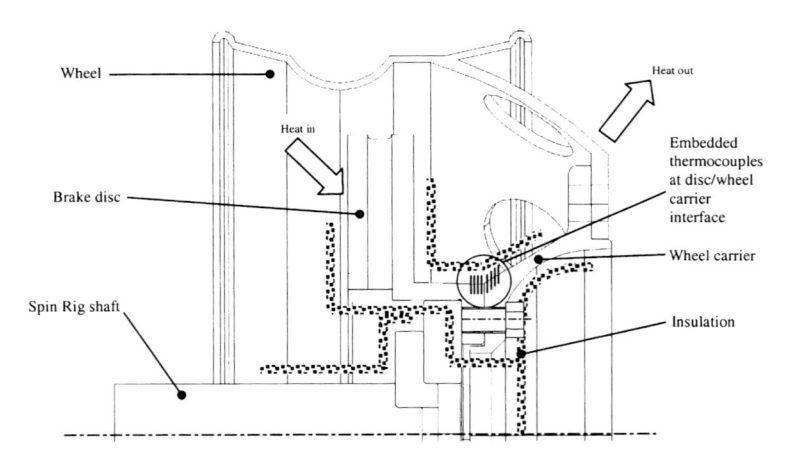

Fig. 3 Thermal contact resistance experiment set up

Figure 4 shows the TCR in the form of a heat transfer coefficient (HTC) for a standard interface condition. For comparison reasons, all heat dissipation results (conduction, radiation and convection) are presented as HTCs in W/m^{2}K. The temperature measurements are taken in two distinctive areas, between the bolts and at the proximity of the bolts. The average contact pressure (total bolt clamp force divided by the contact area) varied between 10 and 50 MN/m^2, the maximum value corresponding to the nominal bolt tightening torque. The HTC increases with the increase in contact pressure and the relationship was found to be linear (Figure 4). Also, the HTC is lower at the locations between the bolts than at close proximity to the bolts (the local contact pressures are higher in this area).

The values of HTC are very high, between 5000 and 10000 W/m^{2}K, for the nominal bolt torque. Since the conduction is speed independent, conductive heat dissipation can be a significant component of brake disc cooling, particularly at low vehicle speeds. To use this advantage, high temperature difference between the disc and wheel carrier is necessary. The effect was also observed by Sheridan [1] for a passenger car fitted with aluminium wheels.

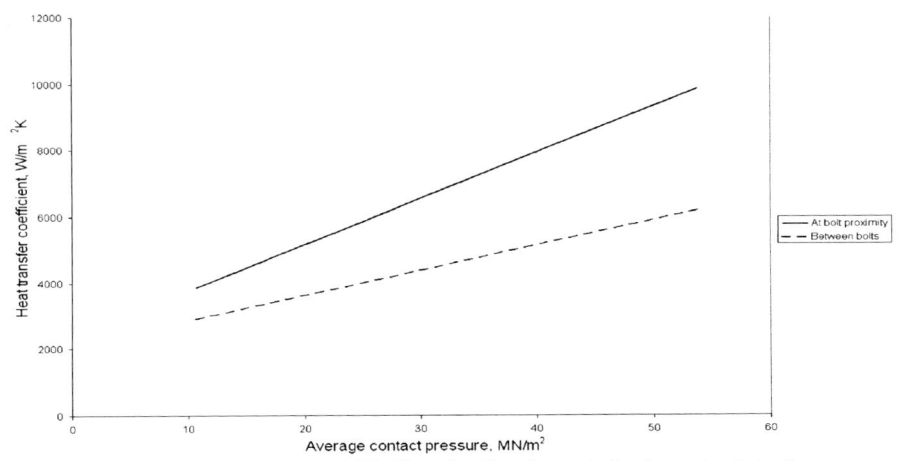

Fig. 4 TCR in the form of HTC at the disc flange/wheel carrier interface

4 STUDIES OF RADIATIVE HEAT DISSIPATION

Radiative heat dissipation is determined by the emissivity and temperature of the component surface. Emissivity values for brake discs found in the literature vary quite considerably. Limpert [2] has recommended an emissivity value of 0.55 for cast iron discs. Noyes and Vickers [3] assumed all emissivities to be 0.8 with a background temperature of 38°C. From thermal imaging work carried out on a brake dynamometer, Grieve et al [4] used an emissvity value of 0.4 for cast iron and MMC rotors. Eisengräber et al [5] showed that the surface emissivity needed to be constantly corrected during the braking application, the measured values varying between 0.15 and 0.9 in different dynamometer tests. During a drag test, emissivity values increased from 0.4 to 0.7.

The authors of this paper have measured the emissivity of the CV disc by comparing thermocouple measurements with infrared sensor readings and adjusting the emissivity to match the temperature values. An emissivity of 0.2 was found for a new, machined brake disc surface at temperatures 20 - 200°C. For a corroded disc surface, the emissivity value was much higher, over 0.9 for the wide temperature range between 20 and 600°C.

Based on referenced emissivity values, Figure 5 shows the calculated HTC values for the disc surface throughout the disc operating temperature range. The dashed line shows the possible HTCs for a brake disc during a drag brake application as per Eisengräber et al [5]. The curves show that significant radiative heat dissipation occurs even at low temperatures, at 100°C the radiative HTC is about 4 W/m^{2}K for an emissivity of 0.55. At 400°C radiative heat dissipation becomes very significant, the HTC values vary between 6 W/m^{2}K for the lowest emissivity (0.2) and 27 W/m^{2}K for the highest emissivity considered (0.9). At 600°C, the HTCs vary between 12 and 52 W/m^{2}K, for the lowest and highest emissivity values.

The results illustrate the importance of correct emissivity values for reliable brake cooling prediction. Furthermore, the effect of radiation being reflected back to the disc by the wheel carrier, wheel and dust shield must also be considered in the analysis of brake disc heat dissipation.

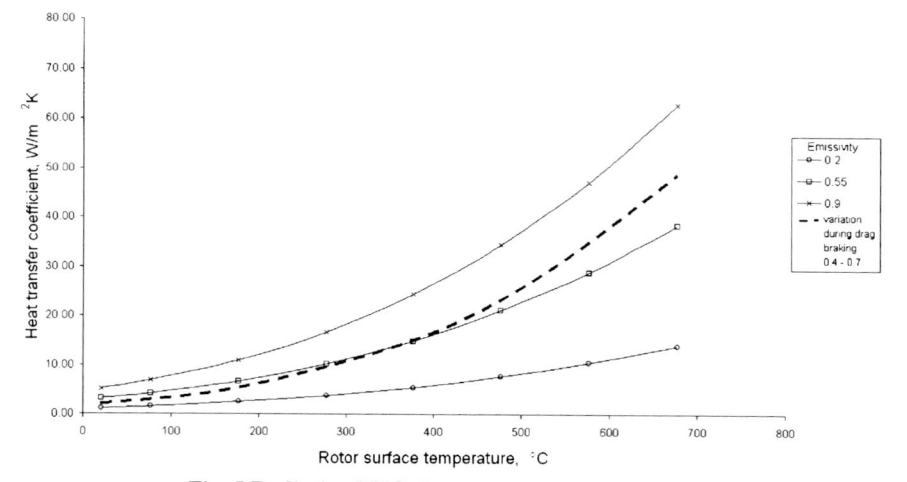

Fig. 5 Radiative HTC change with disc temperature

5 EXPERIMENTAL STUDIES OF CONVECTIVE HEAT DISSIPATION

Experimental studies of convective heat dissipation have been performed on the Spin Rig, shown in Figure 2. Using electric heaters, the disc was first heated to around 200°C. Then, in the cooling phase, the heating was switched off and temperatures measured at different rotational speeds, from 40 to 450 min^{-1}, corresponding to normal vehicle operating speeds. The tests were performed in still air, for the disc only and the wheel assembly, which included: the disc, wheel carrier and wheel (the arrangement shown in Figure 3, but with insulation removed). The result repeatability was excellent, which confirmed the potential advantages the Spin Rig offers in thermal brake analysis.

In Figure 6, two cooling curves are shown, for 40 and 450 min^{-1}, for the disc only (without the wheel carrier and the wheel), from the initial temperature of just over 160°C. As expected, the cooling is much more efficient for the higher rotational speed. Since the other two modes of heat dissipation, conduction and radiation, are speed independent, it is possible to calculate convective heat losses. The HTCs have been calculated and are discussed in the following chapter, with the CFD results.

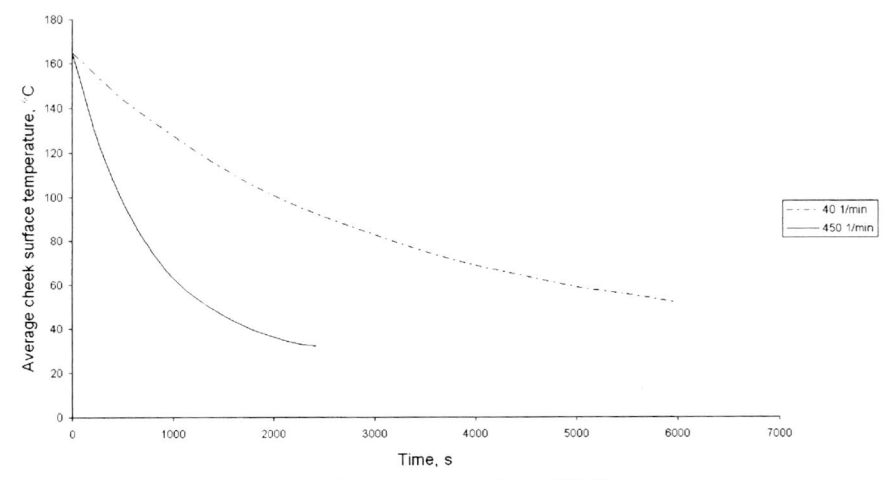

Fig. 6 Cooling curves of the CV disc

6 ANALYSES OF HEAT DISSIPATION USING COMPUTATIONAL FLUID DYNAMICS (CFD)

For the CFD analyses, the SDRC I-DEAS package was used, the ESC module for the flow analysis and the TGM module for the transient thermal analysis. Figure 7 shows a half section temperature contour map of the fluid surrounding the rotating disc mounted to the Spin Rig shaft, for the steady state conditions. The disc is rotating at 450 min^{-1} and its temperature is 600°C. The ambient air temperature was set at 20°C. It can be seen that the analysis has been performed for the 'disc only', the wheel assembly is not included. The wheel assembly would inevitably influence the temperature contours, reducing disc cooling. The work including the wheel assembly is currently underway.

Figure 7 shows substantial difference in the air temperature and flow pattern between the inboard and outboard side of the disc. The average air temperature is higher on the outboard side (hat side) of the disc than the inboard side. The air temperature in the vane channel shows approximately linear increase, from around 50°C at the entry to just over 100°C at the vane exit.

The CFD analyses provide other output results (air speed distribution, HTC distribution, etc.), the most interesting one for this research being average HTC distribution. A number of CFD analyses have been performed for two values of disc temperature, 100 and 600°C and different disc rotational speeds. The values of average convective HTCs are shown in Figure 8, together with the convective HTCs calculated from Spin Rig temperature measurements at 100°C (as discussed in section 5).

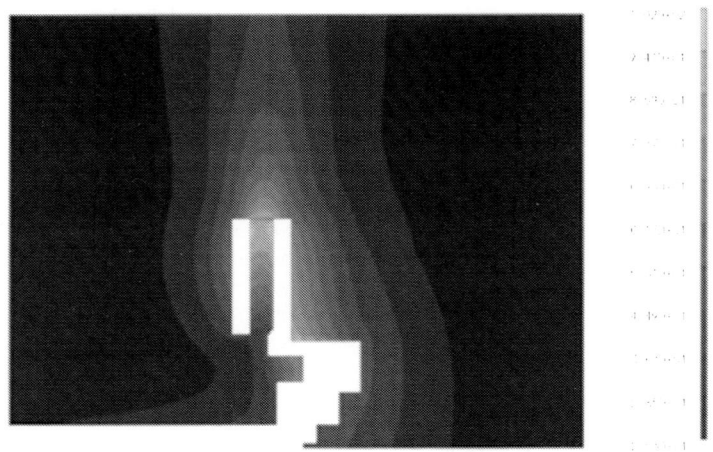

Fig. 7 Air temperature for a 600°C CV disc rotating at 450 min^{-1}

Experimental values show good agreement with CFD results, particularly at mid range rotational speeds. The differences increase at the low and high ends of the speed range, the CFD analyses predicting HTCs about 10% higher at maximum vehicle speed. To achieve greater accuracy and include the complete wheel assembly, further research is currently being conducted. The results are also in general agreement with data published by other authors ([2], [6], [7]).

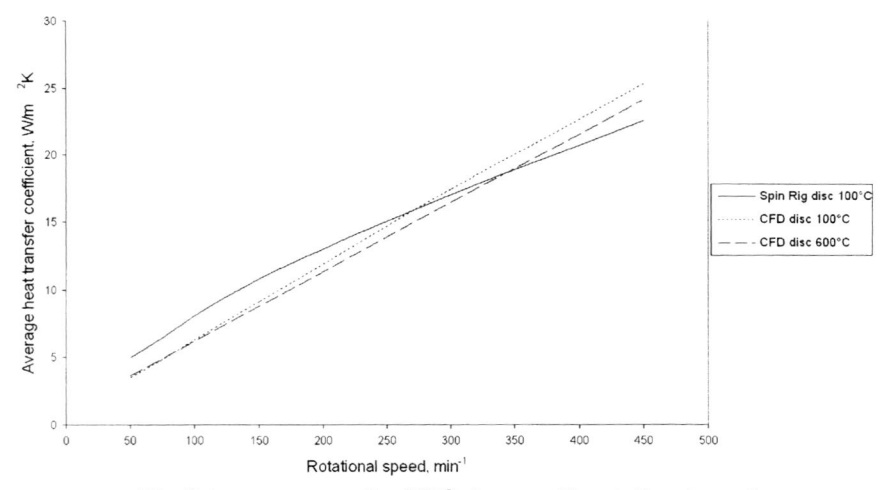

Fig. 8 Average convective HTC change with rotational speed

7 APPLICATION AND DISCUSSION OF RESULTS

Based on previous research (discussed here and in [12]), a finite element brake model was created and drag brake application modelled, as defined by ECE R 13 (type 2, 30 km/h on 6% incline). Appropriate boundary conditions were included and convection modelled as analysed, for 'disc only' in still air. Radiation was modelled assuming change of the emissivity with temperature [5], as shown in Figure 5. Since the wheel was not included, reflective radiation has not been modelled.

The thermal FE modelling of drag braking was performed until the disc surface reached 600°C. The results of total heat dissipation and contribution of individual modes are shown in Figure 9. The results show that the total heat dissipation from the disc at 600°C during the drag braking (rotational speed of 150 min^{-1}) is about 11.5 kW. The contributions of the individual modes of heat dissipation are (approximately): conduction 2 kW (18%), convection 4.5 kW (39%) and radiation 5 kW (43%). It is interesting to note that more heat is dissipated by radiation than convection. The results have been expanded for the range of rotational speeds, from the very low 40 min^{-1} (8 km/h) to the maximum vehicle speed at 450 min^{-1} (90 km/h).

Obviously, conduction and radiation are speed independent, and the ratio of individual contributions changes with speed. At the minimum speed considered, convection contributes by only about 18% to the total heat dissipation. However, at maximum speed, convection is the dominant cooling mode, accounting for 57% of dissipated heat.

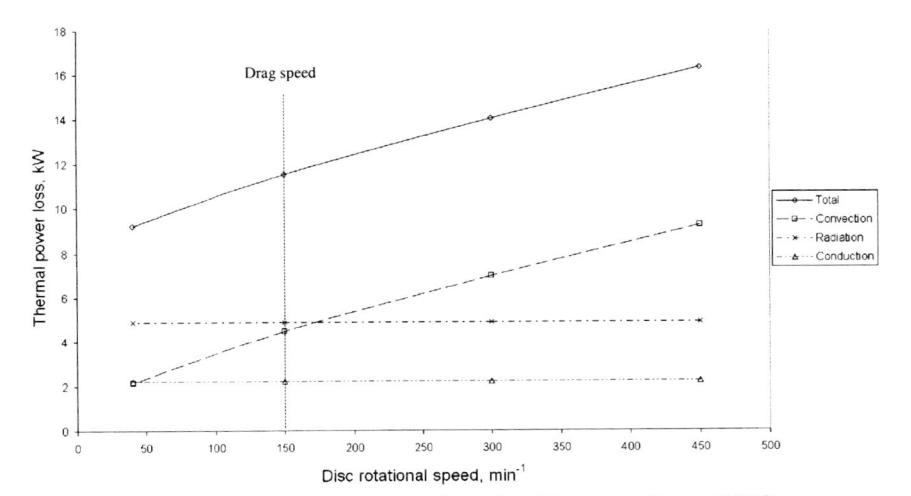

Fig. 9 Heat dissipation by each mode of heat transfer at 600°C

8 CONCLUSIONS

The experimental results have shown that the Spin Rig is a very valuable piece of equipment, providing accurate results for studying all modes of heat dissipation. Further improvement would require a more powerful brake disc heating system.

Conductive studies at the disc/wheel carrier interface showed considerable thermal contact resistance variation between the areas 'in the proximity of bolts' and 'between the bolts'. For the nominal bolt tightening torque, the values of HTC vary between 5000 and 10000 W/m^2K, for the 'standard' slightly corroded contact surfaces. These values are very high, and can be further increased by treatment of contact surfaces. However, a relatively small contact area and low temperature differences between the brake disc flange and the wheel carrier limit the overall gain in heat dissipation.

Radiative heat dissipation is defined by disc temperature, the emissivity of the surfaces and the amount of heat radiated (reflected) back to the disc from the surrounding components. The data and analyses suggest substantial emissivity rise with temperature, which further increases the contribution of this mode of heat transfer at high temperatures. Unfortunately, the heat radiated back can be substantial, therefore the design and surface condition of surrounding components must be very carefully chosen. Radiation remains to be the cooling mode that 'ultimately' prevents disc overheating at high temperatures.

Convective heat dissipation is determined by the airflow in and around the disc. Convective cooling is strongly influenced (increased) by rotational speed and is slightly reduced with temperature rise. Furthermore, cross flow, although not specifically studied in presented research, requires particular attention due to substantial influence. Unfortunately, brake designers are often unable to influence overall vehicle aerodynamics, which must ensure sufficient supply of cool air to ensure maximum benefits of convective cooling. The Spin Rig and CFD results showed good agreement and further work is underway to ensure the best possible solutions for convection, the most dominant cooling mechanism in the majority of driving conditions.

ACKNOWLEDGEMENT

The work presented in this paper was made possible by the assistance obtained from the EPSRC (Engineering and Physical Sciences Research Council), ArvinMeritor (UK) and Brunel University, Department of Mechanical Engineering. The authors are grateful for this help.

REFERENCES

1. **Sheridan, D. C., Kutchey, J. A. and Samie, F.** (1988) Approaches to the thermal modeling of disc brakes. SAE, 880256.
2. **Limpert, R.** (1975) Cooling analysis of disc brake rotors. SAE, 751014.
3. **Noyes, R. N. and Vickers, P. T.** (1969) Prediction of surface temperatures in passenger

car disc brakes. SAE, 690457.
4. **Grieve, D. G., et al.** (1998) Design of a light weight automotive brake disc using finite element and Taguchi techniques. Proc. Instn. Mech. Engrs., Vol. 212, Part D.
5. **Eisengräber, R., et al.** (1999) Comparison of different methods for the determination of the fiction temperature of disc brakes. SAE, 1999-01-0138.
6. **Newcomb, T. P.** (1979) Thermal aspects of railway braking. Proc. Instn. Mech. Engrs., C154/79.
7. **Fukano, A. and Matsui H.** (1986) Development of the disc-brake design method using computer simulation of heat phenomena. SAE, 860634.
8. **Daudi, A. R., Narain, M.** (2000) CAE prediction and experimental verification of maximum temperature of cool running 72 curve fin brake rotor design, International conference on brakes 2000. Leeds, UK.
9. **Morgan, S. and Dennis, R. W.** (1972). A theoretical prediction of brake temperatures and a comparison with experimental data. SAE, 720090.
10. **Newcomb, T. P., Millner, N.** (1965-66) Cooling rates of brake drums and discs. Proc. Instn. Mech. Engrs., Vol. 180, Pt. 2A, No. 6.
11. **Sisson, A. E.** (1978) Thermal analysis of vented brake rotors. SAE, 780352.
12. **Tirovic, M. and Voller, G.** (2002) Optimisation of heat dissipation from commercial vehicle brakes. FISITA 2002 World Automotive Congress, Helsinki, Finland.

Interface temperatures in friction braking

H S QI, K NOOR, and **A J DAY**
Department of Mechanical and Medical Engineering, University of Bradford, UK

Synopsis
Results and analysis from investigations into the behaviour of the interfacial layer (Tribo-layer) at the friction interface of a brake friction pair (resin bonded composite friction material and cast iron rotor) are presented in which the disc/pad interface temperature has been measured using thermocouple methods. Using a designed experiment approach, the interface temperature is shown to be affected by factors including the number of braking applications, the friction coefficient, sliding speed, braking load and friction material. The time-dependent nature of the Tribo-layer formation and the real contact area distribution are shown to be causes of variation in interface temperatures in friction braking. The work extends the scientific understanding of interface contact and temperature during friction braking.

1. INTRODUCTION

Friction brakes are required to transform large amounts of kinetic energy to heat at the contact surfaces between the disc and the pad. The temperature distribution in the friction pair caused by this heat is a complex phenomenon which affects the braking performance directly, and has been investigated by many researchers over many years [1 – 5].

1.1 Multiple levels of contact and local interface temperature

It has been found that measured interface temperatures are generally greater than those which are predicted by single contact models at very low Peclet numbers, due to interacting effects of the multiple heat sources [3]. Using a nominal contact area in an approximate single contact temperature expression allows an average surface temperature rise due to the multiple heat sources to be derived. A similar approach has been employed in surface temperature calculation at high Peclet numbers for temperature-wear maps. In the multiple contact case, which includes most sliding situations involving bodies of finite thickness, there is an additional surface temperature rise (macro-level) besides the localized flash surface temperature found at the contacting asperities (micro-level). Such macro-level temperature increases can affect localized regions over the entire nominal contact area and could be considered to be the nominal contact temperature since these are the regions where sliding contact actually occurs, usually represented as a "hot spot". The temperature at the hot spots can be well above the bulk temperature of the contacting bodies, and, being a nominal contact temperature rise due to frictional heating, is significant in many dry sliding systems, yet is seldom considered in surface temperature calculations. The temperature rise due to local contact at a hot spot has been analysed [3]. At and near the real contact region, a sharp, non-linear temperature drop is caused by the 'small scale' heat flow restriction, which can be modelled as a moving heat source over a semi-infinite medium. The strength of the moving source is the frictional heat entering the body over the real area of contact. Further away from the contact interface a temperature drop is caused by the 'large scale' heat flow restriction. Fourier's heat conduction law governs that temperature drop, with the heat flux in this case being the frictional heat divided by the entire nominal area swept by the moving source. The total, or maximum local, contact temperature was expressed as

$$T_{local} = \Delta T_{local} + T_{nominal} = \Delta T_{local} + \Delta T_{nominal} + T_{background} \qquad (1)$$

This shows that local temperature is always higher than the nominal temperature and the background temperature.

1.2 Local interface temperature measurement

It would be desirable to make interface temperature measurements during actual friction braking tests. However, measuring the interface temperature of friction pair is a difficult task. Several methods have been used to monitor the disc/pad interface temperature [6 - 8], and these can be categorized into two groups: non-contact measurement including methods such as optical and infrared measurement, and contact measurement including methods such as thermocouple and temperature sensitive material coating (or paint). For example, the embedded thermocouple method has been used to investigate thermoelastic instability in an automotive disc brake system experimentally under drag braking conditions [6]. The onset of instability was clearly identifiable through the observation of non-uniformities of temperature measured. The disadvantages of embedded thermocouples include low signal-to-noise ratio due to steep temperature gradients in the pad, and relatively poor time response. A high definition thermal imaging system (Agema Infrared Systems) was used by Allied Signal to assist in the development of new friction materials for brake pads in the automotive, aerospace and rail industries [8]. Surface and near-surface temperatures were monitored at various locations in a disc brake during drag-type testing. The recorded transient temperature distributions in the friction pads and infrared photographs of the rotor disc surface both showed that contact at the friction surface was not uniform, with contact areas constantly shifting due to non-uniform thermal expansion and wear. The disadvantage of the infrared method is its sensitivity to surface emissivity, and affected for example by deposits of wear dust.

In the study presented in this paper, an exposed thermocouple method has been used, which is believed to be a more practical method for measuring the disc-pad interface temperature. The friction braking process is very complex, as many factors affect the frictional forces generated. The complexity is further increased when the interaction between disc and pad at both macro and micro level is included. The main objective of this work is to understand which factors affect most significantly friction braking performance, especially local interface temperature. A statistical design of experiments (DoE) approach was used for the effect analysis in this investigation.

2 EXPERIMENTS

2.1 Measurement system

The thermocouple measurement method, as shown in Fig. 1, is based on the assumptions that the quantity measured by the thermocouple is an aggregate ElectroMotive Force (emf) generated at the thermocouple wire tip and the disc interface, and that the emf obtained by this method corresponds to the local disc and pad interface temperatures. Fig. 1a and 1b show the wire thermocouple (WIRE) configurations used; the thermocouple wire used was 1 mm in diameter. To assess the temperature signals obtained by WIRE, an exposed probe thermocouple (PROB) was used at same time. The WIRE and the PROB thermocouples were located within the disc-pad contact zone, as shown in Fig. 1c. In addition, a conventional rubbing thermocouple (RUB) setup, which is commonly used for monitoring disc temperature, and a dummy thermocouples (DUM) were used for system automatic control and failure diagnosis. The RUB and the DUM were located at a distance from the apparent disc and pad contact zone, as shown in Fig.1c. All these thermocouples, i.e. WIRE, PROB, RUB and DUM, were installed on a small sample friction test dynamometer. This test rig is hydraulically actuated and computer controlled, with a sliding speed range from 5 – 16 ms^{-1}, and power dissipation in the range 0 – 10 MWm^{-2}. The rotor is a 125 mm diameter disc, and the friction material test sample is approximately 25 mm square. The signals were acquired and processed by a computer aided data acquisition system.

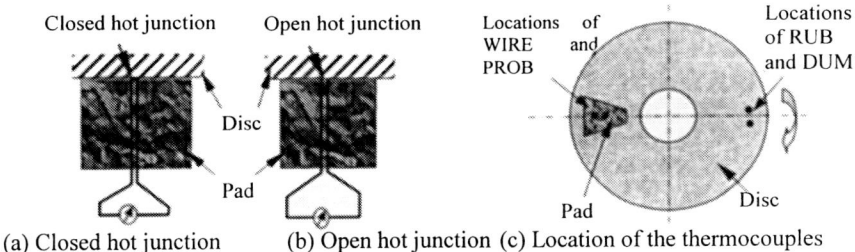

(a) Closed hot junction   (b) Open hot junction   (c) Location of the thermocouples
Fig.1 Thermocouple temperature measurement

2.2 Design of experiments

After a series of preliminary tests, a statistical design of experiments (DoE) was established. To study the effects of braking factors on the braking performance, a four factor, two level full factorial orthogonal design was selected as summarized in Table 1. The main outputs (or responses) from this DoE were the interface temperature, T, (measured by WIRE), and the friction coefficient, μ.

2.2.1 Factors and levels

The control factors and factor levels used in this investigation were:
1. Load applied (A): *150* and *300 N*
2. Speed (B): 6.5 and 16.5 ms^{-1} (*1000* and *2500 Rev/min*)
3. Pad materials (C): *M* and *P*
4. Disc/pad surface conditions (D): *SC1* and *SC2*

The pad materials used were M: a woven asbestos free friction lining, and P: a compression moulded, heavy-duty brake lining for commercial vehicle drum brakes. Disc/pad surface conditions were controlled as $SC1$ and $SC2$. Under the test condition $SC1$, where the number of braking applications is from 1 to 7, the pad and disc surfaces were prepared to the "new" condition before each test, using abrasive paper. Under the test condition $SC2$, where the number of braking applications is from 8 to 14, the pad surface was "used" but the disc surface was prepared to the "new" condition before each test, using abrasive paper. The following conditions were set throughout the experiments: (1) loading time is *20 s*; (2) number of applications for each test is 7; (3) preheating load is *450 N*; (4) preheating temperature is *80 ºC* and (5) cooling temperature is *80 ºC*.

2.2.2 *Experimental design*
An $L_{16}(2^4)$ Latin Square [9] was used for the four factor and two level full factorial experiments as shown in Table 2. The columns in the table represent four factors and six interactions between any two factors in the table. The rows in the table are the 16 tests (T1 to T16) carried out in the experiments.

Table 1 Conditions used in the designed experiments

Factors	Level (-1)	Level (+1)	Assigned Letter
Load (N)	150	300	A
Speed (rpm)	1000	2500	B
Material	M	P	C
Disc/pad surface cond.	SC1	SC2	D

Table 2 $L_{16}(2^4)$ Latin square used in the DoE

Test No.	1 A	2 B	3 A*B	4 C	5 A*C	6 B*C	7 D	8 A*D	9 B*D	10 C*D
1	-1	-1	1	-1	1	1	-1	1	1	1
2	-1	-1	1	-1	1	1	1	-1	-1	-1
3	-1	-1	1	1	-1	-1	1	1	1	-1
4	-1	-1	1	1	-1	-1	1	-1	-1	1
5	-1	1	-1	-1	1	-1	-1	1	-1	1
6	-1	1	-1	-1	1	-1	1	-1	1	-1
7	-1	1	-1	1	-1	1	-1	1	-1	-1
8	-1	1	-1	1	-1	1	1	-1	1	1
9	1	-1	-1	-1	-1	1	-1	-1	1	1
10	1	-1	-1	-1	-1	1	1	1	-1	-1
11	1	-1	-1	1	1	-1	-1	-1	1	-1
12	1	-1	-1	1	1	-1	1	1	-1	1
13	1	1	1	-1	-1	-1	-1	-1	-1	1
14	1	1	1	-1	-1	-1	1	1	1	-1
15	1	1	1	1	1	1	-1	-1	-1	-1
16	1	1	1	1	1	1	1	1	1	1

3 EXPERIMENTAL RESULTS

3.1 Typical temperature signals and reliability assessment of the measuring system
Fig. 2 shows a general picture of the temperature signals obtained during the friction braking tests using the scale test rig. This shows that the signals responded to the preheating periods

(p_i), brake loading periods (a_i), and the cooling periods correctly. The reliability of the measuring system was assessed by analyzing the RUB signals, which has been commonly used for monitoring disc temperature during braking. As shown in Fig.2a, during individual braking application the temperature indicated by RUB increased from 80 °C, which was the preheating temperature and cooling temperature. After each application, the temperature fell steadily to 80 °C and then increased again as the next application started. At the 7th application under the test T7 (-1, +1, +1, -1) in Fig. 1b, for example, the RUB temperature increased from 80 °C to 120 °C and then fell after the braking load is removed. The RUB temperature increased when the braking load or sliding speed increased as shown in Fig. 3a. The temperature increased by about 30°C, for example, when the speed increases from 1000 rpm (T3) to 2500 rpm (T15) in Fig. 3a. The results based on the RUB temperature measurement indicated that the measuring system was reliable and correctly representative of the test conditions and the braking process. Fig. 3b shows that using different temperature sensors, i.e. WIRE, PROB, DUM and RUB, under the same test condition can result in the different temperature outputs. The signals from different methods responded to the friction braking process showing particular characteristics.

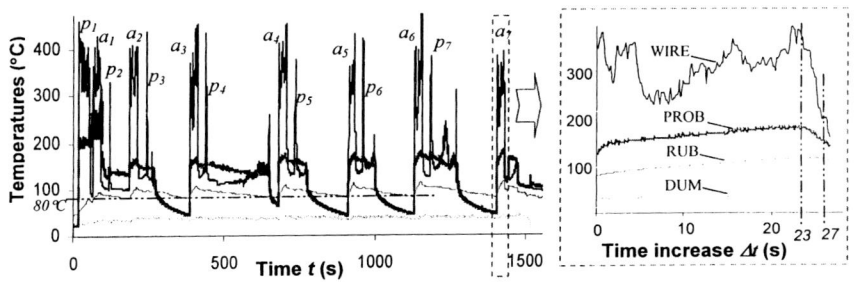

(a) The signals in T7: a_i, p_i the ith application period   (b) In the 7th application period
Fig. 2 A typical signal recorded from the test T7 (-1, +1, +1, -1)
Speed is 1000 rpm (A = -1); Load is 300 N (B = +1);
Material is P (C = +1); Disc/pad surface condition is $SC1$ (D = -1)

3.2 Temperature signal comparisons

3.2.1 Signal response speed

It was found that the temperature signals measured by different measuring methods indicated different behaviour in terms of the signal response speed as shown in Fig. 2b. The temperature signals measured by WIRE and PROB felled immediately at about the 23rd second as the braking load was removed. However, the temperature signal measured by RUB reduced at about the 27th second, which meant a delay of about 4 seconds when the braking load was removed. This was mainly due to RUB being located a distance from the disc and pad contact interface. The WIRE, therefore, is better than the RUB in responding to the braking change between disc and pad during braking process.

3.2.2 Fluctuation of the signals sensed by the WIRE

Fig. 2, and Fig. 3b show that the level of variation of the temperature measured was affected by the measuring methods used. The temperature measured by the WIRE is fluctuated more than those measured by other methods. If it is assumed true that during friction braking the level of contact at local contact area is time dependent, which consequently causes fluctuation in the temperature measured, then this type of information can only sensed by the WIRE.

3.2.3 Magnitude of the temperatures sensed by the WIRE

It was found that the magnitude of the temperature measured was also dependent of the measuring methods used. Under all test conditions, the temperature measured by WIRE was always higher than those measured by other methods under the same test conditions. Under the test T7 (-1, +1, +1, -1) in Fig. 3b, for example, the temperature measured using WIRE was about 225 °C higher than that using RUB. This kind of difference may truly reflect the difference between the disc-pad interface temperature and the disc temperature at some distance outside of the contact zone. The difference between WIRE temperature and PROB temperature is about 150°C. It may be reasonable to assume that WIRE represents the local maximum temperature, T_{local}, the PROB represents the interface nominal temperature, $T_{nominal}$, and the RUB represents the disc background temperature, $T_{background}$, as represented in Equation 1, a relationship of those temperatures. As the result, the temperatures measured by the WIRE were selected to represent the local interface temperatures.

(a) By the RUB under different test conditions (b) By different methods under T7

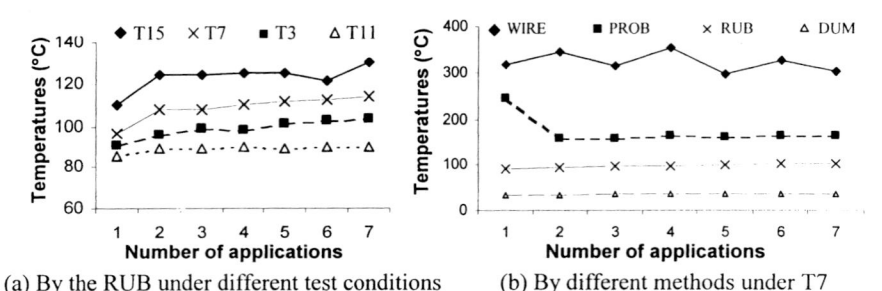

Fig. 3 Temperatures measured under different test conditions with different methods

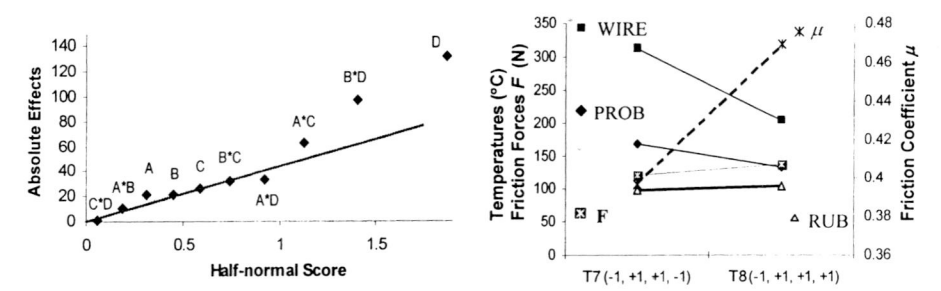

Fig. 4 Daniel plot for temperature   **Fig. 5 Effects of the disc/pad surface conditions**

3.3 Effects analysis

The temperature measured by the WIRE was used for the effect analysis. Based on the DoE, effects plots and Daniel plots were generated. Daniel plots, or half normal plots, are statistical plotting techniques used to analyse experimental data, especially to identify the more significant factors among those investigated [9]. In the Daniel plots in Fig. 4, for example, the absolute values of the temperature contrasts were plotted, so that all the large temperature contrasts appeared on the right-hand side of the graph. The factor D (the disc/pad surface conditions) and the interaction B*D (interaction between the disc/pad surface conditions and the braking load) appeared on the right-hand side in Fig.4. It appears that the disc/pad surface condition, or the number of braking applications, has the strongest effect on the interface temperature, either directly or indirectly (i.e. through interaction with other factors). Daniel plots for friction coefficient also show that the disc/pad surface condition has a strong effect on friction. Further discussion relating to this is shown in the next section.

4 DISCUSSION

4.1 Effects of the disc/pad surface conditions on the interface temperature

A similar observation on the effect of number of braking applications on the friction was reported by Borjesson [10], which showed that friction coefficient was affected by the number of braking applications. And it changed significantly during the first few braking applications. Further analysis of the effects of a number of braking applications on the braking performance indicated that, as shown in Fig. 5 for example, the average local temperatures measured by the WIRE in test condition T7 (i.e. the braking application one to seven) were higher than that in test condition T8 (i.e. the braking application eight to fourteen). In contrast, however, the average background temperatures measured by the RUB in test condition T7 were lower than that in test condition T8. At same time the average friction forces in test condition T7 were lower than that in test condition T8 in Fig. 5. It seems that as the number of application increases, the pad and disc braking friction approach an optimized condition, i.e. generating higher friction torque (or a higher μ) but lower local interface temperature, T_{local}, which is supported by other research [10,11]. To interpret these observations, the change of the contact during the braking has to be studied. It is believed that the Tribo-layer formation and the change of real contact area distribution are the main mechanism in achieving this optimized braking condition.

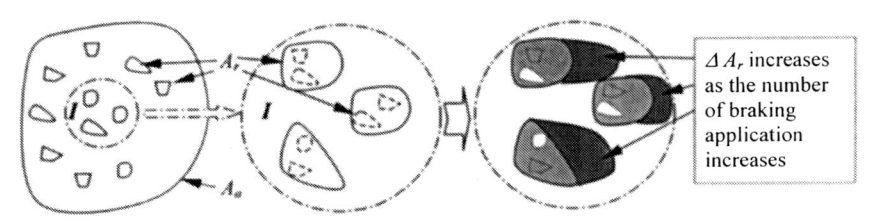

Fig. 6 Shown schematically the wear due to friction causes the increase of the effective contact area A_{ea}

4.2 Effects of Tribo-layer formation and friction braking distance on the interface temperature measured

4.2.1 The change of real contact area distribution and the local interface temperature

The real contact area changes its distribution during braking process for several reasons. When two fresh pad and disc surfaces are in contact under braking load, the pad surface normally only contacts partly, as the two surfaces do not match each other. Wear due to friction will increase the effective contact area A_{ea}. This change will decrease the local maximum temperature as the average normal and friction stress is decreased due to the increase of the effective contact area A_{ea}. Further more, as illustrated in Fig.6, as the number of braking application increases the local real contact area distribution will change at the macro level. Under fresh conditions the pad surface is in an open condition, i.e. the ratio of real contact area A_r and the apparent contact area A_a is small ($\ll 1$). Due to friction and especially the formation of secondary contact plateaux, wear debris fills in the "open" area as described by Eriksson [11], resulting in an increase of the ratio, A_r/A_a. As a result, the local contact stress will decrease, and the local temperature will decrease. The change of the ratio, A_r/A_a, however, cannot explain change of the friction coefficient with the number of the braking applications, as by the friction law, the contact area has no effect on the friction coefficient when $A_r/A_a \ll 1$.

4.2.2 Tribo-layer formation and its effect on the friction coefficient
It is known that the pad/disc contact surfaces will change their characteristics due to friction, wear and other mechanical or chemical interaction at the interface as the friction braking distance increases. The concept of a Tribo-layer or Tribo-film has been used to distinguish the difference between the surface layer and the main body of the pad and disc [11, 12]. As a result, the friction coefficient, which is a function of the friction pair's properties, will change, and here, it changes in a favourable way, i.e. it increases as the number of braking applications increases. One explanation of this phenomena is that as metal wear debris or metal oxide fills in the "open" area on the pad surface by mechanical or chemical means, the affinity between the two contact surfaces will increase. This makes the two contact surfaces more easily to form an adhesive bond (or welding junction). In order to withstand the friction motion a higher share force are required between disc and pad under the same normal load.

4.3 Discussion on the exposed WIRE thermocouple method used
By the law of intermediate metals in thermocouple principle, the algebraic sum of the thermo-electromotive forces (emf) in a circuit composed of any number of dissimilar materials is zero if all the circuit is at a uniform temperature. A third homogeneous material always can be added in a circuit as shown in Fig 7a, with no effect on the net emf of the circuit so long as its extremities are at the same temperature [6, 7]. It follows that any junction whose temperature is uniform and makes good electrical contact does not affect the emf of the thermoelectric circuit regardless of the method employed in forming the junction. The 'open' hot junction configuration used in this investigation is based on this thermocouple principle.

4.3.1 The configuration of the open hot junction
An idealized view of the exposed thermocouple system with "open" and "closed" hot junctions used in this investigation has been given in Fig. 1. The contact between disc and pad during braking is assumed to occur over the hot junction (i.e. at least full contact locally). For the configuration of "open" hot junctions, the disc material bridges the electrical circuit at the hot junction as the material C in Fig 7a. Therefore, during braking, the thermal signal can be recorded as if it were the "closed" configuration. If the disc is removed from contact with pad, however, the electrical circuit is broken. Consequently, the magnitude of the signal reaches infinity as shown in Figure 7b, and the infinity has been treated as zero in the data processing process, as shown in Fig. 7b. Fig. 8 shows the signals obtained by the WIRE (a) with an open hot junction and (b) with a closed hot junction under similar test conditions.

4.3.2 Characteristics of the signals from "open" and "closed" configurations

The results in Fig. 8 shows that the exposed thermocouple technique with the configurations of either closed or open hot junction gives useful data relating to the interaction between the disc and pad during the braking process. The closed hot junction can record the pad surface temperature in braking as well as between braking applications. The open hot junction can detect whether the disc and pad are in contact locally or not. It also can detect whether a Tribo-layer is formed on the disc, based on the assumption that the Tribo-layer formed is a non-conductor.

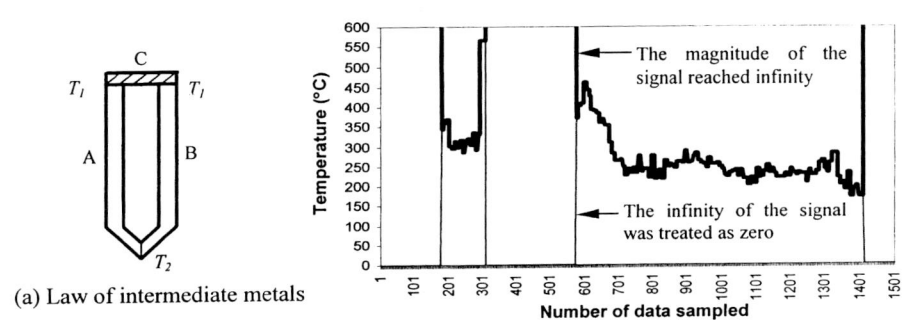

(a) Law of intermediate metals

(b) A signal by the WIRE with an 'open' hot junction

Fig. 7 The exposed thermocouple method with an 'open' hot junction

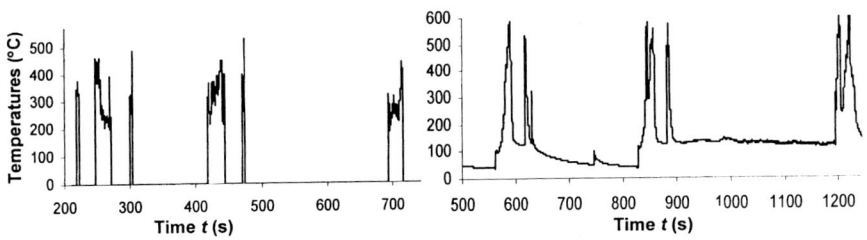

(a) By the WIRE with an open hot junction (b) By the WIRE with a closed hot junction

Fig. 8 Signal comparison

5 CONCLUSIONS

1. Interface temperatures and friction coefficients have been investigated using thermocouple methods and a statistical designed experiment (DoE) approach. Factors including sliding speed, braking force, and friction material type affect the interface temperatures, and, starting from "new" contacting surfaces, the number of braking applications has the strongest effect alone and through interaction with other factors.
2. The real contact area distribution changes as the number of braking applications increases, and this is one of the main reasons for changes in interface temperature with time of rubbing.

3. The existence of a Tribo-layer, which is generated at the disc/pad contact interface in any braking process, is supported by the temperature measurements.
4. The results show that the exposed thermocouple technique with the configurations of either closed or open hot junction gives very useful data on the interaction between the disc and pad during the braking process. The closed hot junction can record the pad surface temperature in braking as well as between braking applications. The open hot junction can detect whether the disc and pad are in contact locally or not, and can also indicate whether a Tribo-layer is formed on the disc, based on an assumption that the Tribo-layer is a non-conductor. The experiments confirm that the exposed wire thermocouple technique can be used effectively and reliably to measure friction interface local temperatures and their variation during braking.

REFERENCES

1. Santini, J. J.; Kennedy, F. E., Experimental investigation of surface temperatures and wear in disk brakes, Lubr Eng, v 31, n 8, 1975, pp. 402-404, 413-417
2. Day, A. J.; Newcomb, T. P., Dissipation of frictional energy from the interface of an annular disc brake, Proceedings of the Institution of Mechanical Engineers, Part D: Transport Engineering v 198, n 11, 1984, pp. 201-209
3. Tian, X. F., and Kennedy, F. E., Contact surface temperature models for finite bodies in dry and boundary lubricated sliding, Transactions of the ASME, Journal of Tribology, 1993, Vol. 115, pp. 411-418
4. Evtushenko, A.A.; Ivanik, E.G.; Konechny, S., Determination of the effective heating depth of the disc brake pad, Trenie i Iznosv 19, no. 3, 1998, pp. 318-322
5. Wollenweber, K.H., and Leiter, R., Function-monitoring brake system: temperature monitoring brake system, proceedings of the Institution of Mechanical Engineers, International Conference on Braking of Road Vehicles, IMechE 1993, C444/049/93, pp. 23-48
6. Dinc, O. S., C., Ettles, M., Calabrese, S. J. and Scarton, H. A., The measurement of Surface Temperature in Dry or Lubricated Sliding', Transactions of the ASME, Journal of Tribology, 1993, Vol. 115, pp. 78-82
7. Shu, H. H. H., Gaylord, E. W. and Hughes, W. F., The relationship between the rubbing interface temperature distribution and dynamic thermocouple temperature, ASME Journal of Basic Eng., Vol. 86, 1964, pp. 417-422
8. Anon, Material development using infra-red thermography, Metallurgia, v 62, 12, (Dec 1995), pp. 409-410
9. Grove, D. M., Davis, T. P., Engineering quality and experimental design, 1992, Longman Group UK Ltd.
10. Borjesson, M., Eriksson, P., The role of friction films in automotive brakes subjected to low contact forces, Proceedings of the Braking of Road Vehicles, IMechE, 1993, vol. C444/026, pp. 259-268
11. Eriksson, M., and Jacobson, S., Tribological surfaces of organic brake pads, Tribology International 33, 2000, pp. 817-827
12. Yu, L, Bahadur, S., An investigation of the transfer film characteristics and the tribological behaviours of polyphenylene sulfide composites in sliding against tool steel, Wear 214, 1998, pp. 245-251

New development of a wet brake system

S NOWAK
Safe Effect Technologies Limited, Balcatta, Australia

SYNOPSIS

The development of a vehicle "wet" braking system referred to as the Solid Rotor Oil Immersed Brake (SROIB) has yielded key innovations and features in comparison with conventional dry friction brake systems.

The wet braking system is fully enclosed and sealed. Rotary disc and pads are of innovative design. To ensure rapid removal of oil from the pad and rotor interface spiroidal grooves of specific geometrical shape and configuration are provided on the faces of the pads and rotor.

To reduce heat ingression from the pads and rotor into the hydraulic piston, PTFE protection boots have been incorporated resulting in a reduced operational temperature.

The technology has to date been applied to braking systems on a range of vehicles, off-road vehicles and industrial applications operating in extreme climatic and harsh environmental conditions.

1. INTRODUCTION

All devices imparting motion are based on friction between interfaces; in fact motion without friction would be difficult if not impossible. Mankind from the inception of motion has endeavoured to understand and use frictional properties to enhance moving devices to his own benefit.

Some of the earlier experiments on friction and study of tribology were carried out by Leonardo DaVinci. One of his well-known conclusions was: **"Frictional resistance is doubled when the load is doubled".** Many of the earlier friction devices were very primitive and were not based on any theoretical knowledge or understanding.

2. KEY OBJECTIVES OF THE SOLID ROTOR OIL IMMERSED BRAKING SYSTEM .

Key objectives of the SROIB system are:
- Improved vehicle safety and reliability performance, including difficult and hazardous environments.

- Reduction of emissions to a minimum i.e. dispersal of harmful brake material particles into the atmosphere.
- Reduced maintenance and down time costs – reduce brake pad or brake shoe wear to minimum on vehicles and industrial applications.
- Easy installation.
- Ability to retrofit the braking system into existing vehicles and machines.

3. GENERAL DESCRIPTION

The main SROIB system elements comprise brake pads, solid rotor, oil bath and outer casing. The callipers are incorporated into the outer casing, with pressure applied to the pads on the side of the rotor, causing them to press against one surface of the rotor disc. This causes axial movement of the splined rotor against the fixed pads on the other side, creating friction braking effect on the rotor and hence wheel. The rotor is in contact with the oil contained within the sealed casing.

Figure 1 shows a diagrammatic layout of the SROIB system for the front wheel of a vehicle and Figure 2 shows a diagrammatic layout of the SROIB system for a rear wheel of a vehicle incorporating a spring applied hydraulically released hand brake.

4. SYSTEM COMPONENTS

4.1 Braking pads

In the beginning, all friction applications were dry, except, of course, when the cart was pulled through a puddle, or it rained, or axle-grease was applied too generously and ran out on the lining.
In the early days:
- Leather was used as a strap or band wrapped around a drum.
- Wood blocks - including bark were used in the early version of calliper brakes.
- Hemp and cotton made into a rope and woven into belting was saturated with liquid drying oil and moulded into shape.

The 1970's were times of great change and new challenges. Governments of developed countries introduced legislation prohibiting the use of asbestos fibres in the manufacture of braking pads or shoes. New technology was evolved and a family of friction materials known as elastomer resilients or "polymerics" emerged. Developments in the area of polymerics are continuing in several directions including "wet" braking application. One of these directions has been the development of friction material with high power absorbing capabilities and excellent heat absorbing characteristics.[1]

A number of automotive brake pads have been qualitatively analysed for composition by scanning electron microscopy and x-ray diffraction spectrometry. Brake pads typically consist of friction material in a resin binder.

The quantity and type of materials that make up the composite depend upon the product application and desired properties. By varying the amounts of each component, characteristics such as mechanical strength and wear resistance can be altered. Other properties such as moulding properties, heat resistance and economy can also be affected by the composition.[1]

From a number of analyses, the following constituents were identified:

- Thermoset resin (Binder)
- Graphite (Wear Reduction Element)
- Barium Sulfate (Mechanical Strength)
- Magnesium Oxide (Friction Modifier)
- Magnesium Aluminium Silicate Mineral (Fibre Base)

Listed below are estimated quantities of each constituent including some of the common constituents that were not identified in the tests:

- Novolak type phenolic resin with hexamethylenetetramine catalyst — 20%
- Graphite (Wear reduction Lubricant) — 10%
- Rubber Powder (Brake Noise Suppressor) — 20%
- Barium Sulfate (Mechanical Strength) — 10%
- Potassium Titanate (Mechanical Strength) — 15%
- Iron Oxide (Friction Modifier) — 5%
- Magnesium Oxide (Friction Modifier) — 3%
- Zirconium Oxide (Friction Modifier) — 2%
- Metal Fibre (Friction Base) — 3-5%
- Cellulose or Ceramic Fibre (Fibre Base) — 10%

This is an initial research phase for brake pads used in "wet" braking systems. Further research will be undertaken to determine physical properties required and this basis will be used in producing trial formulations.

The braking pads used for "wet" braking were dry disc pads, with one fundamental difference; spiroidal grooving of an optimised geometrical configuration were machined on the braking interface. The particular pattern of the grooving was developed to achieve rapid removal of the oil from the brake pad – rotor interface. Figure 3 shows a diagrammatic layout of a spiroidal grooved braking pad.

Measurable pressure of the fluid is generated in the grooves, resulting in a forced circulation of oil away from the rotor. To further increase the rapid removal of the oil PTFE spring-loaded scrapers are located in the front and rear of the pads.

4.2 SROIB oil

The intended purposes of the oil are to reduce the temperature rise of the pads and rotor and to maximise heat transfer to the thermally significant casing, while still enabling the spiroidal-braking pads to maintain sufficient friction at the interface with the rotor. It is critical that the combination of the particular fluid properties, thermal conditions and the interface between the braking pad and the rotor can produce sufficient friction to stop the wheels within effective operational constraints. The introduction of spiroidal grooves on the braking pads and the spring-loaded PTFE scrapers in the design of the braking system achieves the following results:

- Minimise the fluid remaining between the spiroidal braking pad and rotor.
- Prevent hydrodynamic lubrication.
- Achieve a coefficient of friction which is comparable with the coefficient of friction achieved with "dry" pads.

- This scenario calls for a fluid with low viscosity, high boiling point and high thermal conductivity and with sufficient high specific heat to remove the heat generated at the braking pad – rotor interface.

To ascertain the performance of the cooling oil, a four wheel-drive vehicle was fitted with a SROIB braking system on all wheels and field tested on a mine site. After completion of 85,000 km the oil was drained and compared with fresh oil.

The analysis detailed in Table 1 was obtained:

Table 1 : Oil analysis data from field test

Test parameter	New oil	Used oil
Flash Point °C	104	105
Viscosity @ 40 °C (cSt)	4.1	4.25
Iron	<1	1000
Lead	<1	<1
Copper	<1	6
Molybdenum	<1	4
Chromium	<1	<1
Aluminium	<1	8
Nickel	<1	<1
Tin	<1	<1
Silicon	17	59
Sodium	2	5
Zinc	1400	1200
Calcium	3800	3900
Magnesium	21	33
Vanadium	<1	<1
Boron	66	66

The flashpoint is consistent with the low viscosity and indicates the product is very light hydrocarbon oil. The significant additives are zinc and calcium based. The zinc additive commonly present in such products is zinc diethyl dithiophosphate (ZDDP) and it is primarily an anti-wear additive. The oil samples were subjected to a routine test protocol to characterise their physical and chemical properties. The oil stood up well to use with little depletion of its additive package. The analysis of the solids in the oil found the bulk of the material to be iron particles (rust), silica (sand) particles and oil residues.

4.3 Outer casing
The outer casing fulfils two major functions:
- Provides structural connection to resist the torque generated by braking.
- Provides effective sealing of the brake components from dust and foreign material.

In aggressive and corrosive environments, keeping dust and foreign material away from the pad – rotor interface is vital in extending the life of the braking components and maintaining effective operation of the braking system.

In all braking applications it affords a safe braking system with an extended life span of the braking pads and no contamination of the environment.

Internally, damage may also occur if wear particles are produced, and if these particles become entrained in the cooling fluid and recirculate back into the interface between braking pad and rotor.

As a result of prolonged testing the evidence to date shows that wear is minimal and wear particles in the SROIB system settle in the extremities of the casing, minimising any contribution to additional wear.

Three types of casing have been tested:
  a) Cast Iron
  b) Mild and High Tensile Steel
  c) Aluminium

The steel casing has particular application in mining operations and in environments where there is a high likelihood of external damage due to environmental conditions. Similar considerations apply to casings made from cast iron. The use of aluminium casing will be very significant where weight is of the essence and the environment is less dangerous and harsh.

4.4 Rotor

Rotors in all prototypes and field-tested units are made from spheroidal cast iron. Future prototypes and produced units will have rotors produced from a range of materials such as stainless steel, specialised novel materials etc. The innovation introduced is the provision of spiroidal grooves on both faces of the rotor. Results to date indicate an enhanced coefficient of friction. Test will also be conducted using grooves on rotor only and plain braking pads.

4.5 Hydraulic protection boots

Major problems encountered with existing hydraulic braking systems is overheating of the hydraulic fluid. This is caused by the heat permeating from the brake housing via piston and piston seal. A rise in temperature results in the evaporation of brake hydraulic fluid in the small cavity of the piston, affecting efficient operation of the braking system.

The introduction of the PTFE heat protection boots over the piston and the introduction of VITON "O" ring-seals encapsulated in PTFE has resulted in the temperature not exceeding 80 – 85°C. The heat protection boots fit on the standard chrome plated hydraulic brake pistons and instead of square type "O" ring-seals made of NEOPRENE which are prone to swelling when exposed to hydraulic brake fluid, VITON circular "O" ring-seals have been introduced.

5. SYSTEM ENERGY FLOW

Friction devices perform many functions such as the holding, coupling or braking of moving parts or devices. Usually the frictional material (pad or shoe) converts potential/kinetic energy into heat, which, in turn, must be dissipated so that the friction device does not overheat.

The SROIB brake system comprises a sealed casing-housing, brake activators, brake pads/brake shoes with a special pattern of oil dispersing grooves, heat protection device fitted to all pressure pistons and a single or multiple rotors immersed in a bath of oil.

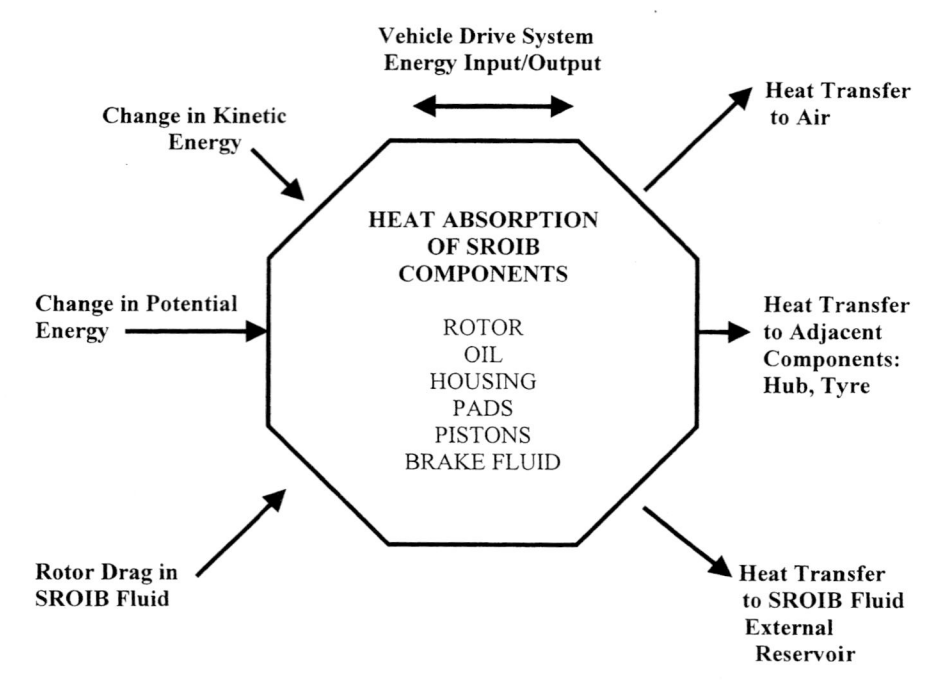

Fig 4 System Energy Flow Diagram

The system Energy Flow Diagram depicted in Fig 4 considers the flow of energy into and out of the braking system.
Energy is applied to the system via the interface between the braking pad and rotating rotor sub-system, where it is converted into heat.

It includes:
- The Kinetic Energy of the vehicle.
- Any height change related to the vehicle's motion.
- Any energy applied by the engine through the drive train.

Drive train energy is ignored in the determination of the highest input scenarios, simulating vehicle freewheeling conditions. This relates directly to the standard performance tests as prescribed by the Australian Design Rules [2].

5.1 Torque capacity
Government authorities administer and validate the element of performance with decelerations well within the specified limits [2].

For a wheel mounted brake;
Torque Transmitted; $T = R_w \times m \times (d + g \sin\theta)/n_w$ (Nm)
where: R_w = Wheel rolling radius (m)
  θ = Angle of decline of surface (deg)
  m = Vehicle mass (kg)
  n_w = Number of wheels
  g = Gravitational acceleration (9.806 m/s^2)
  d = Vehicle deceleration (m/s^2)

Friction generated torque at the pad-rotor interface resists the wheel torque.
Friction Torque; $T_f = R_f \times \mu \times P_b \times A_b$
where: R_f = The average radius of the friction pad from the rotor centre (m)
  μ = Dynamic coefficient of friction (0.11 – 0.15)
  P_b = Brake line pressure (Pa)
  A_b = Area of brake piston(s) (m^2)

6. TESTING AND VALIDATION OF A SINGLE ROTOR OIL IMMERSED BRAKING SYSTEM.

Testing was carried out on a specially designed and constructed dynamometer test rig capable of simulating a range of conditions to validate the performance of a "wet" braking system in compliance with standards prescribed by the Australian Design Rules, Rule 35.00 and 35.01.[2]

The test rig is fabricated from structural steel sections. It has a platform on which is mounted a variable speed electric motor coupled to a shaft on which flywheel is mounted. The shaft is connected to a fixed ratio gearbox. The output shaft of the gearbox houses adaptor for mounting the brake.

Conditions simulated were:
- Road speed.
- Kinetic Energy developed per wheel.

Key objectives of testing:
To determine temperatures on the pad interface, oil temperature in the reservoir, temperature of the braking fluid, temperature of the brake housing and components.
- To determine torque and pressure values during the braking cycle.
- To determine the dynamic coefficient of friction for a range of conditions and brake pad geometry.
- To determine the performance of all innovations.
- Monitor wear rates of pads, and the quality and condition of the cooling oil.
- Monitor performance of a range of prototypes developed i.e. vehicles and industrial application.

Instrumentation and conduct of tests:
- Acquisition and data collection was carried out using a data Taker DT800 using normal mode and burst mode.
- The following outputs of instruments were connected to the data acquisition unit: Force Transducer, Pressure Transducer, Rotary Encoder/Frequency Convertor and Thermocouples (Type K).

- Other input signals to the acquisition unit: Programmable Logic Controller and Variable Speed Motor.

Most of the data was collected in a burst mode at a rate of 625 inputs per second or for a complete braking cycle of 6.0 sec equal to 3750.0 signals.

Graphs of selected samples are illustrated in figures 5 and 6.

Results from the tests are very encouraging and will serve as a benchmark in future development. as well as validating the introduced innovations such as spiroidal grooved pads, spiroidal grooved rotor, heat protection boots etc.

Pressure applied to hydraulic pistons is as on standard vehicles ie. 1200 psi and the brake fluid used is alcohol based liquid commonly used. The dynamic coefficient of friction on wet braking system is between 0.11 – 0.13, this compares very effectively with dynamic coefficient of 0.12- 0.15 on dry braking system. The results also compare extremely well with results obtained by other researchers in the field of wet braking. Main reason for the coefficient of dynamic friction being high is the lower operating temperature achieved with the SROIB braking system.[4]

Another point worthy of note is torque transmitted. By providing spiroidal grooves on the rotor the torque transmitted has increased by 25-30%.

The pad wear based on units in service is minimal and with further research in physical properties of brake pad material and the composition of the cooling fluid it is possible to extend the life span of the braking pads to that of the vehicle or industrial braking device.

7. CONCLUSIONS

The most significant achievements in wet braking to date have been temperature control, braking efficiency, minimisation of braking pad wear establishing that **wet braking is braking into the future** for all types of vehicles and industrial application.

References:
[1]   F.A. Lloyd and M.A. DiPino
      Paper No 800977 presented at International Off-Highway Meeting and Exposition
      Milwaukee, Wisconsin, Sept. 8-11, 1980.
[2]   The Australian Design Rules for Road Vehicles.
      As at Determination 1 & 2 of 2000.
[3]   Fredrick A. Loyd, John N. Anderson and Laurie S. Bowles
      International Off-Highway & Powerplant Congress and Exposition
      Milwaukee, Wisconsin, Sept. 12-15, 1988
[4]   Robert L. Fish and Frederick A. Lloyd
      1975 SAE Off-Highway Vehicle Meeting
      Milwaukee, Wisconsin, Sept. 8-11, 1975

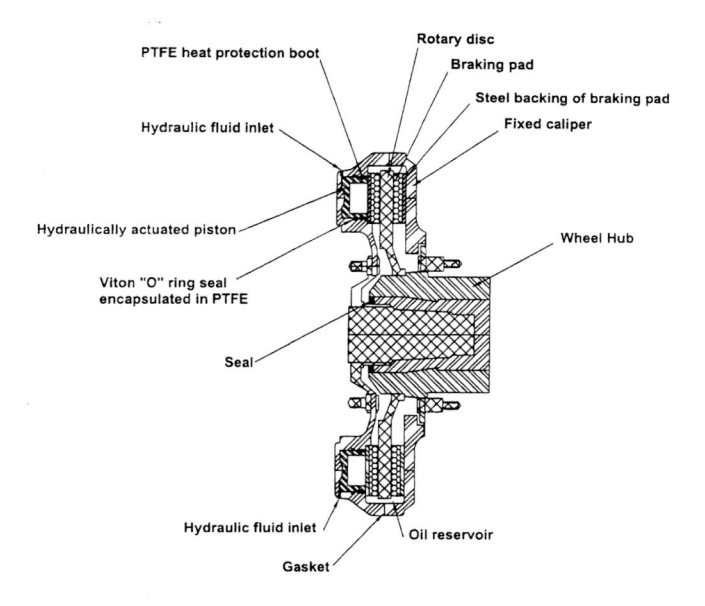

FIG 1 SINGLE ROTOR OIL IMMERSED BRAKE FOR VEHICLE (FRONT)

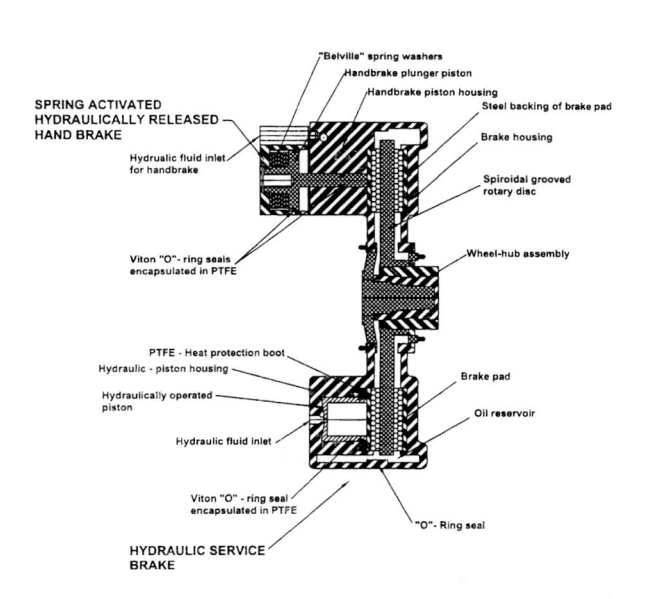

FIG 2 SINGLE ROTOR OIL IMMERSED VEHICLE BRAKE (REAR WHEEL)

Fig 3: Spiroidal grooved pad

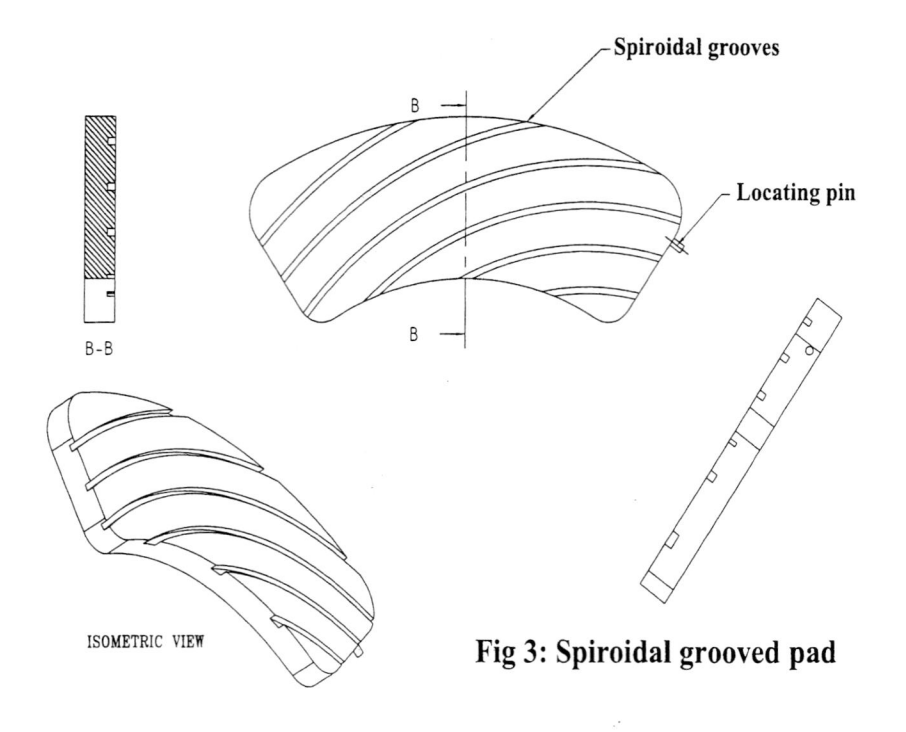

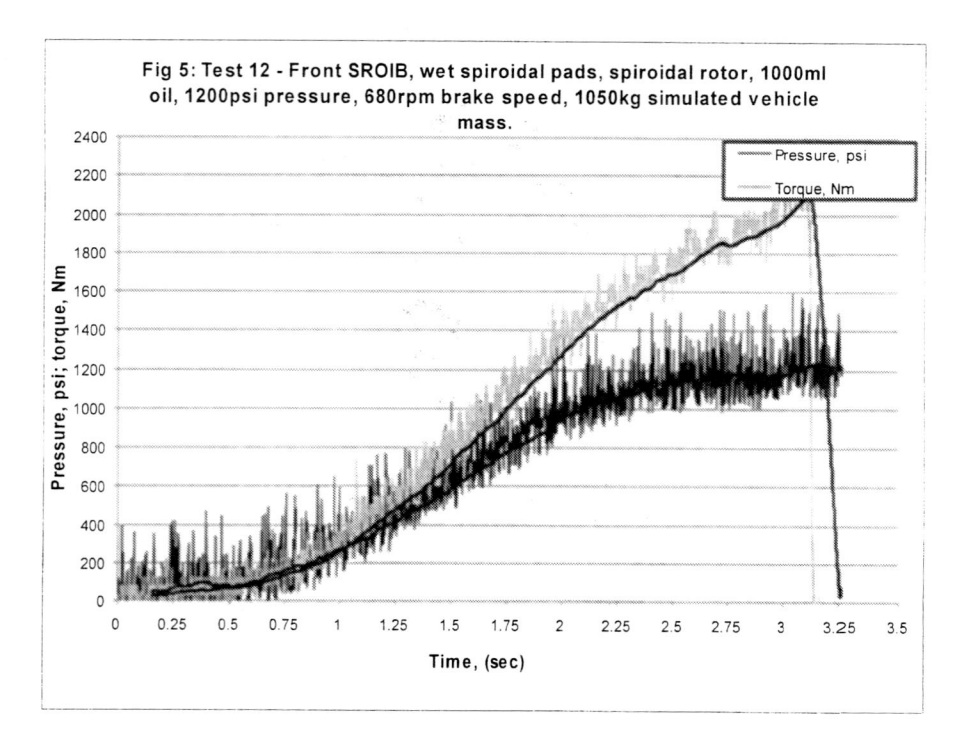

Fig 5: Test 12 - Front SROIB, wet spiroidal pads, spiroidal rotor, 1000ml oil, 1200psi pressure, 680rpm brake speed, 1050kg simulated vehicle mass.

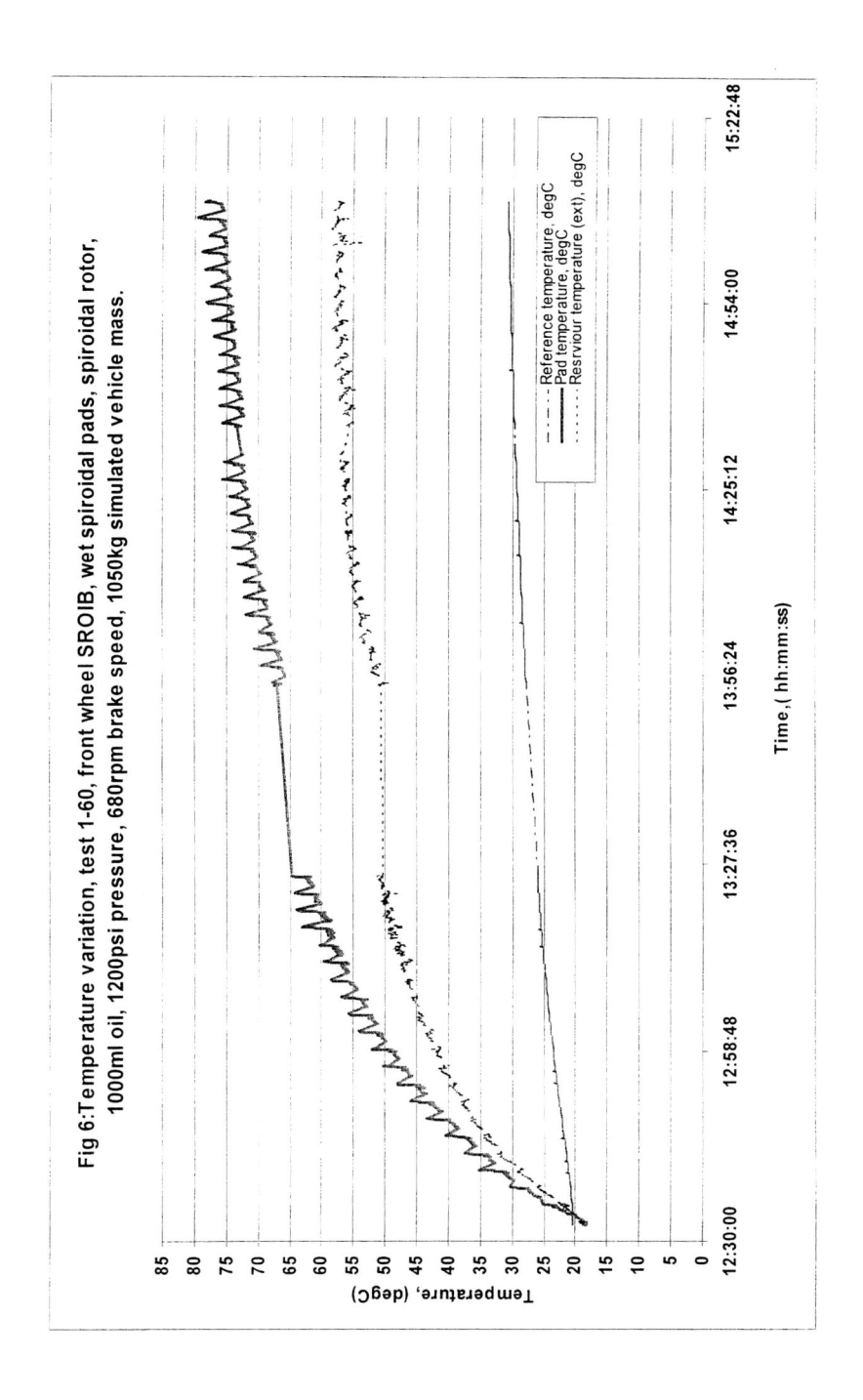

Fig 6: Temperature variation, test 1-60, front wheel SROIB, wet spiroidal pads, spiroidal rotor, 1000ml oil, 1200psi pressure, 680rpm brake speed, 1050kg simulated vehicle mass.

Friction and wear of polymer matrix composite materials for automotive braking industry

P FILIP
Center for Advanced Friction Studies, SIU at Carbondale, Illinois, USA

ABSTRACT

During braking, a polymer matrix brake lining is rubbed against a rotating counterpart (typically a pearlitic gray cast iron), and the energy is dissipated in the related friction process. The performance of friction materials depends to a significant extent on the friction layer, which is generated on the friction surface as a result of complex mechano-chemical interactions. In this research, laboratory materials were formulated, tested, and analyzed with the aim of developing an alternative method for modeling the performance based on an understanding of real phenomena on the friction surface. An analysis of the friction layer was performed by the use of light, scanning, and transmission electron microscopy supported by energy dispersive microanalysis, grazing angle X-ray diffraction, and atomic force microscopy. The performed analysis and synthesis using modified thermodynamic and kinetic relationships allowed the modeling of friction and wear properties.

1 INTRODUCTION

The materials used in the manufacture of brakes are known as friction materials. In contrast to other tribological applications, a relatively high friction coefficient is desirable when using friction materials. A coefficient of friction in the range of 0.3 to 0.7 is normally observed in these materials when rubbed against a proper counterface. A characteristic friction material is a multicomponent polymer matrix composite with a formulation, which is often developed empirically. The high-energy conditions and complex mechano-chemical interaction on the friction surface during braking make it difficult to predict performance based on the properties and formulation of the bulk materials. The tailored design, modeling, and development of polymer matrix friction materials requires an understanding of processes on the friction surface. It has been demonstrated that the friction phenomena depend on the chemistry and topography of the surface, time, which is at disposal for interactions between individual species on the friction surface, pressure, and temperature of the friction surface (1-4). A successful solution of "friction problems" requires an integrated approach (1, 2). Consistent and well-defined properties of friction materials as well as their tailored manufacturing are important from the point of view of economy, transportation safety, and environmental load.

This contribution is devoted to the characterization of the structure and chemistry of the friction layer developed in a laboratory friction test. The structure and chemistry does not follow "classical" thermodynamic and kinetic laws frequently applied in materials research. The demonstrated analyses of the friction layer and following synthesis, using modified thermodynamic and kinetic relationships, represent a newly developed tool and allow the modeling of friction and wear properties. The tailored design and smart testing were accomplished when the newly developed tool was used.

2 EXPERIMENTAL

2.1 Materials
The formulation of investigated model polymer matrix composite (PMC) materials is given in Table 1. The details of fabrication technology are given in (2). PMC samples were rubbed against pearlitic gray iron discs.

Table 1. Formulation of investigated samples.

Material	Content (vol. %)
Brass chips	30
Petroleum coke	30
Vermiculite	5
BaSO$_4$	7
MgO	5
Phenolic Resin	23

2.2 Friction Assessment and Screening Test (FAST)
A relatively simple and sensitive FAST was used (3). A $12.7 \times 12.7 \times 5$ mm^3 sample is pressed against the rotating cast iron disc with a diameter of 300 mm. This drag test does not represent a real braking simulation. The friction force was kept constant and the normal force varied accordingly. The friction coefficient was measured every five seconds, and wear was calculated as weight loss after the test. Different energy dissipation rates within the extreme limits of 232.2 and 464.4 J/s, corresponding to two extreme sliding speeds of 3 and 6 m/s were applied. The duration of individual FAST tests varied between 600 and 5400s, which corresponds to the total amount of energy between 0.14 (600 s at 3 m/s) and 2.51MJ (5400 s at 6 m/s) spent during the friction process.

2.3 Analytical techniques
Surface topography was characterized using a stylus profilometer Mahr (Perthometer PRK, diamond tip radius 5 µm) and an Atomic Force Microscopy (AFM, Topometric and Digital Instruments). Lateral force microscopy was applied to characterize friction on selected areas.
The roughness parameters R_a and SmR_a were calculated in accordance with well-known relationships (5). The microstructure and surface of the investigated samples were analyzed using a light microscope Nikon FX35 (polarized light, analyzer in, retarder plate out) and CCD camera Sony DXC-151, environmental scanning electron microscope Hitachi S2460N (pressure 5Pa; energy dispersive x-ray microanalysis EDAX Noran Voyager System), field emission scanning electron microscope Hitachi S4700 (energy dispersive x-ray microanalysis

EDAX Kevex System), and an analytical TEM/STEM microscope Hitachi H7100FA, 100kV(energy dispersive x-ray microanalysis EDAX Noran Voyager System). X-ray and Grazing Angle diffraction analyses of the bulk and friction surfaces, respectively, were performed using the Rigaku Dmax flash B ($Cu_{K\alpha}$ radiation, Ni filter, 40kV, 30mA), Seifert ID-3000 ($Cu_{K\alpha}$ radiation, Bicron scintillation detector with curved graphite diffracted beam monochromator, 45 kV, 40 mA), Spellman DF3 series ($Cr_{K\alpha}$ radiation, Scintag liquid N_2-cooled Ge detector, 0.25° radial divergence limiting Soller slit, 40 kV, 35 mA), diffractometers, and the Advanced Photon System (synchrotron) (λ= 2.1939 A). Optimas©6.1 image analysis software further modified by (6) was applied for quantitative analyses. Statistical methods were applied in the designing and evaluating of the friction tests and in friction layer analysis (4, 6). Statistically relevant results were used in thermodynamic and kinetic calculations and the related modeling of the performance.

3 RESULTS

3.1 Performance parameters μ and wear

The evolution of the friction coefficient and wear as detected for the two extreme conditions is shown in Figs. 1a and b, respectively.

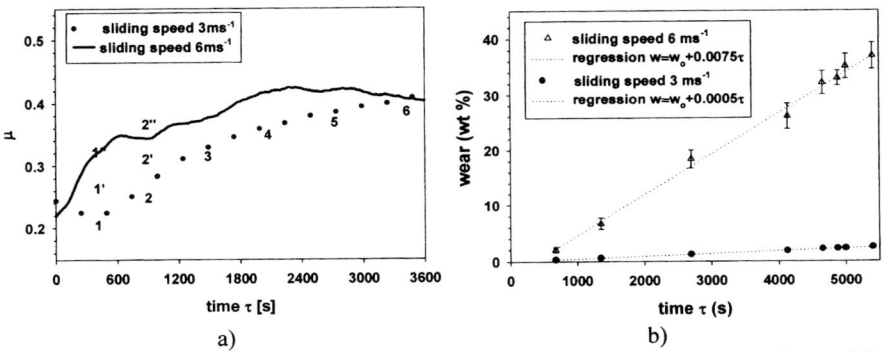

Figure 1. Time dependence of the coefficient of friction (a) and wear (b) as detected for samples tested at two extreme sliding speeds in FAST.

These μ vs. τ plots represent statistically relevant data obtained by kernel smoothing (4). The error bars in Fig. 4b represent standard deviation 1σ. Both the coefficient of friction and wear strongly depend on an applied energy dissipation rate (sliding speed). The selected data obtained for different sliding speeds and different testing times applied are marked by points 1 to 6. The analysis of the friction layer was performed for a limited number of selected data points marked as 1 to 6 and 1" to 6", respectively. The corresponding modeling based on this analysis allowed the prediction (interpolation) of performance (μ and w) at any chosen point (e.g. 1') within limits given by dependencies shown in Figs. 1a and b. Both the coefficient of friction and wear rate could be expressed as a function of friction layer composition and structure. The structure and chemistry of the friction layer also dictates the surface roughness.

3.2 Analysis of the friction layer

The roughness parameters obtained from entire surfaces of PMC samples tested at two extreme sliding speeds are given in Table 2.

Table 2. Average roughness R_a and smoothened average roughness SmR_a of samples tested in FAST.

Sliding speed (m/s)	Parameter (μm)	Testing time (s)			
		600	1200	3600	5400
3	R_a	3.225	2.988	2.545	2.445
	SmR_a	2.113	1.996	1.562	1.560
6	R_a	4.656	3.166	3.282	3.155
	SmR_a	2.231	2.046	1.626	1.455

When smoothened data are compared, the SmR_a values are very similar for two different friction conditions applied. The detected differences in average roughness R_a are related to the fact that the samples tested at higher sliding speeds developed higher waviness due to thermally activated macroscopic deformation (thermoelastic instabilities). The deformation causing waviness can be ascribed to a limited thermal conductivity and heterogeneity of investigated brake lining samples. The detailed microstructure of the surface before FAST is shown in Fig. 2a, and Figs. 2b and c represent the characteristic microstructure of the friction surface developed in FAST at low (3m/s) and high (6m/s) sliding speeds, respectively.

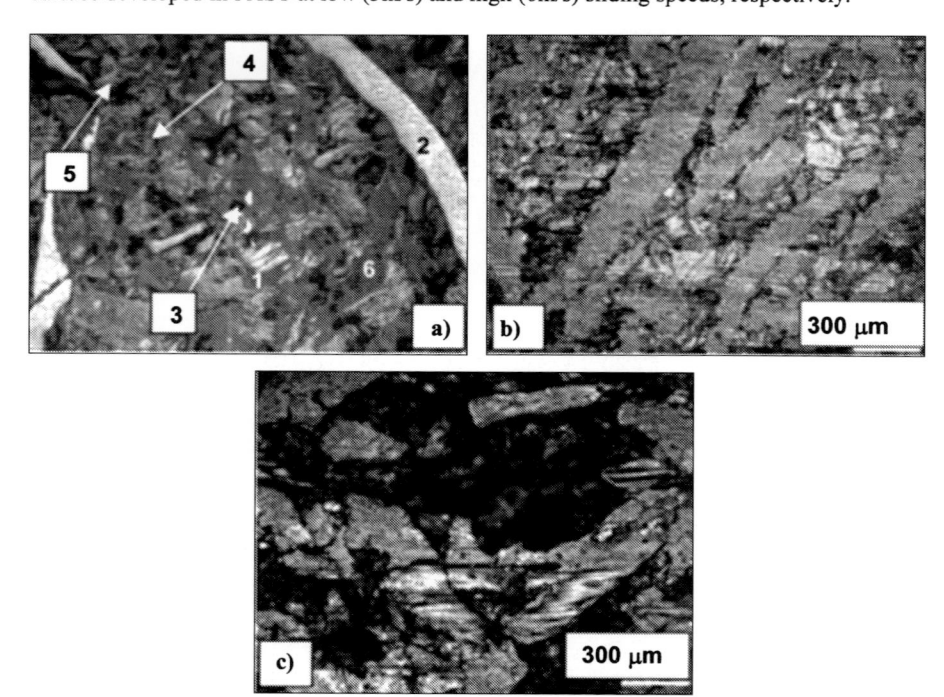

Figure 2. (a) Microstructure of investigated PMC materials before friction test. Coke (1), brass fiber (2), BaSO₄ (3), MgO (4), vermiculite (5) are embedded in a phenolic resin matrix (6). (b) Surface of brake lining samples as detected in Light Microscopy after FAST testing. (b) Low sliding speed, (c) high sliding speed.

It is easily visible that the friction surfaces and contact areas differ considerably at different sliding conditions. A friction layer characteristic for each condition applied develops on the friction surface.

AFM studies revealed that individual parts of the friction layer, differing by the chemistry and structure, exhibit a different level of friction (Fig. 3). The entire value of the coefficient of friction is a function of volume content of the individual phase on the friction surface: $\mu = \Sigma\mu_i$. The value of μ_i depends on the volume content of the i-th component V_i. No simple relation between surface roughness and the value of the coefficient of friction and wear was found in this experiment.

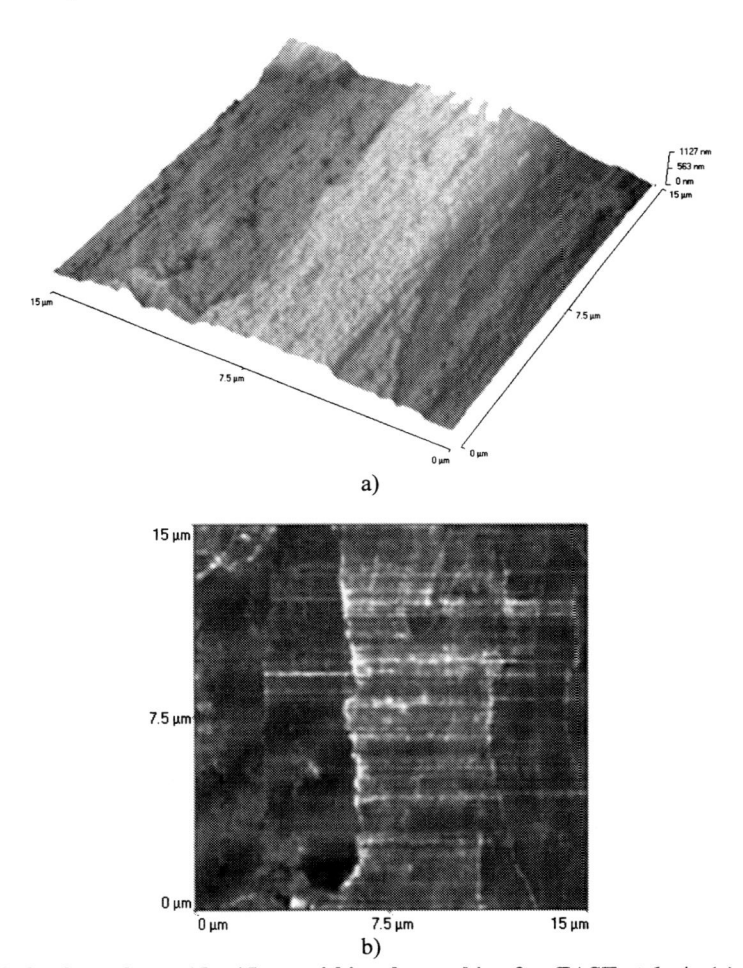

a)

b)

Figure 3. Surface of area 15 × 15 μm within a brass chip after FAST at 6m/s. (a) surface topography, (b) lateral force demonstrating the differences in the coefficient of friction of central and edge regions shown in (a)

SEM images revealing the character of the friction layer as observed on perpendicular cuts with respect to the friction surface are shown in Figs. 4a and b. As clearly seen, the thickness of the friction layer developed on the PMC samples depends on the applied sliding speed. The measured data reflecting development of the friction layer thickness are plotted in Fig. 4c.

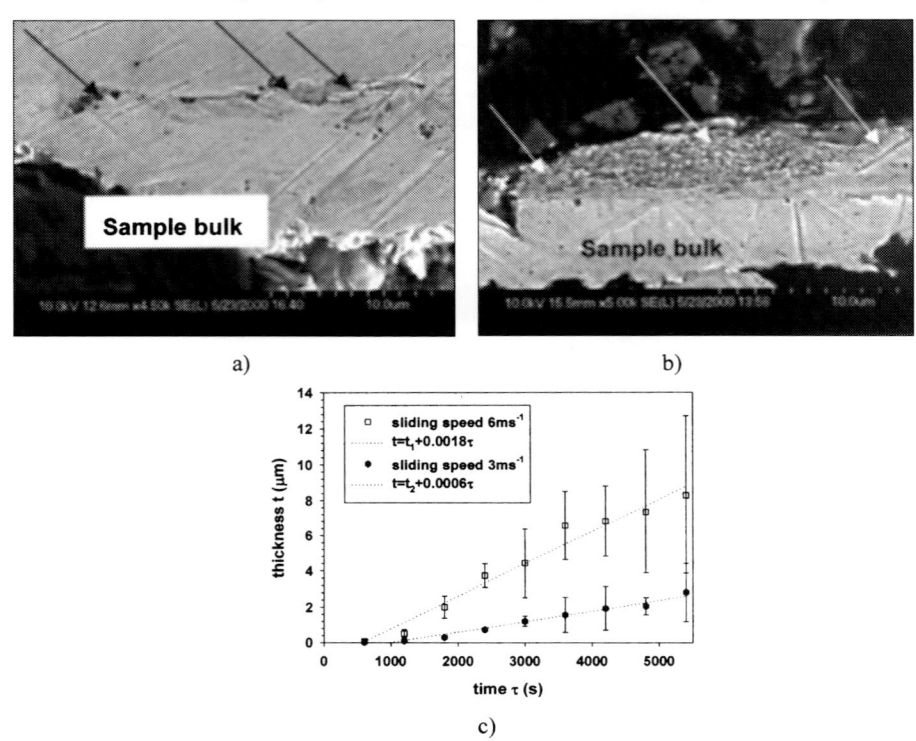

Figure 4. Friction layer (marked by arrows) of the samples tested at (a) the lowest sliding speed 3m/s and (b) the highest sliding speed 6m/s for 5400s. SEM, back-scattered electron images. (c) Thickness of the friction layer developed at two extreme sliding speeds in FAST (measured in SEM).

In contrast with experiments performed for metals and with theoretical works presented in (7-13), no relationship between the friction layer thickness and wear or the value of μ was detected. Instead, it was found that the coefficient of friction and wear rate depend on friction layer composition and structure. As apparent from the grazing angle synchrotron diffraction spectra (Fig. 5), the structure of friction layers developed at different testing conditions differs significantly from the bulk. Cu-oxides, Fe-oxides, MgO, barite, vermiculite, carbon, and organometallic complexes were detected on the friction surface. While the amount and quantity of barite, MgO, and vermiculite remains almost unchanged, the quantity and structure of copper oxides, iron oxides and carbon differ significantly for each of the applied testing conditions. It is clear that the formation of different components of the friction layer results from synergistic effects occurring at the surface. TEM analysis revealed the presence of a crystalline carbonaceous matter at a low sliding speed.

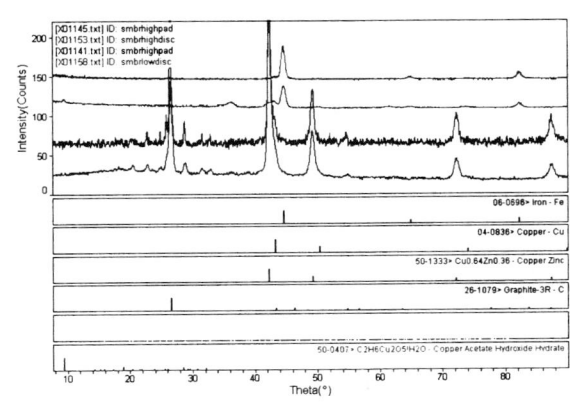

Figure 5. GAXRD spectra as detected for two extreme FAST testing conditions for cast iron disc and brake lining samples (5400s testing time)

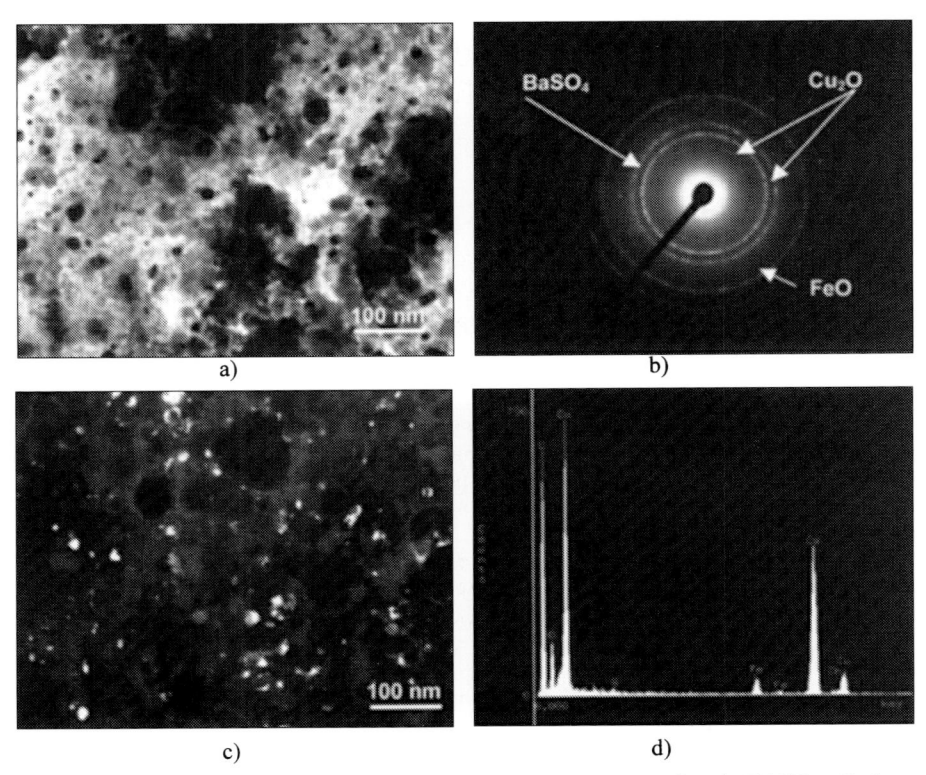

Figure 6. TEM of the friction layer generated on the friction surface in FAST at 6m/s. (a) bright field image, (b) diffraction pattern, (c) dark field image revealing size and distribution of Cu_2O particles, and (d) EDX spectrum from TEM thin foil.

The amount of amorphous glassy carbon increases with an increasing energy dissipation rate, and no crystalline carbon was detected after FAST performed at 6m/s. The development of CuO demonstrated a similar dependence. At higher sliding speeds, CuO transforms to Cu_2O, which was the only copper oxide detected after FAST at 6m/s. Selected TEM images are shown in Figs. 6a to c. Submicron size particles were detected in the friction layer, and it can be expected that kinetic of reactions is extremely fast. The defect density analysis indicated that the elastic strain energy accumulated in the friction layer is low.

An interesting phenomenon was the detected decomposition of Cu65Zn35 brass. Pure copper and brass Cu99Zn1 diluted of Zn were detected on the surface of PMC samples and discs. Iron oxides were transferred to the PMC sample surface from cast iron discs. An example of analysis of the friction debris is shown in Fig. 6d. The significantly lower quantities of friction debris embedded in carbonaceous matter were detected after the FAST testing was performed at lower sliding speeds compared to the amount detected at higher sliding speeds.
The fairly good correlation of XRD (synchrotron), SEM, and TEM analyses allowed a confident description of friction layers developed at individual testing conditions.

3.3 Modeling the performance
The performance parameters are closely related to amounts of species detected in the friction layer. Figure 7 represents an example of the experimentally identified dependence between the coefficient of friction, content of Cu_2O and testing time for three sliding speeds.

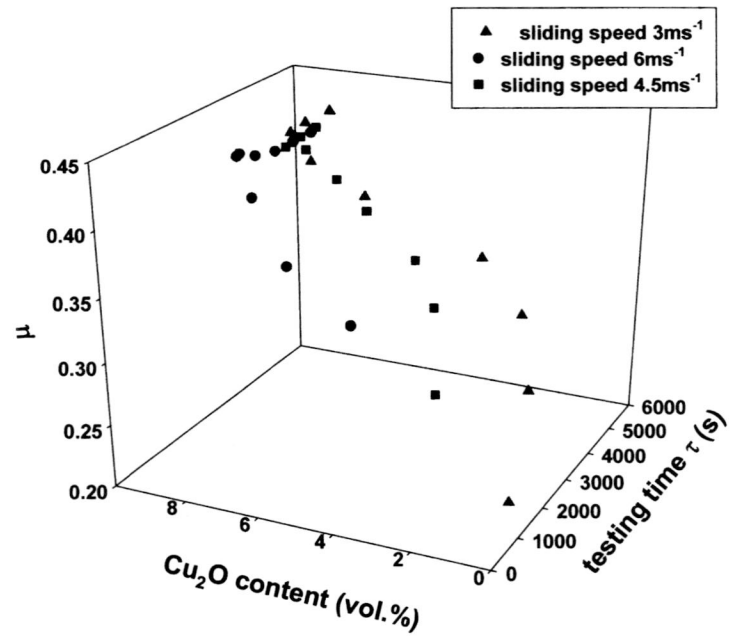

Figure 7. Experimental data, generated in FAST tests for three different testing conditions.

A functional relationship describing the correlation of these three parameters allows modeling of the performance. This method enables the prediction of the value of the coefficient of friction and wear rate. It is necessary to note that the presence of any matter on the friction surface is a result of synergistic effects. All species present in the friction layer simultaneously interact or influence each other interaction. Therefore an analyzed dependence valid for one matter (e.g. Cu_2O) reflects the properties of the entire friction layer developed at given friction conditions. In other words, the characteristic amount of Cu_2O typifies only the analyzed friction layer, and completely different amounts of Cu_2O would be representative for other brake lining formulations tested at the same conditions in FAST. The detected time dependencies of Cu_2O content did not follow the typically observed Kolmogorov-Johnson Mehl-Avrami relationship describing the kinetic of solid-state reactions (14-17). Instead, the following relationship describes the formation of Cu_2O in analyzed system:

$$X = 0.0166 \times v + 8.5 \times (1 - \exp[-0.0019 \times v \times \tau]),$$

Where X is the amount of Cu_2O (vol. %) in the friction layer, v is the applied sliding speed (m/s) and τ (s) represents the friction test duration. Thermodynamic analysis allowed the prediction of the occurrence of different phases. The Gibbs free energy G is a function of temperature T, pressure P and the number of moles of all of species present in the friction layer, i.e.:

$$G = G(T, P, n_i, n_j, n_k, \ldots)$$

where n_i, n_j, n_k, ... are the number of moles of the species i, j, k, ... present on the friction surface. The state of the system is defined only when all of the independent variables are known. Differentiation of the above equation gives:

$$dG = (\partial G/\partial T)_{P, ni, nj, nk, \ldots} dT + (\partial G/\partial P)_{T, ni, nj, nk, \ldots} dP + (\partial G/\partial n_i)_{T, P, nj, nk, \ldots} dn_i + (\partial G/\partial n_j)_{T, P, ni, nk, \ldots} dn_j + \ldots$$

and

$$dG = -SdT + VdP + \Sigma(\partial G/\partial n_i)_{T, P, nj, \ldots} dn_i$$

where $\Sigma(\partial G/\partial n_i)_{T, P, nj, \ldots} dn_i$ is the sum of k terms (one for each of the k species) each of which is determined by the partial differentiation of G with respect to the number of moles of the i-th species at constant T, P, and n_j, where n_j represents the number of moles of every species other than the i-th species. dG varies with time in complex friction situations. The Ellingham diagram calculated for detected oxides and ZnO at static atmospheric conditions did not describe the phases present in the friction layer. Calculations were done numerically based on thermodynamic data obtained from (18). According to "equilibrium classical thermodynamic" the formation of all oxides is feasible in whole temperature range. However, the experimentally detected structure and chemistry of the friction layer is completely different, as the thermodynamic relation has to take into account the surface energy, deformation energy, and corresponding different activities of individual species due to the presence of defects (19, 20). Also, the realistic partial pressure of oxygen, which differs from atmospheric conditions (21 kPa), has to be found. The elements with a different affinity to oxygen tend to oxidize simultaneously, and the realistic partial pressure of oxygen in the friction layer is significantly lower than 21 kPa. The oxygen present is not only "consumed" by Cu, but even more intensely by carbon, Zn and Fe; hence, the partial pressure p_{O2} decreases. This is practically demonstrated by the presence of different oxides at different temperatures. For instance, in the case of Cu_2O at 900°C, the partial pressure of oxygen p_{O2} has to decrease to 1×10^{-7}kPa, and at 600°C, to 1×10^{-17}kPa, for the Cu_2O to be present in the friction layer in equilibrium with CuO. Since Cu_2O was detected at both temperatures, it is obvious that the indicated low partial pressure of oxygen was achieved at the friction surface. Based on the knowledge of the partial pressure of oxygen and the surface area of Cu_2O detected from the TEM experiment, the surface excess energy can be calculated, and the "nonequilibrium" temperature dependence of ΔG for Cu_2O reaction can be obtained (17, 21-24). The change in the Gibbs free energy for reaction $2Cu + 1/2O_2 \rightarrow Cu_2O$ can be expressed as:

$\Delta G = RT \ln [(a_{Cu}^2 \cdot p_{O2}^{1/2})/a_{Cu2O}]$.

Temperature represents another important variable in thermodynamic calculations. Temperature of the friction surface was measured as a dependence of the $d_{(002)}$ interplanar distance of vermiculite. This method is related to dehydradation and the change of crystal lattice of vermiculite, and it allows a reliable and accurate measurement of temperature in the vicinity (upper 30μm layer) of the friction surface (12). A calibration experiment with the investigated samples allowed finding the following relationship for the temperature of the friction surface T_s:

$T_s = 221.9 \times v \times (1 - \exp[-0.0002 \times v \times \tau])$,

where v is the sliding speed (m/s) and τ is testing time (s) applied.

The calculated thermodynamic data based on the above-described analysis of the friction layer confirmed experimental findings (Table 3). Note that if the change in Gibbs free energy for a product phase is positive ($\Delta G \geq 0$), this phase cannot be formed. In accordance with the experiment, these calculations lead to values $\Delta G \geq 0$ at lower temperatures and confirmed the absence of Cu_2O for a low sliding speed and testing times 600 and 1200s. Based on "nonequilibrium" thermodynamic calculations, Cu_2O is only stable above 890°C at the lowest sliding speed 3m/s. The experimental measurements indicate the stability of Cu_2O in temperature interval above 900°C. On the other hand, experimental data show that Cu_2O is stable within the entire temperature interval (room temperature to maximum reached temperature ≈1000°C), when the sliding speed 6m/s is applied. The developed thermodynamic model confirms this, and the Gibbs free energy differences ΔG, calculated for the highest sliding speed, are negative within the entire temperature interval shown. The presence of Cu_2O was detected experimentally in the entire temperature interval at the highest sliding 6m/s. This is a good confirmation of the reliability of the thermodynamic model applied.

Based on the experimentally analyzed presence of different materials on the friction surface, the modified thermodynamic and kinetic relationships were established for the given system. Also temperature can be predicted as a function of time (τ) and energy dissipation rate (sliding speed v). Based on this information, the structure and the chemistry of the friction surface can be predicted and related to the performance (μ and wear) for any selected time and energy conditions. The fairly good correlation between the measured and the calculated (predicted) data is easily visible in Table 4.

Table 3. Calculated changes in Gibbs free energy of Cu_2O present in friction layer of samples tested at two extreme sliding speeds.

Temperature (°C)	Calculated "nonequilibrium" ΔG (J/mol) for sliding speed (m/s)	
	3	6
0	1.56×10^8	-0.2×10^6
100	1.39×10^8	-0.2×10^8
200	1.21×10^8	-0.5×10^8
300	1.04×10^8	-0.6×10^8
400	0.9×10^8	-0.7×10^8
500	0.7×10^8	-0.9×10^8
600	0.5×10^8	-1.1×10^8
700	0.3×10^8	-1.3×10^8
800	0.2×10^8	-1.5×10^8
900	-1.4×10^6	-1.7×10^8
1000	-0.2×10^8	-1.8×10^8

Table 4. Measured data and calculated values of the coefficient of friction (μ) and wear (w) for intermediate sliding speed 4.5ms^{-1}.

Time τ (s)	μ/w [wt. %]		
	Experimental	Calculated	Difference
600	0.31/0.62	0.30/0.55	0.01/0.07
1200	0.36/2.3	0.35/1.8	0.01/0.5
1800	0.38/3.98	0.37/3.65	0.01/0.34
2400	0.40/5.6	0.40/5.2	0/0.4
3000	0.41/7.3	0.41/7.23	0/0.07
3600	0.42/9.01	0.42/9.2	0/-0.18
4200	0.41/10.7	0.42/10.2	-0.01/0.5
4800	0.41/12.4	0.41/11.9	0/0.5
5400	0.41/14.1	0.41/13.8	0/0.3

In contrast with numerous other models, the presented approach requires that selected friction tests at defined conditions (covering the span of potential applications) have to be performed. The most proper parameter controlling the performance of friction materials is the energy dissipation rate (power) related to a friction process (4). Analysis of the friction layer is a critical part of the presented approach, and the following synthesis and modeling is based on these data. The efficiency of friction materials development can be increased significantly when the presented analysis and synthesis involving modified thermodynamic and kinetic laws is applied or at least combined with the frequently used empirical approach (4, 12).

In spite of the fact that it was not presented here, the proposed alternative of analysis, synthesis, and modeling can be applied to other types of friction tests. Based on the data presented here, it can be easily seen that the discussed analysis, synthesis, and modeling based on the understanding of real phenomena on the friction surface also allows the comparison of different friction tests and their modification (development). Identical testing conditions will always lead to the formation of identical products on the friction surface.

Figure 8 illustrates the suggested design strategy with an application of the newly developed analytical tool. The empirical strategy based on formulation, testing, and redesign loop is improved by involving the newly developed analytical tool, which can provide relevant information about the key parameters affecting the performance and efficiency of testing.

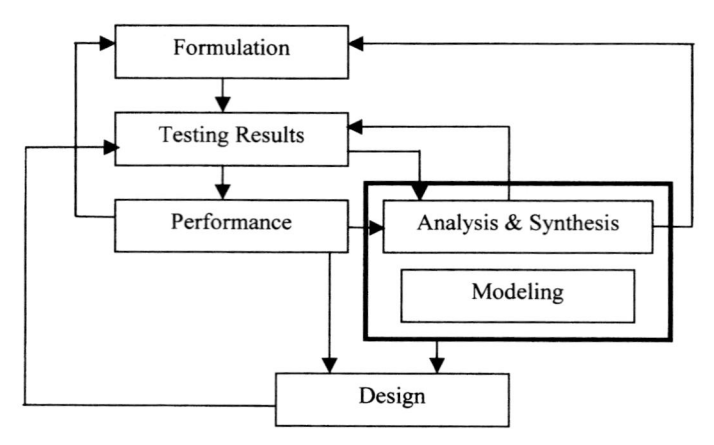

Figure 8. Suggested design strategy for PMC brake lining formulation. Bold framing marks newly developed analytical tool discussed in this paper.

4 CONCLUSIONS

The performance of friction materials depends strongly on the character (structure and chemistry) of the friction layer developed always on the friction surface. The tailored design and smart testing of friction materials may incorporate the developed analytical tool (Fig. 8), which makes the entire process significantly more efficient. The presented analysis and synthesis allows the modeling of performance, tailored design, and testing of friction materials.

The thermodynamic and kinetic of processes on the friction surface differ from those typically observed in materials science. The detected time development does not follow the typically observed Kolmogorov-Johnson Mehl-Avrami relationship describing the kinetic of solid-state reactions. Thermodynamic relationships have to be modified, and realistic surface energy and pressures of oxygen have to be incorporated into models. The proper modifications of kinetic and thermodynamic relationships can be estimated from the analysis of the friction layer following friction testing at different conditions covering the span of the potential application.

5 ACKNOWLEDGEMENT

This research was supported by the CAFS university-industry research program, by NSF (project #EEC9523372) and by ORNL (project # 2000-055). Drs. T. Watkins and J. Bai kindly assisted with the experiments performed at ORNL Oak Ridge, TN, and at BNL Long Island, NY.

6 REFERENCES

(1) P. J. Blau, Friction Science and Technology, New York: Marcel Dekker, 1996.
(2) P. Filip, L. Kovarik, M. Wright, Automotive brake lining characterization, Proc of 15th Annual SAE Brake Colloquium 1997, SAE, Warendale, PA, 1997, pp.319.
(3) A. Anderson, S. Gratch, H. P. Hayes, A new laboratory friction and wear test for the characterization of brake linings, Proc. Of Automotive Engineering Congress 1967, SAE, New York, NY, 1967, pp 1-13.
(4) P. Filip, L. Kovarik, P. Pacholek, J. Pavliska, M. Wright, On-Highway Brake Linings, Area two, Quarterly report to the Center for Advanced Friction studies, , Vol.3 No.3 , 1999.
(5) Surface texture (Surface Roughness, Waviness, and Lay), ASME B46.1-1995 The American Society of Mechanical Engineers, June 14, 1996.
(6) P. Filip K Butson, H. Lorethova, L. Kovarik, M.A. Wright, Stereology and Quantitative Image analysis Quarterly report to the Center for Advanced Friction studies, Vol.4 No.1 , 2000.
(7) J. L. Sullivan, T.F. Quinn, D.M. Rowson, Origins and development of oxidational wear at low ambient temperatures, Wear, 94, (1984), pp.175-191.
(8) D. M. Rowson, Frictional heating and the oxidational theory of wear, Journal of Physics D: Applied physics, 13, (1980), pp.209-219.
(9) J. L. Sullivan, T.F. Quinn, D.M. Rowson, Development in the oxidational theory of wear, Tribology International, 13 vol.4, (1980), pp.153-158.
(10) G. B. Schaffer, P. G. McCormick, Displacement reactions during mechanical alloying, Metallurgical Transactions A, 21A, (1990), pp.2789-2794.
(11) G. B. Schaffer, P. G. McCormick, Reduction of metal oxides by mechanical alloying, Applied Physics letters, 55, 1989, pp.45-46.
(12) P. Filip, M. Kristkova, Quarterly report to the Center for Advanced Friction studies, October 2000.
(13) O. Kubaschewski, B. E. Hopkins, Oxidation of metals and alloys, Butterworths, London, 1967.
(14) A. N. Kolmogorov, Izvestija Akad. Nauk USSR, Serija Fizika, 3, 1937, 355.
(15) M. W. A. Johnson, K.F. Mehl, Trans. Am. Inst. Mining Met. Eng., 135, 1939, 416.
(16) M. Avrami, J. Chem. Phys., 7, 1939, 1103. and 8, 1940, 212 and 9, 1941, 177.
(17) J. W. Christian, The Theory of Transformation in Metals and Alloys, Pergamon Press, Oxford, 1975.
(18) R. A. Robie, B.S. Hemingway, J. R. Fisher, Thermodynamical properties of minerals and related substances at 298.15 K and 1 bar (10^5 Pascals) pressure and at higher temperatures, U.S. Geological Survey Bulletin 1452, United States government printing office, Washington, 1979.
(19) M. Zdujic, C. Jovalekic, Lj. Karanovic, M. Mitric, The ball milling induced transformation of γ - Fe_2O_3 powder in air and oxygen atmosphere, Materials Science and Engineering, A262, (1999), pp.204-213.
(20) M. Zdujic, C. Jovalekic, Lj. Karanovic, M. Mitric, D. Skala, Mechanochemical treatment of γ - Fe_2O_3 powder in air atmosphere, Materials Science and Engineering, A245, (1998), pp.1009-1117.
(21) D. A. Rigney, Effect of structural and chemical changes on sliding friction and wear of ductile material, MRS workshop, Tribology on the 300th anniversary of Amontons' law, 1999.
(22) I. J. Lin, S.Nadiv, Review of the phase transformation and synthesis of inorganic solids obtained by mechanical treatment (Mechanocemical reaction), Materials Science and Engineering, 39, (1979), pp. 193-209.

(23) J. Molgaard, V.K. Srivastava, The activation energy of oxidation in wear, Wear, 41, (1977), pp.263-270.
(24) D. R. Gaskel, Introduction to the thermodynamics of Materials, Taylor &Francis, Washington, third ed., 1984.

The interactions of phenolic polymers in friction materials

B H McCORMICK
Borden Chemicals UK Limited, Penarth, UK

ABSTRACT

Four resins, modified and unmodified, were studied for their effects in friction materials. The cured resin properties were evaluated without filler, for physical properties and friction performance.

The changes that occur in the friction material during service were followed. An interaction between a common filler, barytes, and the resin at service temperatures was identified. Changes in the physical structure of friction dust during service were identified, suggesting an explanation for the performance advantages of friction dust.

The thermal properties of the surface layer altered during friction testing and the formation of a transfer film correlated to noise generation.

1. INTRODUCTION

Phenolic systems have been used in the manufacture friction materials for many years, but the interactions within these materials are still not fully understood. Technology for developing improved friction systems continues to evolve utilising modelling techniques to describe these system. However, models are only as good as the input.

The technology of friction manufacture 'formulating and processing' is the optimisation of the interactions, ultimately to meet the demands of particular application. These can occur within the matrix during manufacture, and subsequently in service due to from the temperatures generated.

An understanding of the interactions in friction materials requires data relating to the type of resin and the other components present. However few studies have concentrated on looking at these interactions from manufacture to final friction performance.

2. CURED PHENOLIC RESINS

Phenolic resins are used as the binders in friction materials to hold together many other materials, without the resin the product could not perform as a friction material. However there have been few studies (1) of the properties and performance of cured unfilled phenolic resins. This may relate to the problems in preparing suitable samples. Phenolic resins are notoriously difficult to mould into void free specimens.

Much time and energy was invested into the task of preparing unfilled cured resins. Using a press fitted with a transducer to monitor the platen displacement it was possible to follow the compaction of the resin. A temperature ramp for the heating cycle allowed the design of a moulding cycle so that a number of cured resin samples could be prepared.

Four different resin systems were evaluated:
- Unmodified high molecular weight resin, regularly used as a binder in the friction industry
- PVME modified resin developed to improve adhesion in friction material formulations
- Nitrile rubber modified resin, used for commercial vehicle lining manufacture
- Polychloroprene modified resin, used in the friction industry and identified as having excellent fade properties

2.1. Mechanical Properties
The cured resin samples were evaluated for their physical properties using ASTM Standards. A range of these properties are shown in the table below.

Table 1. Mechanical properties of phenolic resins

Modifier	Modulus GPa	Standard Deviation	Strength MPa	Standard Deviation	Fail Strain %	Standard Deviation
Unmodified	5.88	0.303	126.3	8.8	2.24	0.18
PVME	5.95	0.01	110.5	11.9	1.97	0.24
Nitrile	5.08	0.07	108.4	7.3	2.15	0.17
Polychloroprene	4.65	0.02	92.4	4.9	1.99	0.11

(Flexural Properties)

Modifier	Yield MPa	Standard Deviation	K_{IC} MPam$^{0.5}$	Standard Deviation	No	Standard Deviation
Unmodified	283.7	2.2	0.70	0.05	46.7	0.6
PVME	266.5	10.6	0.86	0.18	45.4	0.5
Nitrile	235.4	5.8	1.21	0.15	40.7	3.0
Polychloroprene	219.5	3.9	1.15	0.11	39.2	0.8

(Compressive Properties | Fracture Toughness | Vickers Hardness)

These results indicate that the properties of the resin system are affected by the nature of modifiers used. The unmodified resin system shows high flexural strength and modulus but low toughness, whereas the rubber modified resin shows lower strength and modulus properties but increased toughness. Changes in these properties would affect friction performance. Compressibility of a friction material can affect surface contact area, whilst strength of the resin film could affect how fine abrasive particles are held in position during the wear process. There are many ideas concerning the mechanism of noise generation in friction materials and the use of 'softer' resin systems has been proposed. However a study of the cured resin systems without fillers indicates that many physical properties could be interrelated, making the evaluation of changes more difficult. This study has shown a linear relationship between hardness and compressive strength.

2.2. Morphology

Specimens from the cured plaques were metallurgically polished and examined by optical microscopy. Three of the resins (including the nitrile rubber modified resin) showed no obvious phase morphology. The only resin exhibiting features was the polychloroprene modified resin where a dispersion of rubber particles in the size range 1 – 10 μm could be found.

The fracture surfaces were studied by Scanning Electron Microscopy (SEM). The unmodified and PVME modified resins showed flat surfaces with shear bands only where crack growth had occurred on different planes. These are features typical of brittle materials. Although the PVME modified materials had shown higher fracture toughness there was no evidence for crack deflecting mechanisms in the morphology. The polychloroprene modified resin showed disturbances to the fracture path by the distributed particles. Interestingly, the nitrile rubber modified resin showed a highly disturbed surface down to the μm level, indicating the nitrile rubber was evenly dispersed and any potential particle rubber size was below this level. This is direct confirmation that for nitrile rubber there is a significant difference in the rubber dispersion when using nitrile modified resins compared to the addition of rubber into friction material mixes.

2.3. Tribology

The cured resin samples were evaluated for their friction performance using a Mk 1 Denison pin-on-disc machine under constant sliding speed conditions and grey cast iron discs. A range of test conditions was used, varying the sliding speed (between 1 and 10 m/s) and also the load applied (either 25.28N or 50.56N, giving a contact stress of 0.5 or 1 MPa). The resins showed markedly different performance in both friction performance and wear rate.

Looking at the wear rates, all the resins show two wear regimes. At low power dissipations the wear rates are high and an abrasive wear mechanism is occurring, whilst at higher power dissipations thermal wear becomes significant and thermal transformation of the resins gives low wear rates as shown in figure 1.

There are also interesting effects in the temperatures generated during the friction testing, as shown in figure 2. The equilibrium disc temperature for a particular load/sliding speed is affected by the resin. The nitrile rubber modified resin generated the lowest disc temperatures whereas the polychloroprene modified resin generated the highest disc temperatures. This result may indicate one reason why nitrile rubber modified resins have frequently been used for the production of linings for commercial vehicles.

These unfilled resin samples show changes significant variations in friction performance and

Figure 1. Wear rate of phenolic resins

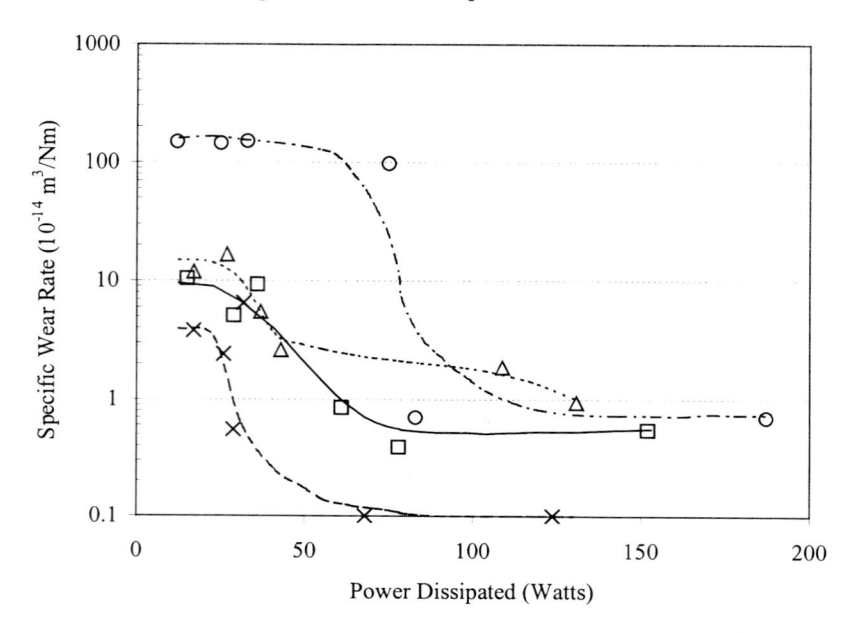

Figure 2. Frictional heating effects

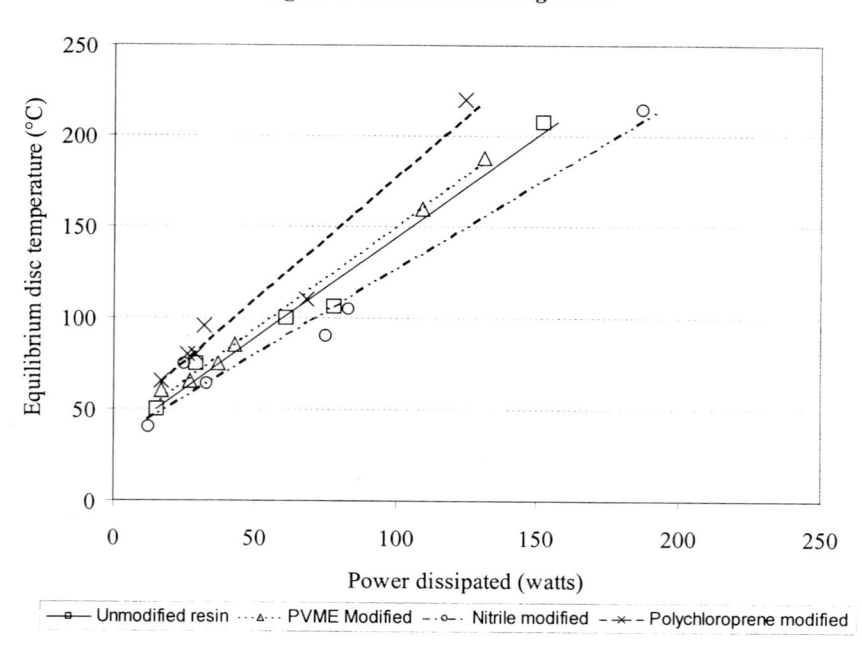

—□— Unmodified resin ⋯△⋯ PVME Modified –·o·– Nitrile modified –×– Polychloroprene modified

it could be expected that significant variations would be found in the friction materials prepared from them.

3. RESIN INTERACTIONS IN FRICTION MATERIALS

Friction materials are frequently complex mixtures of materials but manufacture is often considered to be a simple press and cure operation with little account taken either of interactions occurring within the matrix or of the effect of the curing operation.

Friction material mixes can have very large surface areas, specific surface areas for a number of raw materials are shown below.

Table 2. Surface area of friction material components

Material	Surface Area m^2/g
Rockwool	0.2
Glass Fibre	0.008
Barytes	1.21
Alumina	0.7
Lime	5.44

This can give surface areas for non-asbestos formulations of the order of 3 to 4 m^2/g which then leads to calculations of the resin thickness as $<10^{-7}$m. Based on these figures the resin film thickness in a friction material would only be a few molecules thick. This leads to the inevitable question of how the resin is distributed in the composite. Optical microscopy of the composites did not reveal any insights into the way the resin was distributed. However moving to a technique using laser fluorescence microscopy revealed the location of the resin. It was found that the phenolic resin fluoresced, allowing areas where resin was present to be identified. The images below show the standard optical image (left) and a combined optical and fluorescent image (right) taken from a false coloured image. Where the optical and fluorescent image overlap the result is white or a shade of grey. The areas showing only the optical image are very dark grey or black. Large areas are shades of grey, indicating that the resin has in fact spread widely in the composite.

Figure 3. Distribution of resin

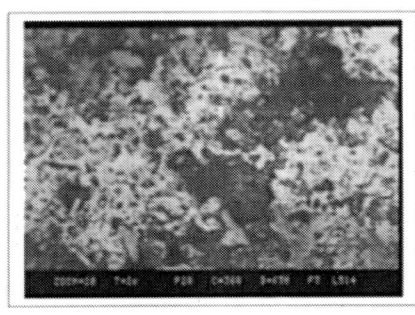

Topographic image Composition image

This is not the only interaction of the resin. Many friction materials contain fibres and there can be significant interactions with certain resins. The glass tow consists of a group of fibres held together to allow processing of the chopped fibre. The images below were again taken using laser fluorescence microscopy, in this case white areas show the presence of resin The image on the left shows the fibre tow (left of centre on the image) can remain unaffected when using an unmodified resin. There was no fluorescence from within the bundle indicating that the resin had not penetrated. In contrast the image on the right shows the effect of using the PVME modified resin, where wetting and penetration of resin into the fibre tow has occurred. The fibres had separated within the tow and fluorescence occurred around the individual fibres confirming the presence of resin. Work was also carried out by video microscopy on a hot stage using the base resin without hexamine. As the resin melted the PVME modified resin was able to undergo capillary movement along the fibre tow. Interestingly only the PVME modified resin exhibited capillary flow.

Figure 4. Wetting of fibre tows

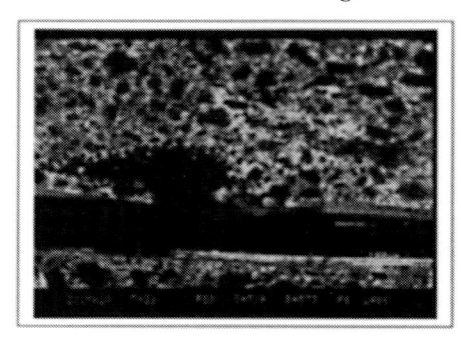

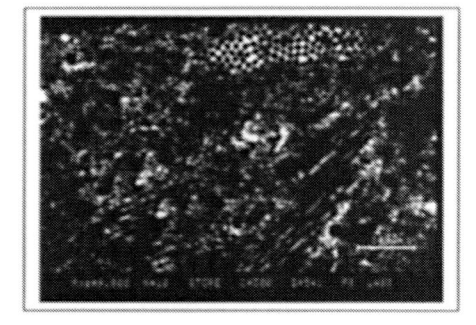

Unmodified resin					PVME Modified resin

4. CHANGES IN FRICTION MATERIALS DURING SERVICE

After production of the friction material it might be assumed that interactions of the resin have ceased and that the only other action of the resin is to form a carbon char during service, but further studies show this is not correct.

4.1. Interaction between resin and barytes

At higher temperatures in service there is a reaction occurring between the resin and one of the major fillers used in friction materials, barytes. During an operation to bake a simple friction material up to temperatures of 1000°C mass spectrometry work was carried out to determine the composition of the volatiles formed. Surprisingly at temperatures around 700°C an excessive production of carbon monoxide and carbon dioxide was found, the mass spectrum trace for the latter is shown below in figure 5.

The friction material being used for this study contained only barytes as a filler, so the reaction responsible for carbon dioxide production was anticipated to be reduction of the barytes. This was confirmed by using X-ray Photoelectron Spectroscopy analysis (XPS) where the barium sulphate was shown to have changed to barium sulphide.

Figure 5. Formation of Carbon Dioxide

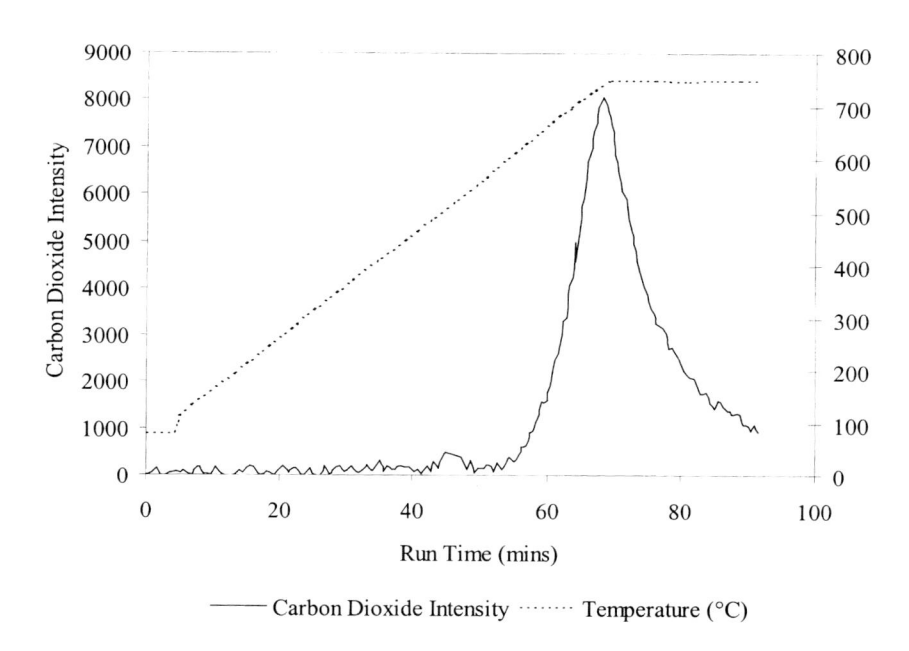

4.2. Friction dust decomposition

CNSL based friction dust has been used for many years as an additive to friction material formulations. However little information has been available to describe how friction dust gives significant improvements in friction material performance (2). Friction performance depends not only on the composition of the friction material but also on the wear surface formed by heat and abrasion in service. The heat generated during service affects much more than the interface wear layer, it has a significant effect on the composition of the body of the friction material behind the wear layer. After carrying a dynamometer wear test with disc temperatures up to 600°C the friction material was cross-sectioned and studied using SEM and optical microscopy. Structural changes were seen in the body of the friction material to a depth of several millimetres after this high temperature brake application. The SEM image is shown in figure 6, with the wear surface at the top.

Near the surface of the friction material there are voids present indicating the loss of friction dust particles. Further into the friction material, i.e. down the temperature gradient, it can be seen that shrinkage of the friction dust particles has occurred, leaving voids around the particles. A similar evaluation of friction materials after different dynamometer test schedules was carried out. After short friction tests carried out with disc temperatures below 300°C, the friction dust was found to be essentially unchanged with particles present up to the friction surface. After a longer friction test involving a fade cycle up to a disc temperature of 500°C a loss of friction dust was found for a small distance into the material, up to 2mm below the surface. Further into the friction material shrinkage of the friction dust had occurred.

Figure 6. Friction material after dynamometer test

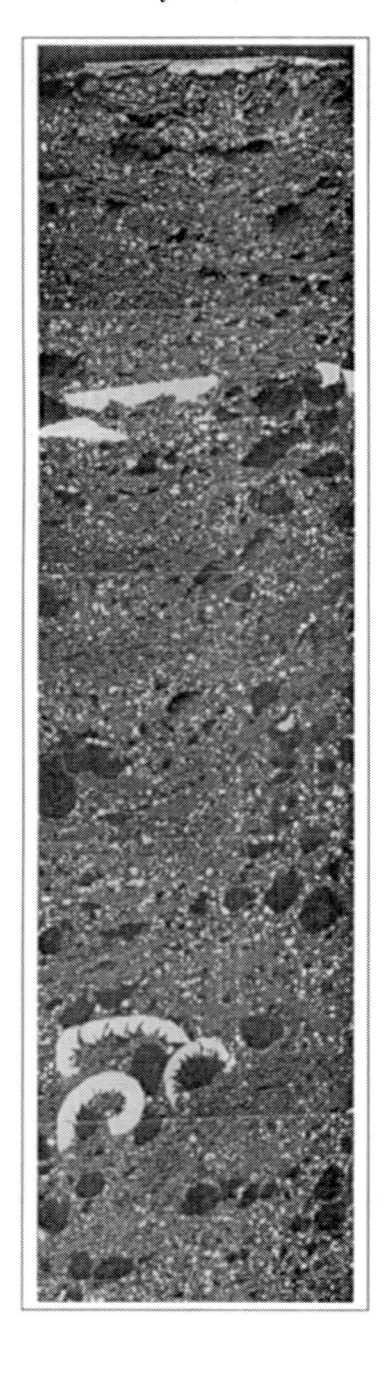

Figure 7. Friction material microstructure

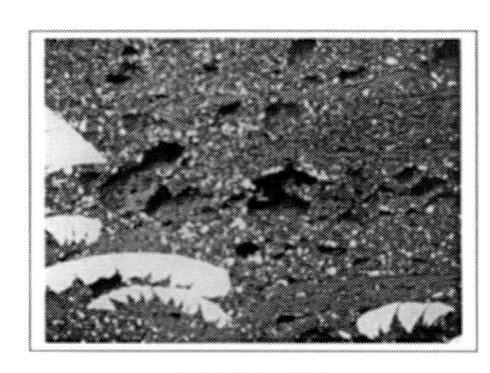

500°C

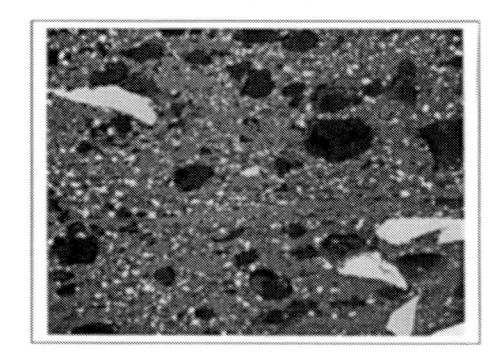

375°C

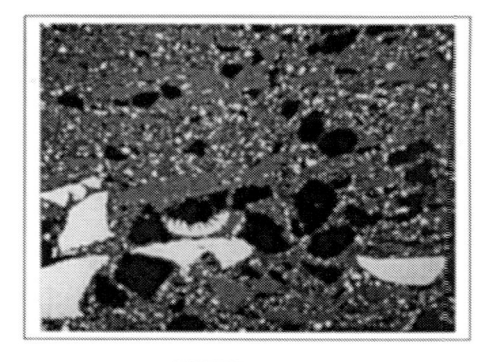

250°C

Friction material specimens were then heated at three different temperatures in an oven, 250°C, 375°C and 500°C. The microstructure of these blocks is shown in figure 7.

The microstructure is essentially unchanged at 250°C, whereas shrinkage of the friction dust occurs at 375°C and by 500°C total loss of the friction dust has occurred. This information also allows an evaluation of the temperatures seen in the body of the friction material in service.

This data indicates that one mechanism for the improvement in friction performance by CNSL friction dusts is the formation of additional porosity during service.

5. WEAR LAYER INTERACTIONS

Changes occur in the friction material as a result of the brake application and a new surface is formed on the friction material. Pin-on disc testing of friction materials showed that differences in behaviour could be seen between modified and unmodified resins.

5.1. Thermal conductivity of wear layers

The equilibrium disc temperature at different power dissipation was studied. The relative order of the heat generation changed from that seen for the neat resins. The results, as shown in figure 8, also indicate that lower disc temperatures were generated for the friction material systems compared to the resin only systems. This suggests that more heat was being dissipated through the friction material.

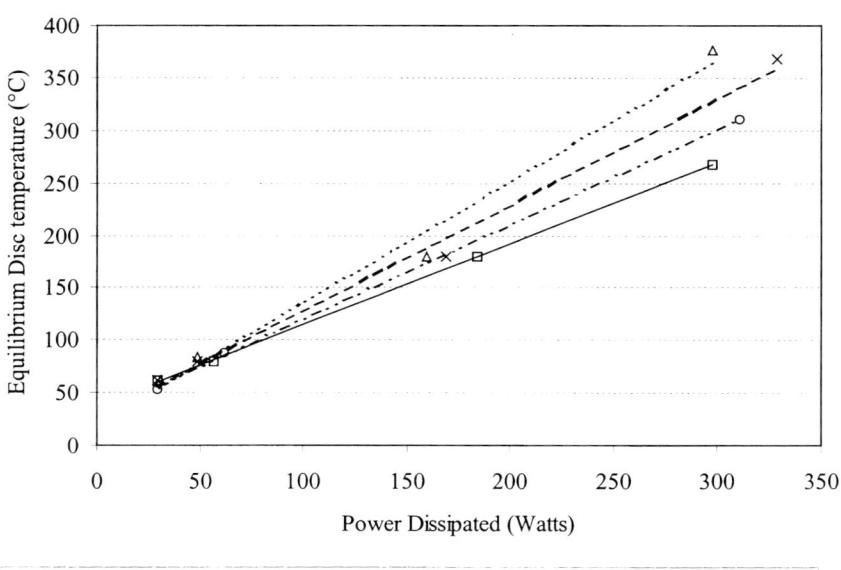

Figure 8. Frictional heating effects

These changes, both in temperature generation and the relative resin order, prompted further questions on the properties of the wear layer. The data suggested that the thermal properties of the system had changed so an investigation of the thermal conductivity of the friction materials was carried out. The technique of Opto Thermal Radiometry (OTTER) was used. This uses a pulsed laser to instantaneously raise the surface temperature, then the heat produced can be dissipated in three ways

- radial conduction within the surface layer
- conduction into the underlying substrate
- radiation

The time decay of the Infra-Red radiation emitted is recorded and this decay curve is matched to a heat flow model to deduce the rates of the various thermal processes in the sample. The heat model used relates a decay constant τ for the thermal process to the radial conduction in the sample.

$$\tau = \alpha^{-2} D^{-1}$$

In the equation α is the absorption coefficient for the radiation from the laser and D is the thermal diffusivity. Since all the friction materials were very similar α is assumed to be constant, so the conductivity is inversely related to the values of D.

D is related to the thermal conductivity:

$$D = K\rho c$$

Where K is the thermal conductivity, ρ is the density and c is the specific heat. Provided that ρ and c are similar then the lifetime can be used as a direct measure of thermal conductivity. The thermal conductivity is inversely proportional to the lifetime.

Tests were carried out to determine the thermal conductivity of the surface of a friction material before friction testing and then after a friction test.

Table 3. Thermal Conductivity by OTTER

Resin	Condition	Decay Constant τ (μs)
Unmodified	Before friction test	111 ± 20
Unmodified	After friction test	166 ± 60
PVME Modified	Before friction test	73 ± 7
PVME modified	After friction test	280 ± 64

The data shows that for the friction tested material the unmodified resin has the lowest τ and hence the highest thermal conductivity. This provides support for the theory that disc temperature can be related to the thermal properties of the wear layer. It was also interesting to note that the friction tested material was much more variable in thermal conductivity than the untested material. This could be an additional reason for hot spots on the frictional material during brake application.

5.2. Composition of wear layers

During the pin-on-disc testing of the friction materials there were a significant number of occasions when noise was generated during the test. These noisy tests tended to occur under conditions of low loads and low sliding speed, this being similar to that noted for noise generation in vehicles. The level of noise generated was monitored by determining the band pressure level at a frequency of 4kHz. This was frequency at which the maximum sound level occurred for this particular test equipment.

A friction test that was generating noise was stopped and the wear surface of the friction material was studied, figure 9. An SEM topographic image showed that the surface was relatively flat and this was also confirmed by carrying out a profile of the surface. The SEM image showed a number of friction dust particles at the surface and these particles were essentially flat. This topographic image (left) was compared to a composition image (right). The friction dust areas remained dark in the composition image indicating that in this area elements of low atomic number were present.

Figure 9. SEM Images of noisy friction material surface

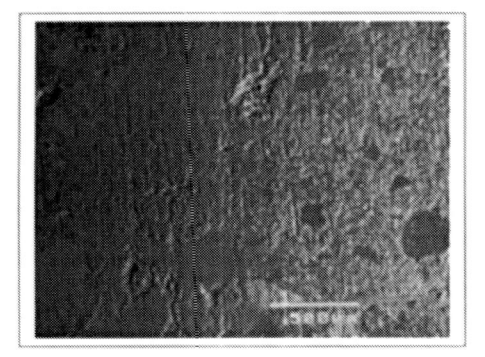

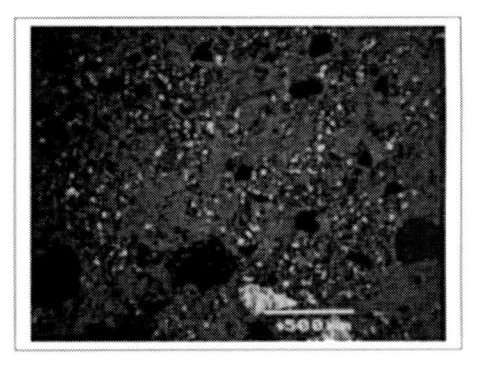

Topographic image   Composition image

The composition of the friction material surface was also evaluated using Energy Dispersive X-ray Analysis (EDAX). A strong signal for Fe was found indicating that a transfer film had been formed from the rotor. However, the fact that the friction dust particles gave a dark image in the SEM composition image showed that the friction dust particles had not been coated with the Fe transfer film.

The sample was reconditioned by running a further friction test to take the temperature up to 240°C. The test became quiet and the wear film was analysed to see if a change had occurred in this layer. The composition of this surface was again determined using EDAX. This time a signal for Ca was more prominent, figure 10. The signal for Fe was very much reduced in size indicating that the transfer film had been destroyed.

Figure 10. Composition of friction material surface

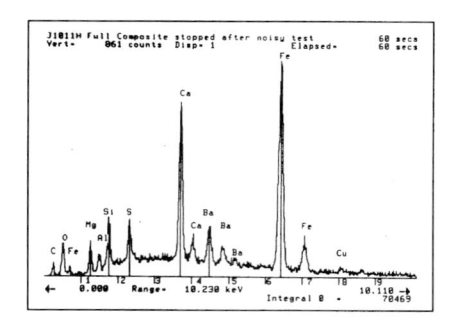

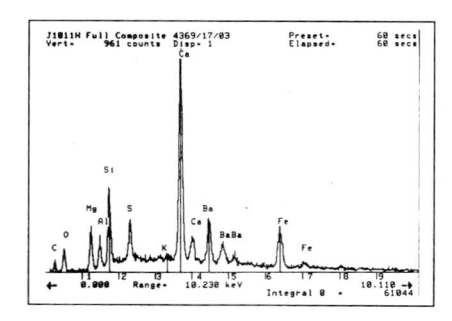

Surface after noisy test Surface after quiet test

SEM images were again taken during this quiet test, figure 11. The topographic image now showed a rougher surface and this was confirmed with a profile of the surface. The SEM image also showed that the friction dust was starting to degrade in a way similar to that shown earlier in the paper. Gaps could now be seen between the friction particles and the matrix.

Figure 11. SEM Images of quiet friction material surface

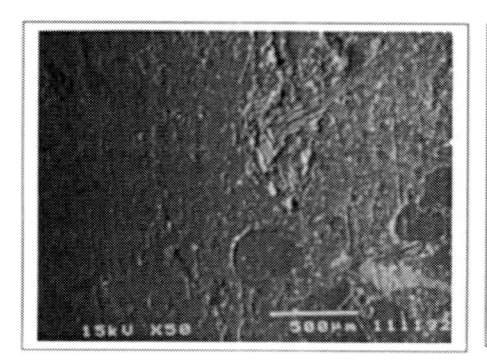

Topographic image Composition image

It is also interesting to compare the transfer film formed in the pin-on-disc work with the film on the surface of a friction material (using the same formulation but with the PVME modified resin) during dynamometer testing. The EDAX result for two materials is shown in figure 12, the two traces are virtually identical.

Figure 12. Analysis of transfer film

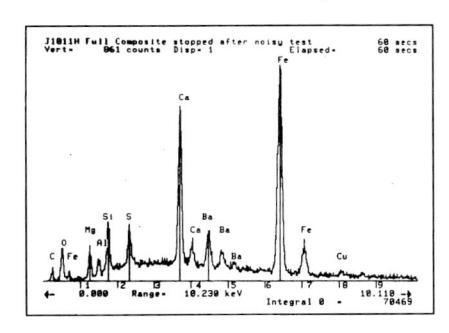

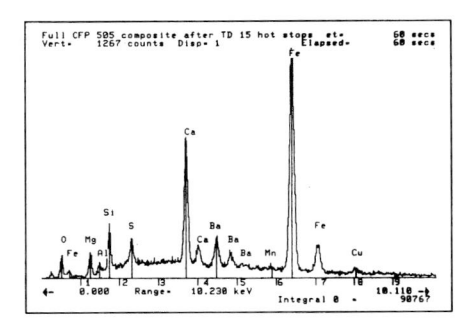

Surface from pin-on-disc   Surface from dynamometer test

Further elemental analysis of this surface layer was carried out and showed a high concentration of oxygen, indicating the transfer film would actually be composed of iron oxide.

6. CONCLUSION

Four different resin systems, modified and unmodified, have been studied. This has given a greater understanding the properties of the resin systems, how the resins interact in friction materials and how they affect friction performance.

It has been found that the resins have different physical properties when cured. These resins also influence the friction performance when evaluated without any filler present. In formulated friction systems further interactions occur under the influence of the temperatures found in service. This study of the changes will help to develop the understanding of the mechanisms surrounding the performance of friction materials and suggest new techniques to help optimise the performance of friction materials in the future.

REFERENCES

1 **LAMBLA, M.** and **VO, V. C.** (1986) Optimization Of Phenolic Resins for Friction Materials. Polymer Composites **7(5)** pp262.

2 **JACKO, M. G.** (1978) Physical and Chemical Changes of Organic Disc Pads in Service. Wear **46** pp 163 - 175

Effect of friction lining additives on interfaces of friction components

R HOLINSKI
Dow Corning GmbH, Wiesbaden, Germany

ABSTRACT

Since many decades additives have been added to friction lining composites in order to improve braking performance and prolong life of friction components. Various metal sulfides are still used in large quantities as friction lining additives. It was found, that lead sulfide and molybdenum disulfide support the formation of a transfer layer on the metal counter part, which reduces wear and stabilizes friction. At higher temperatures these metal sulfides oxidize by tribooxidation.

Surface analysis by SEM and ESCA revealed, that antimony sulfide does not contribute significantly to a formation of a transfer layer. At rotor temperatures higher than 500°C no transfer of composite material was detected anymore on rotor surfaces, rather a chemical reaction took place. Detected was antimony pentasulfide, which contributes excellently to wear protection at high temperatures.

A new additive called "polarized graphite" was found to contribute to a formation of a thin homogeneous transfer layer on the rotor surfaces. These transfer layers reduce brake noise and wear of both friction components and stabilize friction. At rotor temperatures higher than 500°C no transfer of material takes place any more rather a chemical reaction of the additive.

Metallisation of brake pads - transfer of metal flakes from rotors to brake pad surfaces - is initiated by hydrogen diffusion into the rotor surface. It was found that additives containing zinc- and tin compounds form a metal layer on rotor surfaces, which act as a barrier layer for hydrogen, thus preventing hydrogen diffusion into the rotor surface and thereby eliminating metal pick-up.

1. INTRODUCTION

In a friction brake, organic friction components are in sliding contact against the cast iron drum or rotor. Surfaces of friction components change during the braking operation. At the surface of the organic composite a friction - heat - affected layer is formed, which has a different composition and different physical properties than the matrix. Such a layer is generally very wear resistant. On the cast iron surfaces transfer layers are formed (1). Both layers - the friction layer and the transfer layer - control tribological performance of braking such as friction level, fading, vibration, comfort and wear rate (2). Therefore it is of utmost importance to affect the characteristics of these layers in order to achieve desired physical properties.

Empirically it was found by developers of friction composites that the addition of metal sulfide powders to the friction lining formulation improves the braking performance. A number of metal sulfides are added to composites in a powder form in a concentration ranging from 1% to up to 10% by weight (3). Without the use of metal sulfide in lining formulations the composite wear is very high.

Also cast iron surfaces deteriorate under certain conditions and affect the braking procedure in a negative way. Due to decarbonisation of the rotor surface, resistance against shear is reduced. Thin films of metal are sheared from the rotor surfaces and transferred to the composite surface. Wear and noise increase. Additives can assist in elimination of rapid destruction of rotor surfaces by prevention of decarbonisation of rotor surfaces. In a series of investigations it has been studied how additives in friction composites affect the interfaces of components of a friction brake.

2. GROWTH OF TRANSFER LAYERS

The solid lubricant Molybdenum Disulfide is added to various composites like carbon brushes, sinter bearings and plastic components in order to reduce wear rate. It was found that MoS_2 works as an adhesion promoter for transfer layers and build wear resistant layers at interfaces of components in sliding contact. In case the adhesion force from an MoS_2 particle to the cast iron surface is stronger than the cohesive force within the composite then the build-up of a nucleation layer will start. The build-up of such a layer mostly starts with nucleation sites (4). Further growth is taking place around the sites if cohesive strength of the composite is lower than the adhesion force between small sites and the composite particles. A secondary transfer layer is formed on the nucleation layer. This secondary transfer layer has very good cohesive strength because it resists high shear forces and is not rubbed off very easily during braking. Tranfer films have been investigated on a ring-block-tester. A composite block is in sliding contact with a ring. The ring weight has been determined periodically. It was found that the weight gain on the ring was due to the transfer of composite material to the ring surface. After a certain time growth of the transfer layer stopped and film thickness remained constant (5).

Basic investigations were performed on a ring/block test machine. A composite block was pressed against a rotating steel ring with a pressure of 80 MPa. (Fig. 1 left)

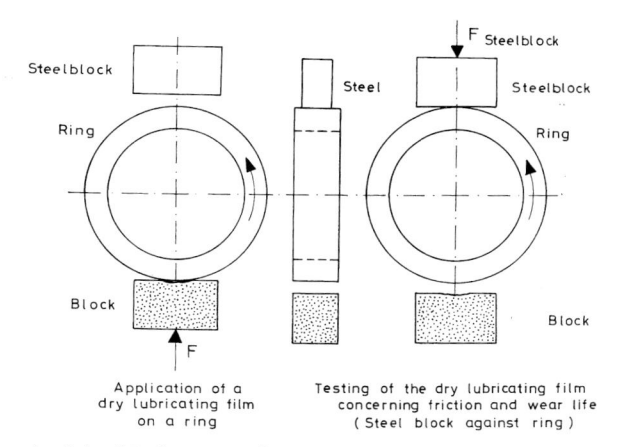

Fig. 1: Sketch of ring/block test specimen
Ring diameter: 35 mm   Speed of rotation: 72 rpm

It was found that initially there is no transfer of material from the composite to the metal surface. During the first 100 revolutions friction and frictional temperature increase (6). Then a continuous transfer of material was observed. After a certain transfer of material the film thickness leveled off; the film stopped growing. After this growth friction and frictional temperature stabilized. Film thickness of a transfer film depended on the type of additive present in the composite formulation. In some cases the film continued to grow and blistered off eventually. A new film started to be formed by transfer of material, which resulted in higher composite wear.

In order to determine adhesive strength of the nucleation layer and cohesive strength on the film, the composite block was removed from the ring and a steel block was pressed against the rotating ring at a pressure of 650 MPa. The friction coefficient depended on the formulation of the transfer film. Some films had a limited life time others had a rather long life time (Fig. 2). Certain additives contribute to adhesion of transfer layers to the substrate and to high shear strength of the layer.

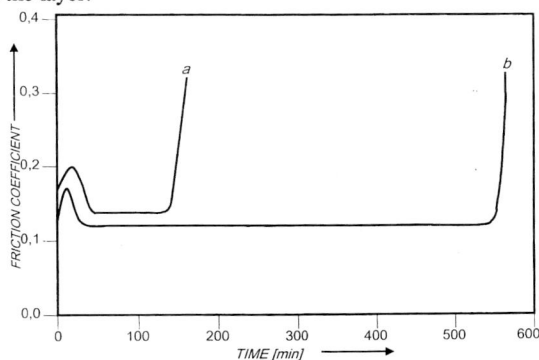

Fig. 2: Duration of transfer films during dry sliding of a ring against a block.

Molybdenum disulfide is known for its good adhesion on metal surfaces. This adhesion can be explained by the polar nature of MoS_2 lamellae. In order to create a solid lubricant with better adhesion than MoS_2, graphite lamellae have been furnished with polarity. Untreated graphite has poor adhesion on metal surfaces because of lack of polarity. Adhesion of polarized graphite on metal substrates was found to be better than that of molybdenum disulfide films. (Fig. 3). Rotor wear was reduced significantly when polarised graphite was added to the composite in place of molybdenum disulfide. Obviously adhesion to the rotor surface and transfer film strength of polarised graphite is superior over MoS_2.

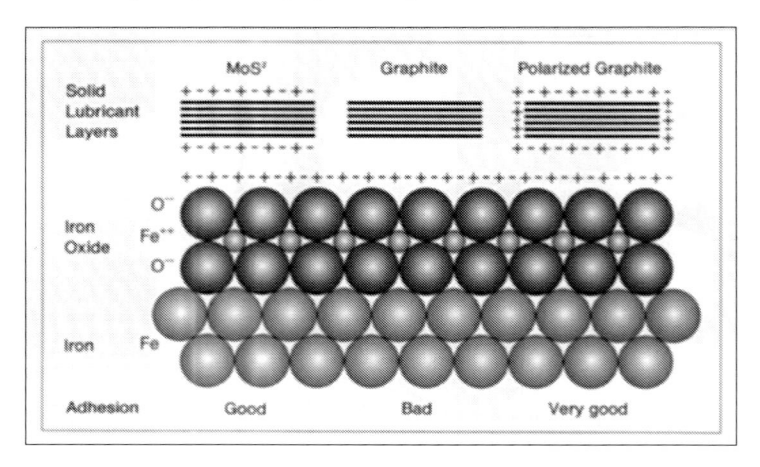

Fig. 3: Model of polar surfaces of lamellar solids on a metal surface.

3. THE TRANSFER LAYER

On a brake lining simulator (3) brake linings consisting of phenolic resin, steel wool, barium sulfate and calcium carbonate have been tested during continuous braking. Surfaces of the cast iron rotor have been investigated by scanning electron microscopy (SEM) and energy dispersive X-ray analysis (EDX). It was found that during the first 100 cycles there is no transfer of the film to the rotor surface. The analysis showed only the elements of the steel like iron, carbon and silicon (Fig. 4). After transfer of material the rotor surface looks different and a transfer layer is visible. Detected are barium, calcium, sulfur and oxygen from the compounds of the friction composite. The transfer layer has a thickness of about 0,5 µm; at a tension of 10 KV of the microprobe the electron beam has a penetration of about 0,5 µm. Only a small iron peak is detected which means the transfer layer up to a thickness of 0,5 µm covers the rotor surface.

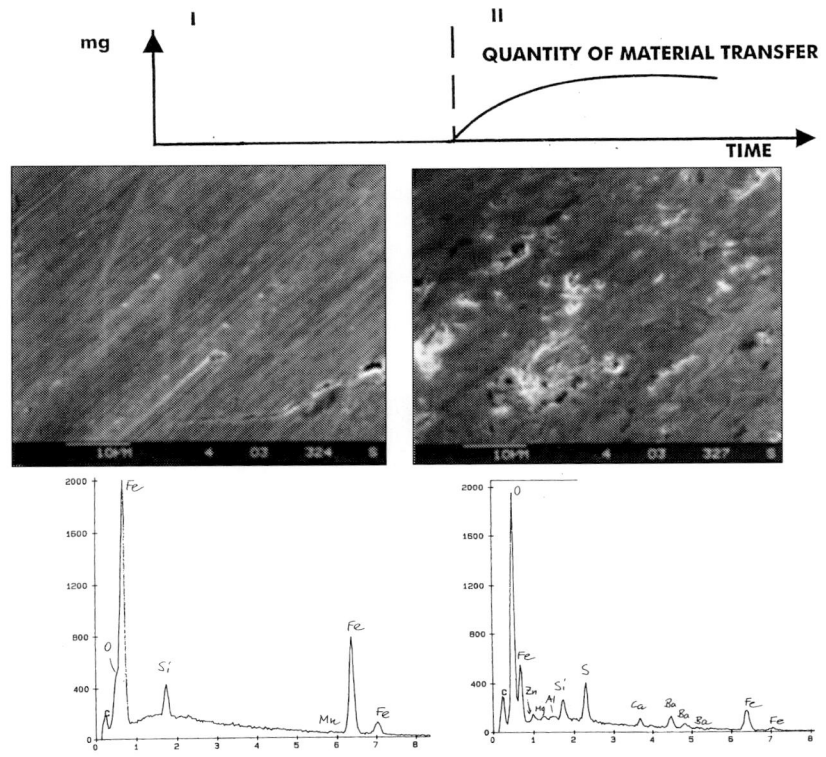

Fig. 4: Top: Transfer of material to rotor surface
Middle: SEM micrographs (1000 x) of rotor surfaces during running-in (left) and after run-in (right)
Bottom: EDX analysis of rotor surfaces

To the brake lining matrix additives for wear protection have been added. Formulation A contained 3% antimony sulfide and 1,5% lead sulfide. Formulation B contained 4,5% polarized graphite. Dynamometer tests at 100 °C showed a pad wear of 0,7 mm for formulation A and 0,4 mm for formulation B. These results say that the transfer layer being formed by the composite containing metal sulfides have poorer wear protection than the polarized graphite containing formulations. Rotor wear was half as high for the formulation containing polarized graphite. This indicated that the transfer layer containing polarized graphite has a high degree of adhesion on the rotor surfaces and high shear resistance.

In field tests with passenger cars it was found, that brake linings containing polarised graphite formed on rotor surfaces homgeneous transfer layers. In particular during low braking pressures rotor wear was reduced significantly; it was reported from the field that differential thickness variation (DTV) of the rotor was eliminated since DTV occurs during low braking pressures.

Brake composites containing antimony sulfide, lead sulfide and polarised graphite were evaluated on a friction lining tester at various frictional temperatures.

Up to a rotor temperature of 400°C the elements of the transfer layer have been detected. At a rotor temperature of 500°C no transfer layer was analysed anymore; the surface of the rotor looked entirely different (Fig. 5). The chemical analysis showed no peaks of barium and calcium anymore which may have been transferred from the friction composite. However, elements detected were antimony, lead, zinc and carbon. Only components from the additive package in the composite determine the nature of the rotor surface.

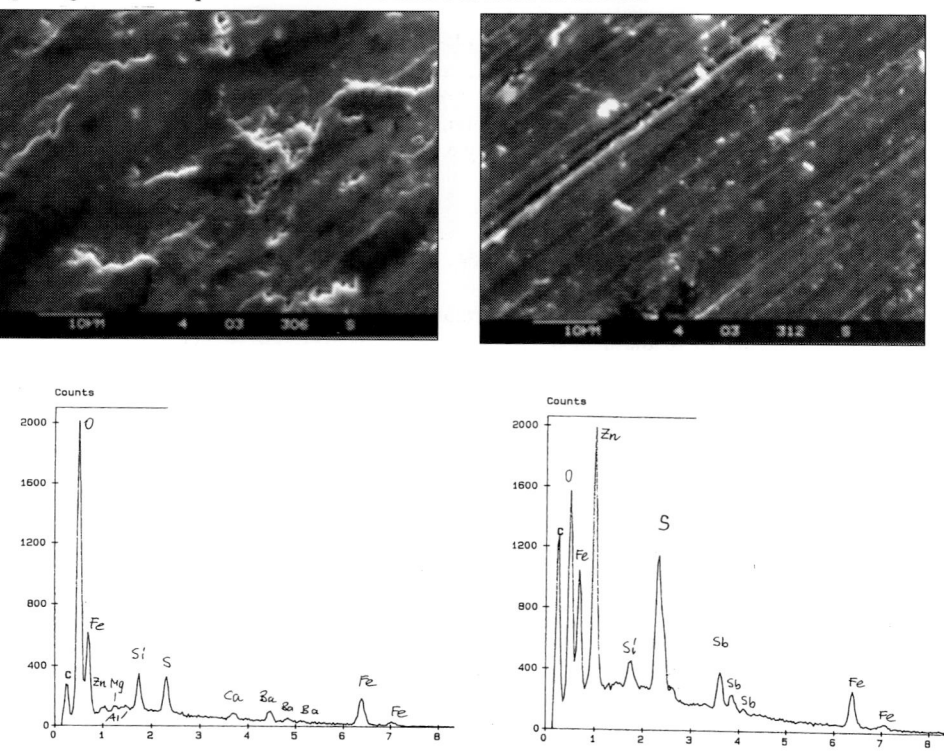

Fig. 5: SEM micrographs (1000 x) (top) and
EDX analysis (10 KV) (bottom) of brake rotor surfaces
at a rotor temperature of 400°C (left) and a rotor temperature of 500°C (right).

At 500°C rotor temperature there is no formation of a transfer layer anymore but the additives of the friction composites react chemically with the rotor surface and build a reaction layer. To gain more information about the reaction layer the rotor surface has been investigated by ESCA (Electron Spectroscopy for Chemical Analysis). By this method, only a very thin surface layer of a thickness of 5 nm is detected. The area next to the wear scar and the wear scar have been analysed. In the wear scar no iron has been found but the elements antimony, lead, zinc and oxygen are present. No barium and calcium has been detected which again confirms that there is no transfer of material from the composite onto the rotor surface. Rather there is a chemical reaction by decomposition of the additives like lead sulfide and antimony sulfide. It was of interest to find out what kind of reaction takes place with antimony trisulfide. Detected was a binding energy for antimony of 530.8 eV at the rotor surface. The binding energy of antimony trisulfide is at 529.4 eV and for antimony trioxide 530.0 eV. For antimony pentaoxide the binding energy of antimony is 530.8 eV. This is an indication that during the braking process at high temperatures, the antimony sulfide has decomposed and reacted to antimony pentaoxide. The reaction layer is very wear resistant and reduces primarily composite wear at high temperatures.

4. METAL PICK-UP OF BRAKE LININGS

Investigated has been a phenomenon called metal pick-up which means that foils of the rotor surface are transferred to the composite surface. This metallisation of the brake composite surface has significant disadvantages of braking concerning wear rate and noise level. The phenomenon of metal pick-up is mostly observed in the springtime when there is a significant amount of water in the environment. It is suggested that during the braking process water is trapped between surfaces of friction components and decomposes to hydrogen and oxygen.

In a number of publications it was reported that hydrogen diffuses into the rotor surface (7). The rotor consists of cast iron that means it contains a lot of cementite (Fe_3C). By diffusion hydrogen enriches in the surface of the rotor and reacts with the cementite to alpha iron and methane. Figure 6 shows the concentration of hydrogen diffusion in the rotor surface. The surface area of the rotor changes to alpha iron and the rest of the rotor remains as cast iron (8). Because of metallurgical stress, thin layers of alpha iron are sheared from the rotor surface and are transferred to the composite surface. The chemical reaction is given in Figure 6.

Tests have been performed with friction linings containing copper, zinc sulfide and tin disulfide. Microprobe analysis of rotor surfaces showed a rather high amount of copper, zinc and tin. This is an indication that at the rotor surface there is a formation of a metal layer consisting of copper, zinc and tin (Fig. 7). In these cases no metal pick-up has been observed since the hydrogen which is generated in the interface during the braking procedure does not diffuse into the rotor surface. Solubility of hydrogen in heavy metals is very poor and therefore this layer acts as a barrier for hydrogen diffusion. This additive package was found to prevent metal pick-up in railroad brakes.

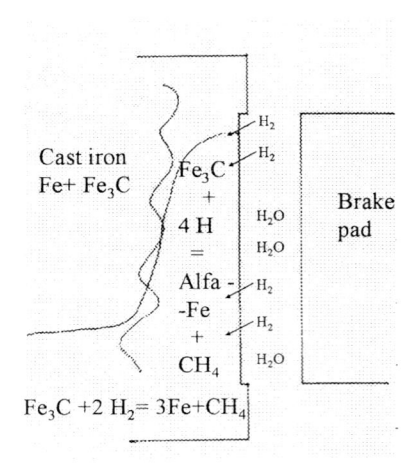

Fig. 6. Model of hydrogen diffusion into the rotor surface

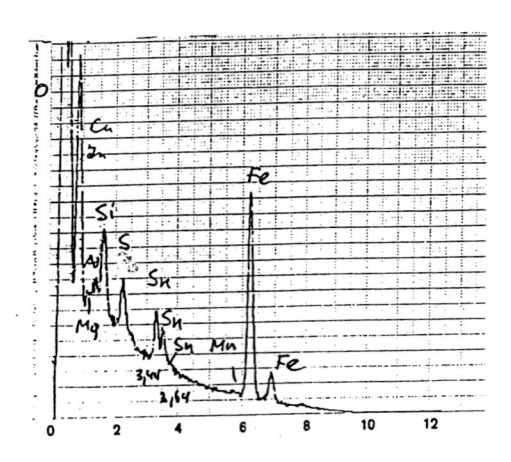

Fig. 7: EDX analysis of rotor surface (10 KV) after brake tests with pads containing Cu, ZnS, and SnS$_2$. Formation of a metal layer : Cu$_x$,Zn$_y$, Sn$_z$

5. CLOSURE

Research work revealed that additives play a significant role in brake composites. During braking thin transfer layers are being formed on metal surfaces of the rotor. Additives - in particular polarized graphite - form very thin transfer layers with good adhesion and good film strength. This results in the reduction of rotor wear and composite wear.

It was found that at rotor temperatures higher than 500°C no material is transferred to rotor surfaces anymore. However, additives react chemically and there is a formation of a new compound on the rotor surface. It was confirmed that antimony trisulfide reacts to antimony pentaoxide at a rotor temperature higher than 500°C. Antimony pentaoxide is very wear resistant and reduces composite wear in particular at high temperatures.

Composites containing metal sulfides and copper form, on rotor surfaces, a metal layer consisting of copper, zinc and tin. This layer acts as a barrier against hydrogen diffusion. No change of metallurgy takes place at rotor surfaces and metal pick-up is prevented. This effect ameliorates high rotor wear and high noise level during braking.

The use of new additives designed for friction linings has increased significantly over the last decade also. It has been confirmed that additives contribute significantly to performance improvement of friction brakes concerning safety of braking and comfort. Life of friction components is also extended.

6. ACKNOWLEDGEMENTS

The SEM micrographs and EDX analysis have been performed by Dr. Hermann Klingele of the Dr. Klingele Institute in Munich, Germany.

7. LITERATURE

1. Jacko M.G.: Physical and chemical changes of organic disc pads in service - Wear, 46 (1978) page 163

2. Wirth A. and Whitaker R.: A Study of transfer films, chemistry and its influence on friction coefficient. Presented at the International Conference of Frontiers of Tribology. Stratford-upon-Avon, U.K., April 91

3. Holinski, R.: A Review of the Contribution of Solid Lubricants to Performance Improvement of Friction Linings. Presented at the 3rd ASLE International Solid Lubricant Conference in Denver, Colorado , August 5-6, 1984.

4. Brendle M., Turgis P. and Lamouri S.: A General Approach to Discontinous Transfer Films: The Respective Role of Mechanical and Physico-Chemical Interactions. STLI Volume 39 (1996), 1, 157-165

5. Industrial Lubrication and Tribology: Volume 53, Number 2, 2001, pp.61-65 MCB University Press, ISSN 0036-8792

6. Langlade C., Fayeulle S. and Olier R.: New insights into adhesion and lubricating properties of graphite-based transfer films, Wear 172 (1994) p. 85-92.

7. Scieska S.F.: A Study of Tribological Phenomena in Friction Couple: Brake Composite Material - Steel. ASLE Transactions, Volume 25, 3, p. 337-345.

8. Polyakov, A.A., Garnukov D.N.: Physical-chemical mechanics of materials. 1969, vol.5, No. 2, p. 197.

Index

B

Barton, D C 53–74
Beveridge, C 25–42
Bingham, C M 175–184
Brookfield, D J 101–108
Brooks, P C 53–74

C

Cao, Q 43–52, 101–108
Cartmell, M P 43–52

D

Day, A J 285–296, 319–328
Dieckmann, T 223–234

E

Ettemeyer, A 109–120
Evans, R 109–120

F

Fieldhouse, J D 25–42, 85–100
Filip, P 341–354
Fudge, C M 3–14

G

Gerum, E 261–270
Gibbens, P 307–318
Grünberg, H 211–222

H

Harju, R 299–306
Hartmann, K-H 211–222
Holinski, R 369–378
Howe, D 175–184

I

Ioannidis, P 53–74

J

James, S 101–108

K

Klaps, J 285–296
Kohrs, C 245–260
Krupka, R 109–120

L

Lamperty, M U 197–208
Leiter, R 145–162

M

Markovic, T 271–284
Marwitz, H 245–260
Matsunaga, T 129–144
McCormick, B H 355–368
Mendelson, A J 123–128
Miyazaki, T 129–144
Morris, R 307–318
Mottershead, J E 43–52, 101–108

N

Nakamura, E 129–144
Nishiwaki, M 53–74
Niwa, S 129–144
Noor, K 319–328
Nowak, S 329–340

O

Obuchi, Y 185–196
Otomo, A 129–144
Ouyang, H 43–52, 101–108

P

Povel, R 245–260

Q

Qi, H S 319–328

R

Rumold, W 75–84

S

Sakai, A 129–144
Schofield, N 175–184
Shimada, M 185–196
Shimura, A 129–144
Slevin, E J 15–24
Smales, H 15–24
Soga, M 185–196
Steel, W 25–42
Stenman, A 223–234
Swift, R A 75–84

T

Talbot, C 25–42, 85–100
Tanaka, Y 129–144
Thompson, J K 3–14

Thompson, P 163–174
Tirovic, M 307–318
Treyde, T 43–52

V

Voller, G 307–318
von Glasner, E-C 245–260

W

Walker, A M 197–208
Walz, T 109–120
Wilkins, S 197–208
Williams, A R 235–244

Advances in Vehicle Design

By John Fenton

This book is an introduction to vehicle and body systems. It provides readers with an insight into analytical methods given in a wide variety of published sources such as technical journals, conference papers, and proceedings of engineering institutions. A comprehensive list of references is provided.

John Fenton distils and presents the best of this research and industry practice into an easily digestible, highly illustrated, and accessible form. Drawing on the available information, the author provides a well-structured and vital reference source for all automotive engineers.

Key features:

● Heavily illustrated ● Hundreds of examples ● Comprehensively indexed ● Concise synthesis of available information ● Written by an established authority ● A ready reference source

Contents include:

Materials and construction advances; structure and safety; powertrain/chassis systems; electrical and electronic systems; vehicle development; systems development: powertrain/chassis; system developments: body structure/systems; references; index.

1 86058 181 1   234x156mm
Hardback   190 pages   1999   **£64.00**

Orders and enquiries to:
Sales and Marketing Department
Professional Engineering Publishing Limited
Northgate Avenue, Bury St Edmunds, Suffolk IP32 6BW, UK
Tel: +44 (0) 1284 724384   **Fax:** +44 (0) 1284 718692
Email: orders@pepublishing.com
Website: www.pepublishing.com

Professional Engineering Publishing

Add 10% for delivery outside the UK

Statistics for Engine Optimization
Edited by S P Edwards, D M Grove, and H P Wynn

In a series of specially commissioned articles, this book demonstrates how statistically designed experiments can make a major contribution to meeting existing and future demands in engine development.

Engine development teams are facing increasing demands from industry, legislators, and customers for lower cost engines and reduced vehicle exhaust emissions combined with improvements in specific vehicle performance and refinement. These necessitate the design, analysis, and testing of an ever wider range of options, while decreasing the process costs and/or duration.

Topics covered include:
Design of experiments; Modelling techniques; Response surface methods; Multi-stage models; Emulating computer models; Bayesian methods; Optimization; Genetic algorithms; On-line optimization; Robust engineering design.

Contents
● Editors' introduction: statistics for engine optimization ● The role of statistics in the engine development process ● Issues arising from statistical engine testing ● Practical implementation of design of experiments in engine development ● Statistical modelling of engine systems ● Applying design of experiments to the optimization of heavy-duty diesel engine operating parameters ● Engine mapping: a two-stage regression approach based on spark sweeps ● The application of an automatic calibration optimization tool to direct injection diesels ● An investigation of the utilization of genetic programming techniques for response curve modelling ● Using neural networks in the characterization and manipulation of engine data ● Empirical modelling of diesel engine performance for robust engineering design ● Improved engine design and development processes ● Subject Index.

1 86058 201 X   234x156mm
Hardback   208 pages   2000   **£49.00**

Orders and enquiries to:
Sales and Marketing Department
Professional Engineering Publishing Limited
Northgate Avenue, Bury St Edmunds, Suffolk IP32 6BW, UK
Tel: +44 (0) 1284 724384 **Fax:** +44 (0) 1284 718692
Email: orders@pepublishing.com
Website: www.pepublishing.com

Professional Engineering Publishing

Add 10% for delivery outside the UK

Integrated Powertrains and their Control

Edited by N D Vaughan
University of Bath

Integrated Powertrains and their Control presents the latest advances in powertrain technology. Specific applications are discussed, along with ideas as to how concept, design, modelling, and simulation are put into practice, to meet the latest/current legal requirements. The book will prove a valuable source of information to all engineers and researchers working on powertrains, fuel, control technologies, or emissions, and related fields. Engineers will find this a useful volume as will vehicle manufacturers and their suppliers and consultants.

The editor, Dr N D Vaughan is a senior lecturer in the Department of Mechanical Engineering at the University of Bath. His principal interests are in the application of control techniques to mechanical systems, the design, of electro-hydraulic control valves, and the study of continuously variable vehicle transmissions.

Contents include: Introduction to advances in powertrain technology; Control of an integrated IVT powertrain; Driveability control of the ZI® powertrain; Performance of integrated engine – CVT control, considering powertrain loss and CVT response lag.

1 86058 334 2   234x156mm
Hardcover   110 pages   April 2001   **£49.00**

Orders and enquiries to:
Sales and Marketing Department
Professional Engineering Publishing Limited
Northgate Avenue, Bury St Edmunds, Suffolk IP32 6BW, UK
Tel: +44 (0) 1284 724384   **Fax:** +44 (0) 1284 718692
Email: orders@pepublishing.com
Website: www.pepublishing.com

Overseas customers please add 10% for delivery